Lecture Notes in Computer Science 8728

Commenced Publication in 1973
Founding and Former Series Editors:
Gerhard Goos, Juris Hartmanis, and Jan van Leeuwen

Editorial Board

Matt Duckham   Edzer Pebesma
Kathleen Stewart   Andrew U. Frank (Eds.)

Geographic Information Science

8th International Conference, GIScience 2014
Vienna, Austria, September 24-26, 2014
Proceedings

 Springer

Volume Editors

Matt Duckham
The University of Melbourne, VIC, Australia
E-mail: matt@duckham.org

Edzer Pebesma
University of Münster, Germany
E-mail: edzer.pebesma@uni-muenster.de

Kathleen Stewart
The University of Iowa, Iowa City, IA, USA
E-mail: kathleen-stewart@uiowa.edu

Andrew U. Frank
Vienna University of Technology, Austria
E-mail: frank@geoinfo.tuwien.ac.at

ISSN 0302-9743   e-ISSN 1611-3349
ISBN 978-3-319-11592-4   e-ISBN 978-3-319-11593-1
DOI 10.1007/978-3-319-11593-1
Springer Cham Heidelberg New York Dordrecht London

Library of Congress Control Number: 2014948637

LNCS Sublibrary: SL 3 – Information Systems and Application, incl. Internet/Web and HCI

Typesetting: Camera-ready by author, data conversion by Scientific Publishing Services, Chennai, India

Printed on acid-free paper

Springer is part of Springer Science+Business Media (www.springer.com)

Preface

The 2014 8th International Conference on Geographic Information Science continued the highly successful GIScience series of conferences. GIScience regularly brings together more than 200 international participants from academia, industry, and government organizations to discuss and advance the state-of-the-art in the field of geographic information science. Since 2004, the biennial conference has alternated between locations in North America and Europe. Following a highly successful GIScience 2012 in the heart of the American Midwest, at Colombus, Ohio, GIScience 2014 was located in the heart of Europe, hosted by Vienna University of Technology, Austria.

Since its inception in 2000, the biennial GIScience conferences have adopted a twin-track program, soliciting the submission of the latest work in progress, through the extended abstracts track (up to 1500 words); and the highest quality completed research, through the full papers track (up to 15 pages). The full papers contained in these proceedings are complemented by the separate extended abstract proceedings distributed at the conference.

The twin-track program is ideally suited to the diversity of disciplines that converge on GIScience, which include (but are not limited to) geography, cognitive science, computer science, engineering, information science, linguistics, mathematics, philosophy, psychology, social science, and (geo)statistics. The combination of full papers and extended abstracts has a proven record of delivering an exciting conference program that is both fast-moving and high quality.

For GIScience 2014, 84 full paper submissions were received. Each paper was thoroughly reviewed by at least three independent members of the international Program Committee. Based on these reviews, supplemented by careful metareviews from the program chairs, 23 papers were selected for presentation, corresponding to an acceptance rate of 27%. 155 further extended abstracts submissions were received, 52 (30%) of which were accepted for short oral presentations.

The accepted full papers provide a snapshot of the breadth of active research topics in the vibrant and maturing field of GIScience. Typically for conferences in the GIScience series, the accepted full papers showcase a mix of basic research connected with the latest hot topics (such as user-generated content, linked data, big data, and text-based navigation systems) as well as important advances in more long-established topics (such as spatial algorithms, qualitative spatial reasoning, spatial analysis, spatial cognition, geovisualization, and geo-ontologies).

This year, however, the excitement of the conference was mingled with great sadness following the tragic deaths of two treasured members of the GIScience Conference community: Prof. Peter Fisher on 20 May 2014 and Prof. Carolyn Merry on 3 June 2014. Pete and Carolyn were both active, serving GIScience Program Committee members, with Pete having been a Program Committee member since the very first GIScience conference in 2000. Pete and Carolyn will be greatly missed, and their lives are remembered in brief tributes from the GIScience Program Committee in the following pages of these proceedings. Indeed, the geographic information science community more broadly will remember in 2014 the lives of Roger Easton, Cliff Kottman, Doug Nebert, and Roger Tomlinson, four other influential contributors to the field who sadly died this year.

The contributions of such influential figures continue to be reflected in the many contributions of others in the field, including those who helped make GIScience 2014 a success. The conference could not have happened without the work of the local organizers Gerhard Navratil and Eva Maria Holy. We would also like to thank Paolo Fogliaroni for his tireless work and support not only as the workshop and tutorial chair but also for assistance with the conference web site. Our thanks go to Thomas Linton at the Vienna University of Technology who provided technical support. We would also like to thank Ross Purves for his time spent sharing his experience with the EasyChair reviewing process, helping to maintain continuity across the different GIScience conference years. Of course we are deeply grateful to the GIScience Program Committee for their considered and thorough reviews, as well as those additional reviewers who also contributed their expertise. We would like to thank all the authors who contributed to the conference by submitting papers and extended abstracts. Most importantly, we would like to thank all those who came to GIScience, presenters and delegates, without whose contributions there would be no conference.

August 2014

Matt Duckham
Edzer Pebesma
Kathleen Stewart
Andrew Frank

Prof. Peter Fisher

"When Pete served as the external opponent on my PhD exam in Stockholm, Sweden, he completely transformed what could have been a nerve-wracking ordeal into one of the most significant academic conversations and learning experiences of my career. Through this conversation I also think he managed to show many of my colleagues the science behind the GIS they mostly saw as a tool."

Ola Ahlqvist

"Pete's work on fuzzy viewsheds was inspirational to me at the beginning of my career, as it was my first introduction to an approach to framing spatial uncertainty of spatial features in continuous space. He was one of the senior scholars I was most excited to meet at the NCGIA International Young Scholars Summer Institute in 1995. Despite relatively infrequent interactions, I count Pete Fisher as an intellectual hero and mentor, and suspect that I am not alone in doing so."

Dan Brown

"I first observed Peter Fisher as a red-bearded newcomer at AUTO-CARTO London, bringing a similar articulate critique of the then-popular "Expert System" to AUTO-CARTO 8. Here was a kindred spirit with no fear of opposing received opinion. We could discuss random topics such as the etymology of the Arabic terms applied to sand dunes. I was present in the deep amphitheatre in Utrecht (AGILE 2011) as Peter handed over the editorship of IJGIS, exhibiting a deft mastery of the range of the field as well as a time for others to take charge. I did not know that event would be my last observation of the trajectory of a dedicated scholar, and a key contributor to a community of scholarship in GIScience."

Nick Chrisman

"I first met Pete in the fall of 2001, at a project meeting of a newly started European framework project. Being in the first months of my PhD, how could I not have been impressed by this strong red-bearded English man who also happened to be a lead researcher in the field I was getting into? I came to know Pete better through the years of this project and quickly realized the wonderful man he was, accessible to students, generous of his time, and passionate about his research field. Pete was an inspiration for me and for many others. Thank you for sharing your time and passion Pete, you will stay well alive in our memories!"

Rodolphe Devillers

"The first time I met Pete in person was in 2005 in his office at Leicester University. I explained to him my tentative research proposal for a PhD that I hoped to start after finishing my Masters at Leicester. It was not a very good research proposal. I remember Pete was supportive and generous with his time; kind and gentle in his manner; at the same time as being unambiguous and direct in his criticism of my idea. I count myself lucky to have had the opportunity many more times over the subsequent years to be grateful for Pete's admirable capacity for combining clear and critical scientific thinking with simple human kindness."

Matt Duckham

"I'm writing this on the train from Leicestershire to London. I remember working on the VFC project with Pete and we used to have meetings with Dave Unwin and Jonathan Raper at Birkbeck. We would take this train down to London and all chat excitedly and noisily about the project and the work we had done. I remember being quite proud of the fact that we were all arguing passionately about the difference between ambiguity and vagueness (or something). Pete would do so at great volume, and only now do I realise just how much this activity must have disturbed the regular commuters. Since the VFC days the rail companies have established a QuietZone on each train - which is where I am now sitting, peacefully reflecting in silence. Pete was never a QuietZone. Rather he was an IdeasZone and always the focus of noisy activity. I do wonder whether his animated booming discussion of uncertainty had an impact on the rail companies?"

Jason Dykes

"Pete Fisher's fundamental geovisualization research has inspired me and many like-minded colleagues since the early 90s in significant ways. His trail-blazing research on uncertainty visualization, animation, sonification, and virtual reality continues to drive our research programs in cartography, geovisualization, and geovisual analytics, for example, under the umbrella of the International Cartographic Association's Geovisualization and Cognitive Visualization Commissions. "

Sara Fabrikant

"Pete is a very sad loss, we need more 'larger than life' people in academia, and he was certainly that. I very much enjoyed his obvious passion, joy, laughter(!) and intensity, as well as enjoying his research. He was a wonderful journal editor too, for IJGIS. He also had the cojones to publish an article by a disreputable colleague of mine (Prof. McNoleg), for which we are both very grateful."

Mark Gahegan

"Pete was a familiar and often audible figure at GIS conferences. There are many occasions when on hearing that distinctive laugh, I would think 'Oh good, he's here!' and gravitate towards the oracle, in the sure knowledge of challenging, insightful, interesting, sometimes rude and always entertaining conversation."

Chris Jones

"Peter Fisher created an exquisitely fine balance between family and academia, always ready to enjoy both, and to mix both to the full. Academically he will be long remembered for his marathon contribution as editor of IJGIS and the various activities that spun off that commitment, including the Classics from IJGIS book of 2007. Among his many ground breaking papers are the series on simulating uncertainty in various GIS operations, and the series on fuzzy sets, heaps, and fuzzy-fuzzy sets. He was instrumental in establishing GIS as a specialty at Kent State, building on the GIS tradition at Leicester, and later putting City University on the GIS map. Pete could always be trusted to have a novel and interesting perspective on whatever topic he chose to pursue, and we can only imagine what other great ideas might have emerged in his presentations, publications, and conversations if his life had not been so tragically cut short."

Michael F. Goodchild

"I worked closely with Pete for nearly twenty years, initially as the Western Pacific Regional Editor of IJGIS. It was a strange relationship as we met very infrequently but corresponded frequently by email. I do recall a visit that he, Jill and family made to Canberra however. Pete spent ages poring over my bookshelves, grunting approvingly when he came across P.G. Wodehouse. I asked if he wanted to borrow something and he replied that bookshelves were the best way of getting to know someone. Their contents were very revealing! I still don't know what he made of my eclectic collection. P.G. Wodehouse sits alongside trashy Sci-Fi, crime fiction and oddments like G.K. Chesterton and Compton McKenzie. When I stayed with him, I slept in a room with extensive, and an equally eclectic collection. It might explain how we worked so happily together for so long."

Brian Lees

"Unsure what to do with a Geography degree, in 1984 I applied for a 6 month research post in 'automated cartography' at Kingston University where Pete was working with Graeme Wilkinson. The work progressed well, and three years later I had a passport to Academia—a land full of fruity folk. 30 years on and I'm still having fun, except in June of this year GIScience lost one of its fruitier fruits. I will always remember that quizzical face, and the bounding laughter!"

William Mackaness

"I was a very junior faculty member when I met Pete. He breezed into a workshop I was attending with a big swoosh and a trail of eager grad students. He said only

'Yes, I'll give you that; you nailed the critics of GISscience.' He was permanently imprinted on me from that day on."

Nadine Schuurman

"I met first Pete in the early 90s, when among other things we talked about founding the GISRUK series of conferences and were partners on a European project on spatial uncertainty. I will remember Pete as a kind, generous, fun-loving guy, with a huge laugh and a sharp mind. He was a wonderful successor to Terry Coppock as editor of IJGIS. I'll miss him."

Mike Worboys

Prof. Carolyn Merry

A brief but heartfelt salute to Carolyn Merry, 1950–2014

- a distinguished teacher and researcher in the field of GIScience
- former Professor and Chair of the Department of Civil, Environmental and Geodetic Engineering at The Ohio State University
- helping connect GIScience to Civil Engineering
- former president of ASPRS, UCGIS, and ASCE

But also ...

- a dedicated, loyal, and warm person;
  - always interacting with the very highest integrity;
    - approachable, and supportive, especially to junior faculty;
      - an inspiring role model;
        - a great force for good.

We are all better for having known and worked with her!

Dawn Wright, Nina Lam, Jeremy Mennis, Tom Cova, and Francis Harvey

Organization

General Chair

Andrew U. Frank   Vienna University of Technology, Austria

Program Chairs

Matt Duckham   University of Melbourne, Australia
Kathleen Stewart   University of Iowa, USA
Edzer Pebesma   University of Münster, Germany

Workshop & Tutorial Chair

Paolo Fogliaroni   Vienna University of Technology, Austria

Local Organizers

Gerhard Navratil   Vienna University of Technology, Austria
Eva Maria Holy   Vienna University of Technology, Austria

Technical Support

Thomas Linton   Vienna University of Technology, Austria

Program Committee

Ola Ahlqvist   Ohio State University, USA
Luc Anselin   Arizona State University, USA
Marc Armstrong   University of Iowa, USA
Kate Beard-Tisdale   University of Maine, USA
Scott Bell   University of Saskatchewan, Canada
Itzhak Benenson   Tel Aviv University, Israel
David Bennett   University of Iowa, USA
Michela Bertolotto   University College Dublin, Ireland
Ling Bian   University at Buffalo, USA
Thomas Bittner   SUNY at Buffalo, USA
Dan Brown   University of Michigan, USA
Dirk Burghardt   Technical University of Dresden, Germany

Barbara Buttenfield	University of Colorado Boulder, USA
Gilberto Camara	INPE, Brazil
Adrijana Car	University of Salzburg, Austria and German University of Technology in Oman, Oman
Nicholas Chrisman	RMIT University, Australia
Christophe Claramunt	Naval Academy Research Institute (NARI), France
Keith Clarke	UC Santa Barbara, USA
Eliseo Clementini	University of L'Aquila, Italy
Tom Cova	University of Utah, USA
Isabel Cruz	University of Illinois at Chicago, USA
Leila de Floriani	University of Genova, Italy
Rodolphe Devillers	Memorial University of Newfoundland, Canada
Juergen Doellner	Hasso Plattner Institut, Potsdam, Germany
Jason Dykes	City University London, UK
Max J. Egenhofer	University of Maine, USA
Sara Irina Fabrikant	University of Zürich, Switzerland
Peter Fisher	University of Leicester, UK
Christian Freksa	University of Bremen, Germany
Mark Gahegan	University of Auckland, New Zealand
Rina Ghose	University of Wisconsin-Milwaukee, USA
Peng Gong	Tsinghua University, China
Michael Goodchild	UC Santa Barbara, USA
Ian Gregory	Lancaster University, UK
Dan Griffith	University of Texas at Dallas, USA
Joachim Gudmundsson	University of Sydney, Australia
Diansheng Guo	University of South Carolina, USA
Muki Haklay	University College London, UK
Lars Harrie	Lund University, Sweden
Francis Harvey	University of Minnesota, USA
Gerard Heuvelink	Wageningen University, The Netherlands
Hartwig Hochmair	University of Florida, USA
Piotr Jankowski	San Diego State University, USA
Krzysztof Janowicz	University of California at Santa Barbara, USA
Bin Jiang	University of Gävle, Sweden
Christopher Jones	Cardiff University, UK
Derek Karssenberg	Utrecht University, The Netherlands
Tomi Kauppinen	Aalto University, Finland
Marinos Kavouras	NTUA, Greece
Maggi Kelly	UC Berkeley, USA
Peter Kiefer	ETH Zurich, Switzerland
Alexander Klippel	Pennsylvania State University, USA
Menno-Jan Kraak	University of Twente/ITC, The Netherlands
Werner Kuhn	UC Santa Barbara, USA
Mei-po Kwan	University of Illinois at Urbana-Champaign, USA

Phaedon Kyriakidis	UC Santa Barbara, USA
Nina Lam	Louisiana State University, USA
Brian Lees	University of New South Wales, Australia
Ron Li	Ohio State University, USA
Xiang Li	East China Normal University, China
Hui Lin	Chinese University of Hong Kong, China
Lin Liu	University of Cincinnati, USA
Yu Liu	Peking University, China
Amy Lobben	University of Oregon, USA
Alan MacEachren	Pennsylvania State University, USA
William Mackaness	University of Edinburgh, UK
Jeremy Mennis	Temple University, USA
Carolyn Merry	Ohio State University, USA
Harvey Miller	Ohio State University, USA
Alan Murray	Arizona State University, USA
Tomoki Nakaya	Ritsumeikan University, Japan
Atsuyuki Okabe	University of Tokyo, Japan
Antonio Paez	McMaster University, Canada
Dimitris Papadias	UST at Hong Kong, China
Karin Pfeffer	University of Amsterdam, The Netherlands
Dieter Pfoser	George Mason University, USA
Alenka Poplin	HafenCity Universität Hamburg, Germany
Lilian Pun	Hong Kong Polytechnic University, China
Ross Purves	University of Zürich, Switzerland
Martin Raubal	ETH Zürich, Switzerland
Tumasch Reichenbacher	University of Zürich, Switzerland
Femke Reitsma	University of Canterbury, UK
Maria Andrea Rodriguez-Tastets	Universidad de Concepción, Chile
Anne Ruas	Institut Français des Sciences et Technologies des Transports (IFSTTAR), France
Christoph Schlieder	University of Bamberg, Germany
Falko Schmid	University of Bremen, Germany
Nadine Schuurman	Simon Fraser University, Canada
Shih-Lung Shaw	University of Tennessee, USA
Takeshi Shirabe	Royal Institute of Technology Sweden, Sweden
Bettina Speckmann	Eindhoven University of Technology, The Netherlands
Emmanuel Stefanakis	University of New Brunswick, Canada
Paul Sutton	University of Denver, USA
Jean-Claude Thill	University of North Carolina at Charlotte, USA
Sabine Timpf	University of Augsburg, Germany
Paul Torrens	University of Maryland, USA
Ming-Hsiang Tsou	San Diego State University, USA

Marc van Kreveld	Utrecht University, The Netherlands
Monica Wachowicz	University of New Brunswick, Canada
Jan Oliver Wallgrün	Pennsylvania State University, USA
Shaowen Wang	University of Illinois, USA
Robert Weibel	University of Zürich, Switzerland
John Wilson	University of Southern California, USA
Stephan Winter	The University of Melbourne, Australia
Michael Worboys	University of Greenwich, UK
Dawn Wright	ESRI, USA
Ningchuan Xiao	Ohio State University, USA
Yichun Xie	Eastern Michigan University, USA
Chaowei Yang	George Mason University, USA
Bailang Yu	East China Normal University, China
May Yuan	University of Oklahoma, USA

Additional Reviewers

Roger Bivand, Norway	Mingyuan Hu, Hong Kong
Olivier Bonin, France	David Jonietz, Germany
Min Chen, Hong Kong	George Kellaris, China
Matthew Dube, USA	Patrick Laube, Switzerland
Junchuan Fan, USA	Joshua Lewis, USA
Riccardo Fellegara, Italy	Xuecao Li, China
Eric Fung, China	Silvia Nittel, USA
Jon Goergen, USA	Mark Padgham, Germany
Venkat Raghavan Ganesh, USA	Christoph Stasch, Germany
Peng Gao, USA	Lu Tan, China
Stefan Hahmann, Germany	Maria Vasardani, Austria
Pierre Hallot, USA	Emily White, USA
Paul Hiemstra, The Netherlands	

Table of Contents

Information Visualization

Spatial Analysis

User-Generated Content

Semantics and Models

Wayfinding and Navigation

Spatial Algorithms

Spatial Relations

Map Schematization with Circular Arcs[*]

Thomas C. van Dijk[1], Arthur van Goethem[2], Jan-Henrik Haunert[3]
Wouter Meulemans[2], and Bettina Speckmann[2]

[1] Universität Würzburg, Würzburg, Germany
[2] Technical University Eindhoven, Eindhoven, The Netherlands
[3] Universität Osnabrück, Osnabrück, Germany

Abstract. We present an algorithm to compute schematic maps with circular arcs. Our algorithm iteratively replaces two consecutive arcs with a single arc to reduce the complexity of the output map and thus to increase its level of abstraction. Our main contribution is a method for replacing arcs that meet at high-degree vertices. This allows us to greatly reduce the output complexity, even for dense networks. We experimentally evaluate the effectiveness of our algorithm in three scenarios: territorial outlines, road networks, and metro maps. For the latter, we combine our approach with an algorithm to more evenly distribute stations. Our experiments show that our algorithm produces high-quality results for territorial outlines and metro maps. However, the lack of caricature (exaggeration of typical features) makes it less useful for road networks.

1 Introduction

Maps are a common and intuitive way of communicating and exploring information with a geographic component. In many cases exact geographic details are not required to convey the primary information. For thematic maps exact details in the base map may even distract from or obscure the thematic overlay. Consequently, there has been a continuous interest in schematic maps (e.g., [1–4]). A schematic map is typically highly abstract and stylized, maintaining only those features that support the message of the map. There exist a wide variety of schematic maps, including metro maps and chorematic diagrams [5].

Most automated methods to create schematic maps have focused on straight-line schematization, often with an orientation restriction [6–8] (e.g., admitting only horizontal, vertical and diagonal lines). In contrast, manually drawn schematic maps often use curves. It can be desirable to have a good continuation [9] of line features, to strengthen their representation. For example, it may be desirable for metro lines to continue smoothly at interchanges or for shorelines to span multiple countries (see Fig. 1).

Results and Organization. In Section 2 we present a new iterative and topologically correct schematization algorithm using circular arcs. This algorithm maintains good

---

[*] A. van Goethem and B. Speckmann are supported by the Netherlands Organisation for Scientific Research (NWO) under project no. 612.001.102 (AvG) and no. 639.023.208 (BS). T. van Dijk is supported by DFG-grant Ha5451/3-1. Collaboration started at Dagstuhl seminar 13151 "Drawing graphs and maps with curves."

M. Duckham et al. (Eds.): GIScience 2014, LNCS 8728, pp. 1–17, 2014.

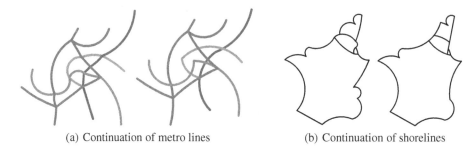

(a) Continuation of metro lines   (b) Continuation of shorelines

Fig. 1. At points where multiple curves meet, continuity may improve schematization

geometric correlation with the geographic input. It allows vertices of degree three or higher to be shifted, creating arcs that continue "through" such vertices. We also show how to tailor our algorithm specifically to metro maps. In Section 3 we present experimental results for three different scenarios: territorial outlines, road networks and metro networks. Our approach yields high-quality results for both territorial outlines and metro maps. It appears less suitable for road networks. A lack of *caricature* (exaggeration of typical features) interferes with the subconscious link between road type and shape. We also discuss the effect of the different algorithmic features on the maximal complexity reduction. In Section 4 we reflect on the implications of our design decisions.

Related Work. Automated schematization has mostly restricted itself to representations with orientation-restricted line segments. There is a large number of results for the schematization of networks (e.g. [1, 3, 7, 8]) or even single lines (e.g. [4, 10, 11]). In contrast, only a few recent algorithms exist that explicitly aim to schematize outlines [6, 12]. The orientation-restricted style can enhance the visual clarity of a map as it promotes parallelism and the continuation of edges at high-degree vertices. Ti and Li [13] discuss the use of strokes and network distortion to further improve the usability of the schematization. However, Roberts [14] recently showed that manually drawn metro maps with curves are more efficient and effective than the long-standing octilinear designs.

Curves are important in manual cartography [15]. There are also several automated approaches to schematize an outline [16, 17] or a subdivision of outlines [18, 19] with circular arcs. However, both methods for subdivisions cannot move vertices of degree three or higher, although doing so is beneficial for subdivisions and crucial in dense metro networks. Fink *et al.* [20] use Bézier curves to draw metro maps. They are able to move high-degree vertices and aim to prevent abrupt turns of metro lines. However, Bézier curves inherently admit more freedom than circular arcs, resulting in a less strict schematic style (similar to the difference between simplification and orientation-restricted schematization).

2 Curved Schematization Algorithm

Our schematization algorithm iteratively replaces two neighboring circular arcs with a single arc while ensuring correct topology. This approach is similar to the one proposed by Van Goethem *et al.* [19] for territorial outlines. However, we present some significant improvements which make our algorithm more suitable for generic networks. Most importantly, we show how to move vertices of degree three and higher, to reduce the number of arcs of a schematization while improving the overall quality. Furthermore, in Section 2.4 we introduce some specific improvements geared towards metro maps.

2.1 Preliminaries

A *network* is a planar straight-line embedding of a graph in $\mathbb{R}^2$ which may represent various types of information such as metro lines or territorial outlines (subdivisions). The edges of a network are circular arcs (line segments are degenerate circular arcs). The edges meet at vertices. The *degree* of a vertex is its number of incident edges. We refer to vertices of degree three or higher as *junctions*. The *complexity* of a network N is its number of edges.

We require that the schematization N is *topologically equivalent* to the input network I. A schematization is topologically equivalent if there is a continuous function transforming I to N where at all times edges intersect only at vertices. This implies that N is planar, the order of incident edges around each vertex is maintained and that adjacencies are preserved.

2.2 Main Algorithm

We describe an algorithm that computes a circular-arc schematization for a given network I. To this end, it maintains a network N; initially, N is a copy of I and consists only of straight edges (line segments). To create the schematization, two edges in N are replaced by a single edge (an *operation*). This reduces complexity and introduces circular arcs. We maintain as invariant that N is topologically equivalent to the input network I. Below, we provide details for the various steps of our algorithm; an overview is given in Algorithm 1.

Stroke Partition. The main innovation of our algorithm is its ability to deal with junctions. We allow the conceptual removal of a junction, joining two incident edges into a single edge. The junction is then implicitly represented as the intersection of edges. To decide which edges may be joined by such an operation, we partition the network into *strokes*. A stroke is a "natural" path through the network, continuing relatively smoothly at junctions. Strokes may correspond to through roads (e.g., [21]) or to coastlines spanning multiple territories.

To compute a stroke partition, we proceed as follows. We first assign each edge to a unique stroke. Let E_v be the set of incoming edges for a vertex v of degree two or higher. For any pair of edges $e, f \in E_v$ we compute the angular deviation. The angular deviation at v equals 180 degrees minus the minimum angle between e and f.

Algorithm 1. COMPUTECURVEDSCHEMATIZATION(N, k)

Input: Network N and desired complexity k
Output: Topologically equivalent circular-arc schematization of N

1: Partition N into strokes
2: Compute all operations O
3: **while** N has more than k edges and O contains an admissible operation **do**
4:     Execute the admissible operation $o \in O$ with the lowest cost
5:     Remove all operations involving an arc replaced by o
6:     Update admissibility of remaining operations
7:     Create new operations involving the edge introduced by o
8: **return** N

We repeatedly combine the strokes of the pair of edges in E_v with the lowest angular deviation and remove them from E_v. The stroke partition can be computed in $O(d_i^2 \log d_i)$, where d_i is the maximum vertex-degree in the network.

Operations. We now define a set of *operations* O that can be executed on the network. An operation removes some vertex v at which two consecutive edges of a single stroke S meet. For now, assume that v has degree 2: no other strokes pass through v. Let u and w be the other endpoints of the two edges in S that are incident to v. To maintain topology, a new edge should be inserted connecting u and w. Thus, an operation replaces two consecutive edge with a single edge.

We call an operation *admissible* if it maintains the correct topology. To ensure a topologically equivalent result, the algorithm performs only admissible operations. As it depends on how we deal with junctions, we further discuss admissibility later in this section.

The *cost* of a replacement indicates the dissimilarity between the replacement edge and the represented input edges. We quantify this dissimilarity with the *Fréchet distance* [22]. For a single edge in N that represents n edges of the input, this measure can be computed in $O(n \log n)$ time [22]. By using this measure to weigh the operations, we maintain a geometric correlation between the resulting schematization and the geographic input.

The replacement with lowest cost for a given pair of edges may be inadmissible. To allow for more flexibility, we add three replacements to the set of operations O. To this end, we create a discrete set of candidate replacements using arcs of different radii and add the three replacements with lowest score. We generate these candidate replacements using angles of $i \cdot \frac{\pi}{k}$ with respect to line $\overline{uw}$ with $-k < i < k$ for some parameter k; we used $k = 20$ in our experiments.

Junctions. Above, we assumed that operations remove only vertices of degree 2. We now introduce operations for two edges that meet at a junction. The examples in Fig. 1, illustrating the gain of schematizing across junctions, have been generated with our algorithm without and with these additional operations.

Let v be a junction on stroke S. When replacing the edges in S incident to v, we must ensure that the junction remains. This constrains the replacement edge. We keep track of these constraints by marking v as a *virtual vertex* on the replacement edge. Note that

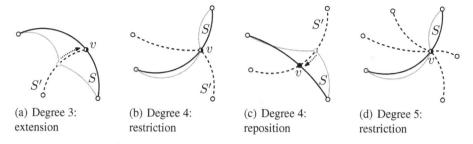

(a) Degree 3:   (b) Degree 4:   (c) Degree 4:   (d) Degree 5:
extension   restriction   reposition   restriction

Fig. 2. Dealing with junctions. Regular vertices are white dots, virtual vertices are black. Replaced edges are indicated in gray, edges of other strokes are dashed.

v is a virtual vertex only for strokes in which is has been "removed"; for other strokes, it remains a *regular vertex*. There are four cases that constrain the possible replacement arc. As v is a junction, it is included in at least two strokes; let S' denote a second stroke that includes v. Refer to Fig. 2.

(a) v has degree 3. Hence, v is an endpoint of S' and we extend or shorten the edge of S' such that its endpoint lies on the replacement edge.
(b) v has degree 4 and is a regular vertex on S'. To maintain the degree-4 vertex, we constrain the replacement edge to pass through v: only a single replacement candidate remains.
(c) v has degree 4 and is a virtual vertex on S'. We can reposition the virtual vertex of S' along its arc. This admits flexibility in the replacement edge and thus we use the same approach as for a degree-2 vertex.
(d) v has degree 5 or more. As in case (b), we constrain the replacement edge to pass through v.

An edge e in the network may have any number of virtual vertices along its boundary. These virtual vertices can constrain further replacements involving e. Let z be a virtual vertex on e. If z originates from a replacement of case (a) or (c), then its incident arc can be extended or z can be repositioned. Otherwise, any operation replacing e must maintain the position of z, thus limiting the possible replacements. If an operation is limited in this way by two or more virtual vertices, a replacement is only possible if these vertices are cocircular with the endpoints of the replacement arc.

Computing Admissibility. An operation is admissible if it maintains the correct topology. To decide on admissibility, we proceed as follows. Let e_1 and e_2 be the two replaced edges and let v be the vertex at which these edges meet. Let u and w be the other vertex of e_1 and e_2 respectively. Finally, let e_r denote the replacement edge. An *involved virtual vertex* is a virtual vertex along e_1 or e_2. The *involved vertices* are all involved virtual vertices as well as u, v and w. A *virtually involved edge* is an edge that is incident to an involved virtual vertex; this does not include e_1 and e_2. The set of *uninvolved edges* contains all edges that are not e_1 or e_2 or an involved virtual edge. To compute admissibility, we check the following conditions; if any of these is not satisfied, the operation is not admissible. These conditions are illustrated in Fig. 3.

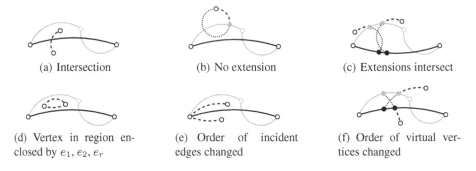

(a) Intersection   (b) No extension   (c) Extensions intersect

(d) Vertex in region en-   (e) Order   of   incident   (f) Order of virtual ver-
closed by e_1, e_2, e_r edges changed   tices changed

Fig. 3. Examples of topology violations of an inadmissible operation. e_r given in black; e_1 and e_2 in gray. Other edges are dashed; extensions are dotted.

(a) There are no intersections between e_r and any uninvolved edge.
(b) Each involved virtual edge connects to or can be extended to e_r.
(c) Extensions of involved virtual edges do not intersect.
(d) There are no vertices in the regions enclosed by e_1, e_2 and e_r.
(e) The order of incident edges around each involved vertex is maintained.
(f) The order of virtual vertices along e_r and along the virtually involved edges is maintained.

To quickly update admissibility, we maintain for each operation a set of edges that cause inadmissibility: an operation is admissible if this set is empty.

Cycles to Circles. When the start and end vertex of an operation are the same, the definitions for the operations can become degenerate. We introduce an additional operation for this case. Let v be the vertex being removed by the operation and u be the other endpoint of both edges. We define a set A of *anchor points* that the replacement circle needs to intersect. A consists of all junctions along both edges, possibly including u or v, that cannot be moved. If A has three or more points all points are required to be co-circular, uniquely defining a replacement circle; otherwise no replacement is possible. If A has two or less points, we extend A by up to two elements by adding u and, if required, v. We regularly sample tangents angles around the connecting chord defined by these two points, defining a set of possible replacement circles.

2.3   Analysis

We now analyze the asymptotic execution time of our schematization algorithm. Let n denote the number of vertices in the input and h the number of junctions. We first determine the time required to compute a single candidate operation. To compute the cost, we compute the Fréchet distance between a single arc and a polygonal line of at most $O(n)$ vertices in $O(n \log n)$ time [22]. For admissibility, we test whether the candidate arc intersects any other arc, requiring $O(n)$ time. If the arcs being replaced contained virtual vertices, then the connected arcs might need to be extended and could start to intersect. There are at most $O(h)$ of these arcs, which are ordered along the replaced arcs.

We need to test only intersections between arcs that are adjacent in this order which takes $O(h)$ time. In addition, for up to four arcs (the first and last on either side of the replacement pair), we check whether they intersect any other arc, taking $O(n)$ time. Thus, computing the cost and admissibility of a single operation takes $O(n \log n)$ time.

At initialization of the algorithm we compute all operations. Since each candidate represents only two edges, we compute the Fréchet distance in $O(1)$ time and thus the initialization takes $O(n^2)$ time. Performing an operation may change the geometry or admissibility of other operations. The geometry of at most $O(h)$ edges changes. Hence, we compute at most $O(h)$ new operations in $O(hn \log n)$ total time. Moreover, we remove the old edges from the sets of edges causing inadmissibility and insert the new edges where necessary. This also takes $O(hn \log n)$ time. Thus, the complete algorithm runs in $O(n^2 h \log n)$ time.

Solution Tree. A sample of the solutions generated by our algorithm for different complexities is shown in Fig. 4. While the complexity of the map in Fig. 4(b) is less than half of the input complexity, the effect of schematization is not very noticeable. On the other hand, the map in Fig. 4(h) is highly schematized, but is also geometrically heavily distorted. The results in between make a trade-off between schematization and geometric accuracy. The optimal trade-off for a schematic map depends on its size, content and purpose.

Since the optimal number of arcs may not be clear a priori, it is desirable to be able to interactively explore schematizations with different complexity. To this end, we maintain not only the current schematic network N, but also some additional structure

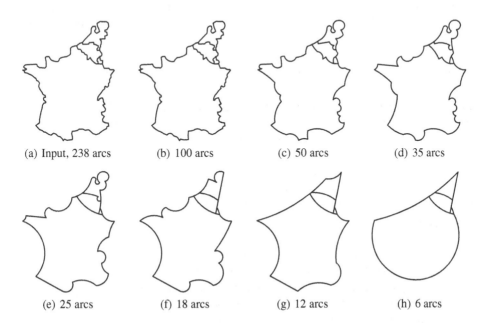

(a) Input, 238 arcs	(b) 100 arcs	(c) 50 arcs	(d) 35 arcs
(e) 25 arcs	(f) 18 arcs	(g) 12 arcs	(h) 6 arcs

Fig. 4. Sample of the possible schematizations for a network that represents Belgium, France, Luxembourg and the Netherlands

that allows us to efficiently recover other intermediate solutions. Two simple options for this structure are an operation list or an operation tree. Both approaches use $O(n)$ space and take $O(nh)$ time to recover an intermediate solution. We use a tree as the expected runtime is lower for schematizations of low complexity. Original arcs are stored in leaves and parent nodes contain changes made by an operation. With every node the current complexity is stored.

2.4   Extensions for Metro Maps

We describe four extensions which are geared towards metro-map construction.

Distributing Stations. A typical metro network consists of a city center that has many highly-connected stations and some lines that go to suburban areas with fewer stations. Keeping stations near their geographic position causes problems due to the high-density city center. We want to distribute the stations more evenly across the drawing, which increases the scale of the center and decreases the scale of the suburbs: this improves readability [23] and helps the schematization. Distributing the stations necessarily distorts the geography of the network; to retain as much geography as possible, we use minimum-distortion focus maps [24]. We set the desired scale factor of a station v to $1 + c \cdot k_v$, where c is a constant and k_v is the number of stations within a disc of some radius d around v. We chose c and d such that the assigned scale factors range approximately up to 2. Note that this sets all scale factors to at least 1.

Stroke Partition Revisited. For metro maps, it is desirable for a single line to continue smoothly at an interchange, even if the angular deviation is high in the original geography. Hence, we change the stroke partition to a two-step process. In the first step, we aggregate adjacent metro connections having the same set of metro lines into preliminary strokes. If a metro line has multiple branches, we aggregate based on the smallest angular deviation. In the second step, we aggregate the preliminary strokes using angular deviation as before.

Interchanges. In the final metro map, interchanges are drawn as circles with some given radius r. Hence, the lines at the interchange do not have to intersect in a single unique point: we may admit some leeway depending on r. In particular, we maintain for an interchange a smallest enclosing disk of the intersections of the incident lines. Topological errors within the disk are allowed, since drawing the disk hides them. We constrain this flexibility by maintaining the order of the incident lines around the boundary of the disk. Moreover, we bound the maximum radius of the smallest enclosing disk by r.

Rendering. To visualize a schematic metro map, we draw the metro lines in appropriate colors. Where multiple metro lines connect two stations, we draw parallel metro lines. The use of circular arcs makes this comparatively simple: we use concentric arcs in such a situation, with slightly varying radius. The ordering of such parallel connections is a problem in itself (see e.g. [25]), one we do not consider here. Though algorithms exist, we manually set the order in our results. Aside from visualizing the metro lines, a metro map should also have vertices for each station. During rendering, we hence reinsert the degree-2 vertices removed during schematization by distributing them evenly along the appropriate edge.

3   Results

We consider three different use cases: territorial outlines, road networks, and metro maps. Our algorithm produces high-quality results for both territorial outlines and metro maps. For road networks, a high complexity reduction can be obtained though the resulting maps do not convey their message well.

Territorial Outlines. A territorial outline represents administrative and geographic boundaries. Examples include country and province borders, or shorelines. Territorial outlines are typically low-density networks and contain a low number of junctions. Often, features can be combined through multiple junctions, see, for example, the combined shoreline of France, Belgium and the Netherlands in Fig. 5(a). Such features should be recognized and schematized as a single arc, since otherwise the saliency of junctions incorrectly increases.

Fig. 5(b) shows the result of our algorithm for a map of the European Union. Note that many of the coastline features have been replaced by single arcs. By schematizing across junctions, we reinforce the importance of the geographic features. For comparison, in Fig. 5(c) the result of the circular-arc schematization method by Van Goethem *et al.* [19] is shown. Here, junctions are fixed and no operations are performed on edges that meet at such a vertex. As a consequence, the junctions become more important in the final map, which is often undesirable. It also prevents the schematization from reaching the same low complexity, as vertices that are close to each other require small details to be maintained.

In contrast to our method, the algorithm by Van Goethem *et al.* [19] maintains the area of each country. This ensures that countries maintain their relative size, which is not guaranteed in our method (e.g. compare Luxembourg, Denmark and the Netherlands). Also, due to the shifting of junctions, some borders can become very short, such as the boundary between Switzerland and Austria. Though in theory the result is topologically equivalent, drawing the actual geometry with nonzero-width lines causes very small elements to be obscured. We can extend the admissibility of operations to include a

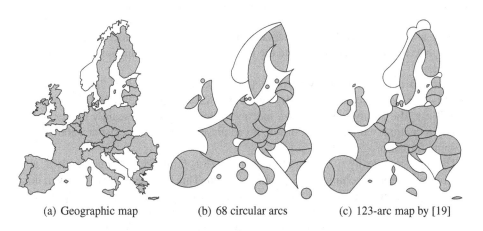

(a) Geographic map   (b) 68 circular arcs   (c) 123-arc map by [19]

Fig. 5. Schematizing the borders of the European Union, augmented with shorelines

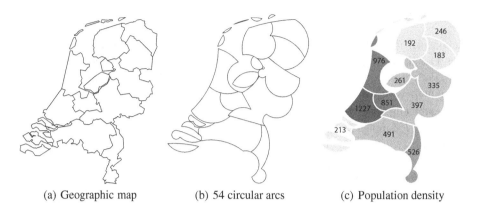

(a) Geographic map   (b) 54 circular arcs   (c) Population density

Fig. 6. Schematizing the province boundaries in Netherlands

check for a minimum distance between affected boundaries. The output then adheres to these minimal distances, assuming the input also does.

Fig. 6 shows a schematization of the provinces of the Netherlands. The ability to schematize across junctions helps to capture both the western coastline feature and the typical shape of the north-east border. Even though this is a highly schematized version of the Netherlands, all boundaries are still geometrically reasonably accurate. As most borders between provinces are represented by a single or a few arcs, the result gives a very stylized impression.

Road Networks. Road networks are typically very densely connected with many junctions. In contrast to territorial outlines and metro maps, however, the geometry of a road implicitly correlated to the type of road. A wiggly mountain road that has been "schematized" to a smooth curve could easily be misinterpreted as a highway. To maintain this implicit correlation, a form of caricature is often required when schematizing roads [26]. By exaggerating the features of roads, the association to their respective road types is maintained. Due to our geometric approach, our algorithm is unlikely to produce a road map that has a "schematic appearance" in this sense. Regardless, we investigate road networks to evaluate the possible complexity reduction in very dense networks.

The road network around Würzburg, Germany, is shown in Fig. 7(a). Our algorithm is able to reduce the complexity by roughly 82%, from 2965 to 540 circular arcs (Fig. 7(b)). The ability to schematize junctions allows for a significantly higher complexity reduction: without this addition, no admissible operations exist at 1566 edges stopping further progress (Fig. 7(c)).

The high complexity of the input map in Fig. 7(a) limits the schematized look attainable even though complexity is greatly reduced. This high density is, however, not the main problem we encounter when schematizing road networks. We also apply our algorithm on a generalized version of the same road network (see Fig. 7(d)). While the schematization is more pronounced in these maps (see Fig. 7(e)), local roads in the input map have been schematized into single, smooth arcs. As a consequence it

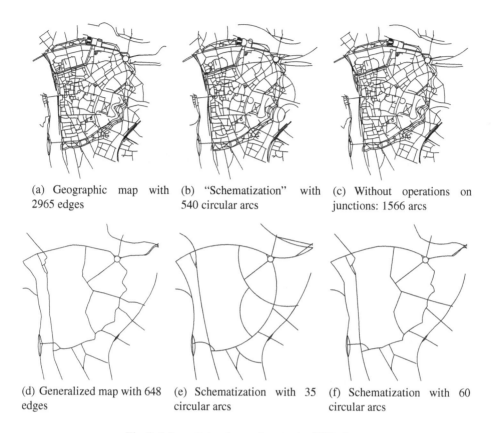

(a) Geographic map with 2965 edges

(b) "Schematization" with 540 circular arcs

(c) Without operations on junctions: 1566 arcs

(d) Generalized map with 648 edges

(e) Schematization with 35 circular arcs

(f) Schematization with 60 circular arcs

Fig. 7. Schematizing the road network of Würzburg

is impossible to distinguish large through roads from local roads without additional knowledge. Schematizations using a higher complexity may give more reasonable results (see Fig. 7(f)), but they require a lower complexity reduction limiting the effectiveness of the schematization.

Metro Maps. The networks of metro maps are usually heavily schematized and have a matching rendering style. As metro maps contain mainly connectivity information, caricature of actual lines is not important. Thus, we do not expect similar recognition problem as with road networks. We computed our results using the extensions described in Section 2.4.

Our results for Vienna are shown in Fig. 8(a)–(c). Distorting the input geometry creates extra space in the center of the map, thus allowing for a better schematization of the network. The circular arcs create a sense of continuity along the metro lines, while giving a very stylized appearance. Fig. 8(d) shows the same network drawn by the algorithm of Fink *et al.* [20]. Their use of Bézier curves leads to a smooth drawing, but has less of a stylized appearance.

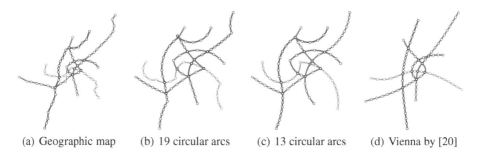

(a) Geographic map   (b) 19 circular arcs   (c) 13 circular arcs   (d) Vienna by [20]

Fig. 8. Schematizing the metro network of Vienna

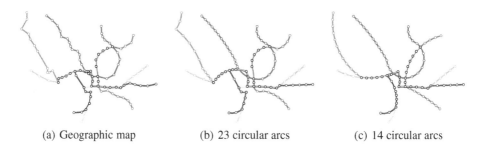

(a) Geographic map   (b) 23 circular arcs   (c) 14 circular arcs

Fig. 9. Schematizing the metro network of Washington, DC

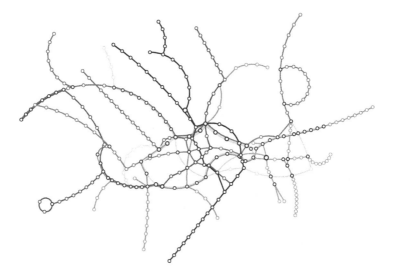

Fig. 10. Schematization of the metro network of London with 70 circular arcs

In Washington (Fig. 9) multiple metro lines connect the same stations. The stroke partition merges connections where the same lines run in parallel and determines continuity of lines at the junctions (interchanges). This increases the continuity of metro lines at junctions.

As a final example, we show the metro map of London computed by our algorithm (Fig. 10). Despite its complexity, the algorithm is able to schematize this network using a comparatively low number of arcs. A number of stations are placed very close to one another and overlap when they are drawn as circles. This reduces the legibility of the map. Future work may investigate ways of appropriately dealing with distance constraints between stations and lines.

Complexity Reduction. Our algorithm has different features, but it is unclear how they affect the minimum complexity that can be obtained. So far we mainly focussed on the ability to schematize across junctions. For metro maps we also introduced automated distortion and the ability to treat interchanges as discs instead of points. Here we briefly investigate the effects on the minimum complexity. Table 1 presents a summary. The additions are cumulative from left to right: the last three columns also include operations on junctions; the last two columns also includes distortion. While the lowest complexity is not necessarily desirable, it gives an indication of the ability of our algorithm to obtain abstract low-complexity schematizations. Many of the figures in this paper use a few arcs more than the minimum, since the lowest attainable complexity may cause undesirable deformations (see Fig. 4(h)).

As expected, the ability to schematize across junctions has a large effect on the complexity. In contrast, both the distortion of the input map and the leeway at interchanges, appear to have only a minor effect on the minimum complexity (if any). However, the complexity of the schematization does not capture all aspects. Distorting the input network, for example, greatly enhances the overall spacing, avoiding visual clutter and increasing legibility.

Table 1. Minimum complexity of schematization achievable with our various additions. The last two additions are used only for metro maps.

Map	Input	Basic	Junctions	Distortion	Interchanges as disks
Europe	1669	105	56		
Netherlands	494	54	42		
Würzburg	2965	1566	540		
Vienna	90	25	12	12	11
Washington	99	22	12	12	12
London	339	128	72	69	67

4   Discussion and Future Work

The experimental results in Section 3 appear satisfactory in most cases, though room for improvement remains. In this section, we review and discuss some of the steps we took to design our algorithm.

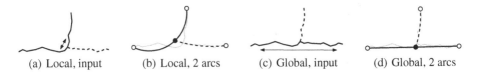

(a) Local, input   (b) Local, 2 arcs   (c) Global, input   (d) Global, 2 arcs

Fig. 11. Different local and global continuity depending on the strokes detected

We use a stroke partition to decide how edges are combined across junctions. This use of strokes has both benefits and drawbacks. Advantages of our partitioning method are its efficiency and simplicity. The main drawback is that the criterion is local and this may negatively impact schematization: local continuity need not correspond to continuity from a global perspective (see Fig. 11). Without a stroke partition we could determine continuation on-the-fly by computing the operations for any pair of edges meeting at a junction. The problem is that once continuation across a junction has been decided it is irreversible. As small operations are executed in order of similarity, it is likely that local geometry still prevails. An advantage of an explicit stroke partition is that we may improve on this independently. Moreover, the continuation may carry information in some networks (e.g. metro networks).

Our algorithm allows for easy integration of both hard constraints and soft constraints. For hard constraints, the admissibility of operations should be modified accordingly. We enforced correct topology, but in addition we could also consider requiring some minimal distance between different strokes. Note that, whereas correct topology is initially guaranteed, this need not be the case for such a minimal distance. In such cases, the algorithm cannot guarantee that the constraint holds for the result; it may even prevent any complexity reduction in areas that violate the constraint initially. This would, for example, also occur when requiring strict smooth continuity at each vertex. To include additional soft constraints, the weighting of operations can be modified. For example, weights can be modified based on the resulting circular arc (e.g. [19]) or on the location of the operation (similar to [27]). We must be careful when combining different soft constraints, to avoid an average solution that is worse than either solution separately. This may occur, for example, if we were to combine a measure of parallelism to other arcs with geometric similarity: the resulting geometry does not exhibit parallelism nor does it represent the geographic situation well.

Our results show that road networks appear unsuitable for purely geometric schematization. This is in line with previous work, stating that road networks require caricature [26]. That this does not cause similar problems for territorial outlines is likely due to the inherent expectations of such geographic boundaries. Country and province borders are mostly not smooth, low-complexity curves. Therefore, observing a low number of circular arcs is a strong indication that the map is schematized. A smooth road on the other hand is not unusual and may cause associations with different road types. Interestingly enough, this assumption for territorial outlines is not always valid: for example, many of the state boundaries in the US are of very low complexity. This raises the question if a higher level of schematization is needed to attain the same perceived level of schematization (see Fig. 12).

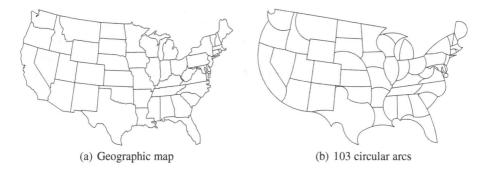

(a) Geographic map   (b) 103 circular arcs

Fig. 12. Schematization of the USA with 103 circular arcs. The complexity of geographic shapes may affect the perception of schematization.

Schematization with circular arcs appears effective in metro maps. Metro maps mainly focus on connectivity and the abstract style of circular arcs fits this purpose well. Maintaining continuity across junctions helps reinforce the structure of the network. The preliminary results from this paper should, however, be further validated in future work. Research into the usability of curved metro maps was recently started [14], but a further study of the effects is required, also for different map types. For metro maps it would also be interesting to see if the property of continuity at junctions can be exploited further. Strokes might be required to always continue smoothly at junctions. This would improve legibility at these positions, be it at the cost of maintaining more degree-2 vertices. The increase in complexity may, however, detract from the visual clarity obtained through extensive schematization. Moreover, as an edge may represent multiple metro lines, it may be infeasible for sufficient lines to continue smoothly at these vertices.

Lastly, we discuss the effect of applying an iterative algorithm. Iteratively selecting the best operation ensures a fast and comparatively simple algorithm. However, it does not guarantee that the obtained result is optimal. Also, it may be rather unstable: a minor change in the input may greatly affect the output. Stability is important for networks that change over time. Though the network changes, the schematic maps should remain similar if possible. Ideally, one would design a stable algorithm that computes an optimal schematic map, e.g. the map with highest similarity given some maximum complexity. However, it is likely that such an algorithm has significantly higher computation time.

5 Conclusion

We presented an algorithm for circular-arc schematization of geographic maps. Our algorithm is able to schematize across junctions (vertices of degree three or higher). Allowing junctions to be shifted makes the schematization highly flexible, which enhances its quality. We did preliminary experiments for three different use cases to test the effectiveness of our algorithm. The results obtained for both territorial outlines and metro maps appear of high quality. Extensive features can be represented with a single arc or just a few and even dense networks can be schematized effectively and efficiently. Though for road networks the complexity can be reduced significantly, the lack of caricature makes the result less effective as a schematization. We also briefly evaluated the potential complexity reduction attainable using the different features of our algorithm. From this evaluation we conclude that the ability to schematize across junctions is the most significant feature that allows for a high complexity reduction.

References

1. Cabello, S., De Berg, M., Van Kreveld, M.: Schematization of networks. Comp. Geometry: Theory & Appl. 30(3), 223–238 (2005)
2. De Chiara, D., Del Fatto, V., Laurini, R., Sebillo, M., Vitiello, G.: A chorem-based approach for visually analyzing spatial data. J. of Visual Languages & Computing 22(3), 173–193 (2011)
3. Merrick, D., Gudmundsson, J.: Path simplification for metro map layout. In: Kaufmann, M., Wagner, D. (eds.) GD 2006. LNCS, vol. 4372, pp. 258–269. Springer, Heidelberg (2007)
4. Neyer, G.: Line simplification with restricted orientations. In: Dehne, F., Gupta, A., Sack, J.-R., Tamassia, R. (eds.) WADS 1999. LNCS, vol. 1663, pp. 13–24. Springer, Heidelberg (1999)
5. Brunet, R.: La composition des modèles dans l'analyse spatiale. Espace géographique 9(4), 253–265 (1980)
6. Buchin, K., Meulemans, W., Speckmann, B.: A new method for subdivision simplification with applications to urban-area generalization. In: Proc. 19th ACM SIGSPATIAL GIS, pp. 261–270 (2011)
7. Nöllenburg, M., Wolff, A.: A mixed-integer program for drawing high-quality metro maps. In: Healy, P., Nikolov, N.S. (eds.) GD 2005. LNCS, vol. 3843, pp. 321–333. Springer, Heidelberg (2006)
8. Stott, J., Rodgers, P., Martinez-Ovando, J., Walker, S.: Automatic metro map layout using multicriteria optimization. IEEE TVCG 17(1), 101–114 (2011)
9. Koffka, K.: Principles of Gestalt psychology. Routledge (2013)
10. Delling, D., Gemsa, A., Nöllenburg, M., Pajor, T.: Path schematization for route sketches. In: Kaplan, H. (ed.) SWAT 2010. LNCS, vol. 6139, pp. 285–296. Springer, Heidelberg (2010)
11. Gemsa, A., Nöllenburg, M., Pajor, T., Rutter, I.: On d-regular schematization of embedded paths. In: Černá, I., Gyimóthy, T., Hromkovič, J., Jefferey, K., Královič, R., Vukolić, M., Wolf, S. (eds.) SOFSEM 2011. LNCS, vol. 6543, pp. 260–271. Springer, Heidelberg (2011)
12. Cicerone, S., Cermignani, M.: Fast and simple approach for polygon schematization. In: Murgante, B., Gervasi, O., Misra, S., Nedjah, N., Rocha, A.M.A.C., Taniar, D., Apduhan, B.O. (eds.) ICCSA 2012, Part I. LNCS, vol. 7333, pp. 267–279. Springer, Heidelberg (2012)
13. Ti, P., Li, Z.: Generation of schematic network maps with automated detection and enlargement of congested areas. Int. J. of Geographical Information Science 28(3), 521–540 (2014)
14. Roberts, M., Newton, E., Lagattolla, F., Hughes, S., Hasler, M.: Objective versus subjective measures of Paris metro map usability: Investigating traditional octolinear versus all-curves schematic maps. Int. J. of Human Computer Studies 71, 363–386 (2013)
15. Reimer, A.W.: Understanding chorematic diagrams: Towards a taxonomy. Cartographic J. 47(4), 330–350 (2010)
16. Drysdale, R., Rote, G., Sturm, A.: Approximation of an open polygonal curve with a minimum number of circular arcs and biarcs. Comp. Geometry: Theory & Appl. 41(1-2), 31–47 (2008)

17. Heimlich, M., Held, M.: Biarc approximation, simplification and smoothing of polygonal curves by means of Voronoi-based tolerance bands. Int. J. of Comp. Geometry & Appl. 18(3), 221–250 (2008)
18. van Goethem, A., Meulemans, W., Reimer, A., Haverkort, H., Speckmann, B.: Topologically safe curved schematisation. Cartographic J. 50(3), 276–285 (2013)
19. van Goethem, A., Meulemans, W., Speckmann, B., Wood, J.: Exploring curved schematization. In: Proc. 7th IEEE PacificVis., pp. 1–8 (2014)
20. Fink, M., Haverkort, H., Nöllenburg, M., Roberts, M., Schuhmann, J., Wolff, A.: Drawing metro maps using Bézier curves. In: Didimo, W., Patrignani, M. (eds.) GD 2012. LNCS, vol. 7704, pp. 463–474. Springer, Heidelberg (2013)
21. Thomson, R., Brooks, R.: Exploiting perceptual grouping for map analysis, understanding and generalization: The case of road and river networks. In: Blostein, D., Kwon, Y.-B. (eds.) GREC 2001. LNCS, vol. 2390, pp. 148–157. Springer, Heidelberg (2002)
22. Rote, G.: Computing the Fréchet distance between piecewise smooth curves. Comp. Geometry: Theory & Appl. 37(3), 162–174 (2007)
23. Merrick, D., Gudmundsson, J.: Increasing the readability of graph drawings with centrality-based scaling. In: Proc. 2006 Asia-Pacific Symp. Info. Vis., pp. 67–76 (2006)
24. van Dijk, T.C., Haunert, J.H.: Interactive focus maps using least-squares optimization. Int. J. of Geographical Information Science (2014), doi:10.1080/13658816.2014.887718
25. Nöllenburg, M.: An improved algorithm for the metro-line crossing minimization problem. In: Eppstein, D., Gansner, E.R. (eds.) GD 2009. LNCS, vol. 5849, pp. 381–392. Springer, Heidelberg (2010)
26. Lecordix, F., Plazanet, C., Lagrange, J.P.: PlaGe: a platform for research in generalization. application to caricature. Geo Informatica 1(2), 161–182 (1996)
27. van Dijk, T.C., Van Goethem, A., Haunert, J.H., Meulemans, W., Speckmann, B.: Accentuating focus map via partial schematization. In: Proc. 21st ACM SIGSPATIAL GIS, pp. 438–441 (2013)

Travel-Time Maps: Linear Cartograms
with Fixed Vertex Locations*

Kevin Buchin[1], Arthur van Goethem[1], Michael Hoffmann[2],
Marc van Kreveld[3], and Bettina Speckmann[1]

[1] Technical University Eindhoven, Eindhoven, The Netherlands
[2] ETH Zürich, Zürich, Switzerland
[3] Utrecht University, Utrecht, The Netherlands

Abstract. Linear cartograms visualize travel times between locations, usually by deforming the underlying map such that Euclidean distance corresponds to travel time. We introduce an alternative model, where the map and the locations remain fixed, but edges are drawn as sinusoid curves. Now the travel time over a road corresponds to the length of the curve. Of course the curves might intersect if not placed carefully. We study the corresponding algorithmic problem and show that suitable placements can be computed efficiently. However, the problem of placing as many curves as possible in an ideal, centered position is NP-hard. We introduce three heuristics to optimize the number of centered curves and show how to create animated visualizations.

1 Introduction

Most people depend on maps for navigation. Regular maps, however, do not ensure that time and distance correlate equally across the map. A village just on the other side of a mountain range might be hours away, whereas a city miles away is only five minutes driving. Temporal conditions, such as traffic jams or roads blocks, can make these effects even more pronounced.

To counteract this, visual cues are used to display (relative) travel times, commonly using colors (see Fig. 1 (a)). Size, however, is a better means to visualize numerical attributes [1]. As an alternative to color, length could also be used to show travel time on stretches of road. As a visual variable, length has a better association to quantity than color. "Relative lengths" of stretches are immediately quantified, whereas "relative colors" do not have a clear interpretation. These observations have led to the development of *linear cartograms* [2–5].

Linear cartograms try to display time more clearly by distorting the base map. Two types of linear cartograms exist: centered and non-centered. The former has a "center" location and only distances to this location correspond to actual travel time. The latter type attempts to have all pairs of locations at travel-time-proportional distances, which

---

* K. Buchin, A. van Goethem, and B. Speckmann are supported by the Netherlands Organisation for Scientific Research (NWO) under project no. 612.001.207 (KB), no. 612.001.102 (AvG), and no. 639.023.208 (BS). M. Hoffmann is partially supported by the ESF EUROCORES programme EuroGIGA, CRP GraDR and SNF Project 20GG21-134306.

M. Duckham et al. (Eds.): GIScience 2014, LNCS 8728, pp. 18–33, 2014.

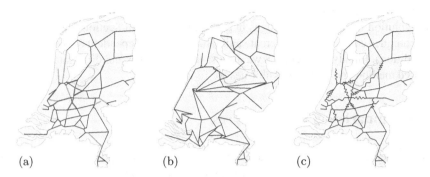

Fig. 1. Expected travel times in the Netherlands during morning rush hour. (a) Color coding. (b) Linear cartogram. (c) Linear cartogram with fixed vertex positions.

is generally not possible without error. By distorting the map, linear cartograms give a more intuitive sense as to what is nearby, in time, on the map. The distortion of the map, however, can make recognition and usage of the cartogram harder [6]. Items on the map might be far removed from their position on the base map, making it hard to find specific items (see Fig. 1 (b), which was computed with the method described in [7]).

We introduce an alternative model that is well suited to visualize travel times on road networks. Instead of distorting the base map, we keep the locations fixed and "distort" only the edges. We do so by using sinusoid curves (see Fig. 1 (c)). Our approach has the advantage that the base map remains undistorted, while length can still be used to quantify and visualize travel time. The resulting maps create a dramatic effect and can also be used in animations, where an increased travel time (delay) is shown by an increased amplitude or frequency. Of course the curves might intersect if not placed carefully. We study the algorithmic problem of generating crossing-free linear cartograms according to our new model.

Related Work. In addition to the results on linear cartograms [2–5], several other papers are also related. Weights of edges (but for very different applications) can also be visualized by the width [8]. When drawing planar graphs with fat edges, the occupation of space by these edges is the main concern. Drawing edges with curves (e.g., [9]) has received considerable attention. Lately, more specifically, the use of circular arcs for edges has received attention (e.g., [10]), in particular the creation of Lombardi drawings [11]. The use of regular sinusoid curves to indicate relative length was recently introduced by Nielsen *et al.* [12] for the visualization of connectivity graphs in genome sequencing. Lastly we note the topic of map labeling, where a suitable position of each label must be found among a set of candidate positions (see [13] for a survey). In particular the edge labeling version studied in [14, 15] is closely related (see Section 4).

Organization. Section 2 explains the model used to find suitable curves, and how to fit cubic Bézier splines. Section 3 discusses the optimal placement of curves, but also shows the restrictions of this approach. In Section 4 we prove that under a mild realistic input assumption, a non-overlapping choice of placement of the curves can be computed in $O(n \log n)$ time, if such a placement exists. We also study the problem of maximizing the centered curve positions under the condition that all curves can be placed, which is NP-hard (Section 5). Section 6 discusses a heuristic approach to compute a solution maximizing the edges in centered position. In Section 7 we show how our techniques can be applied to create animations of time-dependent data such as traffic conditions. Finally, in Section 8 we discuss the advantages and disadvantages of the proposed method and look at possibilities for future work.

2   Preliminaries on Fitting Curves

Problem Setting. We assume a planar graph with a fixed embedding is given, along with the travel times for all edges. We compute a linear cartogram with the same topology and embedding, where all edges are drawn as sinusoid curves whose lengths are proportional to the specified travel times. For each edge e we define a region close to e and draw a curve with the specified length inside that region. To avoid intersections among the curves, we make sure that regions of different edges do not intersect. We consider three possible placements of the regions: above the edge, centered, or below the edge (see Fig. 2). When centered, the curve occupies a diamond-shaped region where the edge is a diagonal of the diamond. In the other two positions, the curve occupies a triangle-shaped region based on the edge. To minimize visual distortion we prefer to place curves in the centered, diamond-shaped position. Reducing the problem solely to the centered positions, however, is overly restrictive, as shown in Section 3.

We call the length of the diagonal of each region that is normal to its edge e the *width* w_e. The width is directly determined by the length of the edge, the associated travel time and the desired frequency. The two triangles and diamond of an edge induce four parts the region could occupy, called *zones* (see Fig. 2 (d)). The vertex of a triangle or diamond that is not part of the edge is called the *apex*. Any edge is associated with four apices, one of each triangle and two of its diamond.

Different Shapes. We represent the distorted edges with C^2-continuous cubic Bézier splines. This type of curve is already commonly present in many maps (e.g., [16]) and a continuous spline maintains continuity of edges. We fit the Bézier spline inside a zone representing the widened edge. This zone can be represented by various shapes. These shapes are not present in the final map and hardly influence the visual appearance (see Fig. 3). However, since space around vertices is limited, tapering shapes, such as triangles, are less likely to intersect at vertices and hence allow for a greater range of feasible solutions. Non-uniform shapes, such as a rectangle in the center of an edge, are also suitable.

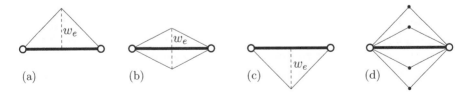

Fig. 2. (a)–(c) The regions with width w_e for an edge e. (d) The zones and apices of e.

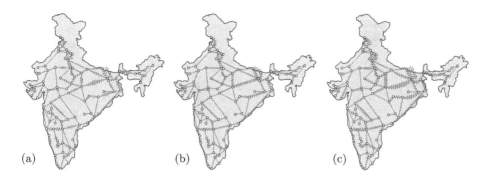

Fig. 3. Different edge shapes for the Indian railroad network. Frequencies are exaggerated for display purposes. (a) Triangles. (b) Rectangles. (c) Ellipsoids.

We study the algorithmic aspects of fitting curves using triangular and diamond-shaped areas. Both the algorithm of Section 4 and the NP-hardness proof of Section 5 also hold for uniform rectangular shapes. Our algorithm does not work for ellipses, however, in this case there is a trivial $O(n^2)$ algorithm.

Relating Width to Curve Length in the Triangle Model. One can fit a C^2-continuous sinusoid cubic Bézier spline such that the length of the fitted curve is (nearly) linearly proportional to the width of the edge. The sinusoid spline starts and ends at the end-points of the edge, so the number of oscillations is a multiple of one half. We let the number of oscillations used depend on the edge length and the specified travel time and we either keep the length of all oscillations (the frequency) or the edge width equal across the network.

Each oscillation is represented by two cubic Bézier curves. The control points of the Bézier curves are evenly distributed along the outer edge of the zone (Fig. 4 (a)) to ensure that the resulting spline uses the full available area. As the control points are equally spaced and the tangents of two consecutive curves are aligned, the connection between consecutive Bézier curves is C^2-continuous. A degenerate case occurs when two curves c_1 and c_2 connect at the apex p (Fig. 4 (b)). To keep the connection C^2-continuous, we select as the connection point the center of the third control point of c_1 and the second of c_2 (Fig. 4 (c)).

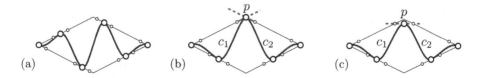

Fig. 4. (a) A sinusoid curve that is fitted to a diamond-shaped area. (b) The spline is not C^2-continuous as the tangents of curve c_1 and c_2 do not line up. (c) The continuity is restored when point p is moved down.

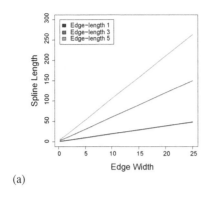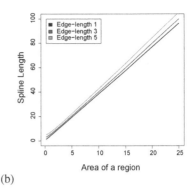

(a)   (b)

Fig. 5. (a) Relation between the width of an edge and the length of the fitted spline for different edge lengths. (b) Relation with the size of the area that is fitted.

The exact length of a cubic Bézier curve cannot be computed by a closed formula, but it can be ϵ-approximated by regularly sampling the curve. In Fig. 5 (a) and 5 (b) the relation between the width, respectively area, of a zone and the length of the fitted Bézier spline is plotted for different edge lengths. We note that the relationship is nearly linear.

3   Optimal Assignment of Centered Curves

To minimize the distortion introduced by lengthening edges, curves should preferably be centered on the edge they represent. Here we explore the effect of placing all curves in their centered region. We assume that edge length and travel time (the desired edge length) are given as input variables. This leaves two variables that can be used to fit curves of the correct length: edge width (amplitude) and curve frequency.

To minimize the number of visual variables in the map, we fix either the frequency or the edge width in the network. This directly determines a function relating the required curve length to the opposing variable. When fitting curves, there should be no overlap between different curves, and thus regions.

By fixing a uniform edge width across the network, the length of a curve is directly related to the number of oscillations on the edge. Using basic geometry we can compute the maximum edge width that still allows a non-overlapping solution. A solution with a maximum width has a minimum frequency, which may be preferable to distinguish oscillations (see Fig. 6 (a)). Any solution with a smaller width can be obtained through scaling.

Instead of fixing the edge width, we can also fix the curve frequency for all edges. This directly implies an edge width for all edges, though a too low frequency may cause overlap between regions. To prevent overlap, we compute the minimum feasible uniform frequency (see Fig. 6 (c)). As frequency maintains the relative edge widths, we can apply similar geometry as before.

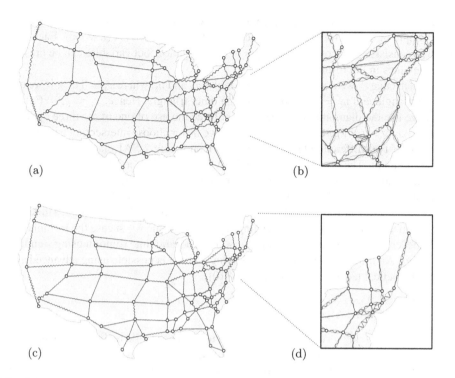

Fig. 6. Expected travel times on the main highway network in the USA. The dense area in the East of the USA overly restricts the results in the rest of the map. Placement regions are indicated in the insets. Bottleneck regions are indicated in red. (a) Using equal edge widths. (b) Using equal frequency.

Critical areas with a high edge density are often restricted to only a small section of the map. The high edge density in combination with a high likelihood of delays, however, can cause the minimum frequency for the entire map to be highly constrained. We can add flexibility to the solution by also using the two outer triangular regions to place the curve, instead of only the centered diamond. This allows solutions with a lower minimum frequency, but changes the problem into an assignment problem which we discuss in the next section.

4   Efficiently Computing Placements of Curves

For each edge we have three candidate regions, a diamond and two triangles, and we must choose one per edge so that the choices do not intersect. Essentially the same problem was studied by Poon *et al.* [14] for rectilinear map labeling. They showed that the problem of selecting non-overlapping regions can be transformed to a 2-satisfiability (2-SAT) instance. In a 2-SAT instance a Boolean formula is given that contains a conjunction of

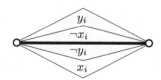

Fig. 7. Assignment of variables to the zones of an edge

disjunctions, where each disjunction consists of two Boolean variables or their negation. An example of such a formula would be $(\neg a \vee b) \wedge (\neg c \vee \neg b) \wedge (a \vee \neg c)$. Each disjunction consists of at most two variables or their negation, but variables can be present multiple times in the formula.

For each edge e_i we use two Boolean variables x_i and y_i, where $x_i = \text{TRUE}$ if we use the one triangle and $y_i = \text{TRUE}$ if we use the other triangle. Hence, $x_i = y_i = \text{FALSE}$ means we use the diamond (see Fig. 7). To enforce the use of one of the three options, we add $(\neg x_i \vee \neg y_i)$. Moreover, to ensure that we never choose intersecting triangles or diamonds of two different edges, we make corresponding 2-SAT clauses representing this requirement. The conjunction of all 2-SAT clauses gives a 2-SAT formula that is satisfiable if and only if there is a choice of regions, one per edge.

Satisfiability of 2-SAT formulas can be tested in time linear in their length [17]. The values of the Boolean variables show which placement to take for each edge. There can be quadratically many pairs of intersecting regions, so in the worst case the formula has quadratic length. This leads to a straightforward quadratic time solution using standard techniques. An improvement to $O(n^{4/3} \text{ polylog } n)$ time is possible using advanced and rather impractical techniques [15] (see also [10]).

4.1 Reduced Time Algorithm

We show that $O(n \log n)$ time can be achieved under a very mild realistic input assumption: for each edge, the apices of its triangles and diamond have angles at least β, for some constant $\beta > 0$. That is, regions may not be arbitrarily wide compared to their edge length. The improvement is based on computing an equivalent 2-SAT formula that has only linear length and that can be constructed in $O(n \log n)$ time. More precisely, the 2-SAT formula has $O(n/\beta^2)$ clauses and is constructed in $O((n \log n)/\beta)$ time. An overview is given in Algorithm 1.

Algorithm 1. `Compute area arrangement`(S, k)

1: Create $2\pi/\beta$ trapezoidal decompositions based on the edges.
2: Mark all zones that are intersected by an edge.
3: Create the arrangement of the valid zones.
4: Detect all overlaps between zones.
5: Generate the corresponding 2-SAT formula.

We first observe that if any edge e intersects any zone of a different edge e', then that zone cannot be used. Hence one or both Boolean variables of e' must be set a certain way to make the formula satisfiable. After finding and removing these zones, we build the arrangement of the remaining zones and show that it has complexity $O(n/\beta^2)$, implying that there will be at most that many clauses in the Boolean formula.

To determine all zones that intersect some edge, we use a number of fixed-direction ray shooting data structures in the set of edges (edges are disjoint since our input is planar). A fixed-direction ray shooting structure is simply a trapezoidal decomposition preprocessed for planar point location; it can be built in $O(n \log n)$ time and supports queries in $O(\log n)$ time. We choose a set D of $\delta = O(1/\beta)$ equal-spaced directions, ensuring that every apex of a triangle or diamond has a direction in D that points to its inside. For each direction in D we build a ray-shooting structure for that direction.

We perform δ ray-shooting queries from each endpoint p of each edge. Whenever the ray hits some edge e, we test if any zone of e contains p, and if so, we remove that zone. Then we perform another $4n$ ray-shooting queries, one for each apex of each edge. The direction of a ray r is toward the defining edge e. If r does not hit e as the first edge, then we can remove that zone of e. Together all ray-shooting queries identify all zones that cannot be used in a solution. This step takes $O((n \log n)/\beta)$ time in total.

We build the arrangement of all remaining zones to find intersecting pairs efficiently [18]. Vertices in the arrangement are either from the remaining zones or intersection points of remaining zones of different edges. The latter give rise to a clause to be included in the 2-SAT formula. We can show with a packing argument that the arrangement of the remaining zones has complexity $O(n/\beta^2)$ (see Section 4.2). Hence, we add at most a linear number of clauses to the 2-SAT formula.

Theorem 1. *Let E be a set of n disjoint edges in the plane, each with an isosceles triangle to one side, an isosceles triangle to the other side, and a diamond with the edge as its diagonal. If the apices of the triangles and diamond have an angle at least $\beta > 0$, then we can find in $O((n \log n)/\beta + n/\beta^2)$ time a choice of a triangle or diamond for each edge in E such that choices of different edges do not intersect.*

4.2 Packing Argument

We assume that all edges have at least an angle β at their apex and that β integrally divides 2π. If this is not the case we can set β to be the largest value smaller than β that does. We show that this requirement restricts the total number of overlaps possible between remaining zones. Recall that no remaining zone intersects an edge.

Lemma 1. *Let set S be a set of isosceles triangles intersecting an isosceles triangle T, where each element $s \in S$ has the same length sides as T. The total area that can be covered by S is bounded by $O(\|e\|^2)$, where $\|e\|$ is the length of a side of T.*

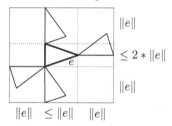

Proof. As all triangles considered are isosceles triangles, we know that if the sides have length $\|e\|$, the length of the base is bounded by $2 * \|e\|$. All points that can be covered by the intersecting triangles of T lie within a rectangle with sides proportional in e (see Fig. 8). Hence, the maximum area covered by intersecting isosceles triangles is bounded by $O(\|e\|^2)$.   □

Fig. 8. The maximum area that can be covered is bounded by $O(\|e\|^2)$

Lemma 2. *The number of remaining, larger isosceles triangles intersecting an isosceles triangle T with sides of length e is bounded by $O(1/\beta^2)$.*

Proof. Define for each intersecting triangle an angle γ corresponding to the counterclockwise angle between its edge when encountered clockwise and the unit vector $(1, 0)$. We partition all intersecting triangles by the angle γ into of a set of intervals $[\alpha, \alpha + \beta]$, where $\alpha \in \{i * \beta : i$ is integral and $0 \le i \le 2\pi/\beta - 1\}$. We bound the number of triangles within one interval by $O(1/\beta)$.

Within an interval, triangles cannot intersect. All triangles intersecting an edge have been filtered out in the first step and their relative direction lies within an interval of width β. All intersecting triangles are at least as large as T. We only look at the top part of the intersecting triangles up to a length e along the sides. The total area covered by all triangles within an interval $[\alpha, \alpha + \beta]$ is at most as large as the area covered by all intersecting triangles. By Lemma 1 this area is bounded by $O(e^2)$. As no triangles overlap and each triangle has a size of at least $\theta(e^2 * \beta)$, there can be at most $O(1/\beta)$ triangles in a partition.

As each partition has $O(1/\beta)$ intersecting triangles and there are $O(1/\beta)$ partitions, the number of triangles intersecting T is bounded by $O(1/\beta^2)$. $\square$

Lemma 3. *When all areas are valid and have at least an angle β at their apex, the number of overlapping pairs of areas is at most $O(n/\beta^2)$.*

Proof. For each pair of overlapping outer areas, we count the overlap towards the smaller triangle. By Lemma 2 it follows that the number of outer areas pairwise overlapping is bounded by $O(1/\beta^2)$ for each outer area. Thus, the total number of outer areas pairwise overlapping is at most $O(n/\beta^2)$. As the number of overlaps for the outer areas is bounded by $O(n/\beta^2)$, the same must hold for intersections with the inner area. The total number of areas overlapping is bounded by $O(n/\beta^2)$. $\square$

5 NP-Hardness

The algorithm in Section 4 computes a valid choice of regions if one exists. It does not, however, have any preference for what region is selected on an edge. Ideally we would want to maximize the number of regions placed in a centered position, the diamonds. Unfortunately, maximizing this number is NP-hard. We first prove that the problem is NP-hard even if all edges have a uniform length. Our construction, however, requires that some edges have a width to length ratio that is strictly larger than 1. We then prove that the problem is NP-hard even if all edges have an arbitrarily small width to length ratio. This, however, requires edges to have non-uniform lengths. Both versions of the problem are reduced from planar maximum 2-SAT [19]. We describe the *gadgets* that we use to encode variables, clauses, and wires between variables and clauses.

Bounded Length to Width Ratio. Fig. 9 illustrates the construction of the first version of the problem. The *variable gadget* consists of four edges whose diamonds intersect in a cycle (see Fig. 9 (a)). A variable gadget has two valid configurations, neither of which uses diamonds (see Fig. 9 (c)). These configurations encode the TRUE and FALSE states of the variable. The number of edges in a variable gadget can be increased to allow more wires to connect.

The *clause gadget* consists of two parallel edges with overlapping diamonds (see Fig. 9 (b)). The two incoming wires represent the two literals used in the clause. If a literal in a clause is FALSE (resp. TRUE), we ensure that the last edge in the wire gadget must choose (resp. need not choose) a triangle region intersecting the clause gadget.

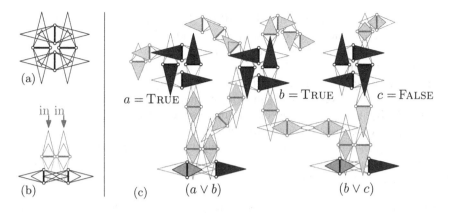

Fig. 9. Gadgets for variables and clauses, and solution for the formula $(a \lor b) \land (b \lor c)$

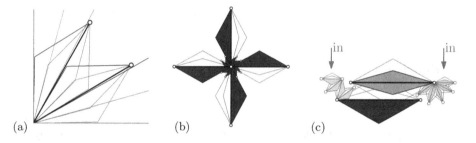

Fig. 10. Gadgets for second version: (a) zoomed in on part of the variable, (b) variable, (c) clause

Consequently, the diamond of the corresponding edge in the clause gadget cannot (resp. can) be selected (see Fig. 9 (c)). If both literals in a clause are TRUE, still only one of the diamonds can be selected as they overlap. Hence, if a clause is satisfied, the clause gadget can select one diamond; otherwise none.

The *wire gadget* consists of a sequence of edges with overlapping regions. Each diamond of an edge intersects the triangles of the previous and next edge. A wire gadget has two valid states: either all diamonds are selected or all triangles pointing from the variable to the clause. We connect each wire to a clause-variable pair such that the center regions of the wire can only be selected if the corresponding literal is TRUE. For each wire gadget that we use, we introduce a *counter-wire gadget*. It has the same length as the wire gadget and solely connects to the opposite assignment of the variable. Therefore, the summed number of diamonds selected in a wire gadget and its counter-wire gadget does not depend on the truth assignment.

Hence, maximizing the number of diamonds in the complete construction (see Fig. 9 (c)) corresponds to maximizing the number of satisfied clauses.

Arbitrary Length to Width Ratio. Now, edges may have an arbitrary length to width ratio, but also different lengths. Our NP-hardness construction is equal to the first version but uses slightly different gadgets.

The *variable gadget* is shown in Fig. 10 (b). Edges are placed around a common vertex, such that consecutive diamonds intersect, while the triangles do not cross the neighboring edges (see Fig. 10 (a)). This is always possible as the angle γ of a triangle is smaller than the sum of the angles of the two involved halves of the neighboring diamonds (2α) (see Fig. 11). By interspersing long edges with short edges in the gadget, we leave space for connecting wires.

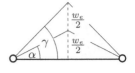

Fig. 11. $2\alpha < \gamma$

The clause gadget is similar to the clause gadget in the first reduction, but rotated by 90 degrees and slightly shifted in opposite directions (see Fig. 10 (c)). As the two edges in the clause gadget are shifted, we can always connect the wires to the clause.

Theorem 2. *Let E be a set of n disjoint edges in the plane, each with an isosceles triangle to one side, an isosceles triangle to the other side, and a diamond with the edge as its diagonal. The problem of choosing a triangle or diamond for each edge in E such that choices of different edges do not intersect and maximizing the number of diamonds chosen is NP-hard.*

6   Heuristics

Exactly computing the optimal setting that maximizes the number of regions in center position is not feasible in polynomial time. Heuristics try to optimize the number of centered edges, but give no optimality guarantee. Thus, they are able to reach polynomial running-times. In this section we present several heuristics that aim to maximize the number of centered edges.

First, note that if an edge intersects a zone of a different edge, this zone can never be part of a solution. We remove these zones by setting the corresponding variables, thus reducing the search space. Edges where the center zone does not intersect any other zone, can safely be selected and are also set. The same holds for clusters of edges that have non-overlapping center regions and only overlap the rest of the instance with their off-center regions.

Second, note that it is never advantageous to set an edge to an off-center position. Hence, we check only the effects of setting an edge to the center position. This may, however, force other edges to be set off-center. We test three heuristics that attempt to minimize the constrains placed on the solution space:

H1 Select the edge that invalidates the fewest regions among other edges.
H2 Select the edge that invalidates the fewest possible center regions.
H3 Select the edge that has the lowest ratio between option 2 and 1.

As selecting an area may have consequences that reach far across the network, we cannot test the result of a selection locally. Instead we use the 2-SAT representation of the input. The variables that correspond to the selected area are set and this information is propagated across the 2-SAT formula. Subsequently simplifying the 2-SAT formula prevents areas from being selected that would lead to invalid results. Using this approach we can guarantee that we always obtain a valid solution for the problem,

Table 1. Number of edges in center position for the different heuristics

Scenario (number of edges)	Optimal	H1	H2	H3
Los Angeles Highway Network (133)	128	128	128	128
Netherlands Highway Network (118)	105	105	105	105
India Railroad Network (148)	143	143	143	143
USA Main Highways (136)	118	118	118	118
Percentage of cases solved optimal (50 cases) - average error				
Random square 11 (120)	100% - 0	66% - 1.2	80% - 1.2	88% - 1.0
Random square 13 (168)	100% - 0	70% - 1.0	76% - 1.0	86% - 1.0
Random square 15 (224)	100% - 0	77% - 1.2	86% - 1.0	93% - 1.3

if one exists. Given the realistic input assumptions discussed in Section 4.1, testing the effect of setting an edge can be done in $O(n^2)$ time, where n is the number of edges. Hence, we obtain an algorithm for all three heuristics that runs in $O(n^3)$ time. As the expected input scenarios are relatively small, this running time is reasonably fast. By comparison, all heuristics solved the test-scenarios within a few seconds, the brute-force approach, however required up to half a day to optimally solve a scenario.

As an informal use case test, we tested our heuristics on three scenarios based on real-life data. These scenarios are relatively easy to solve due to the inherent uneven distribution of dense areas. In Los Angeles, for

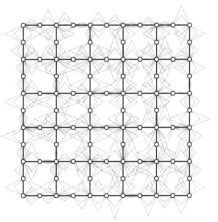

Fig. 12. Random grid-based scenario of size 11 by 11

example, nearly 90% of the center regions do not intersect any other edge. To test the behavior of the heuristic in more complex situations we also compute solutions for several more complex, randomly-generated scenarios. Each random scenario consists of a grid of vertices of size m by m, where all odd columns and rows are connected by edges (see Fig. 12). Edges are attributed a random width in the range $(0, 10)$ and then scaled to the largest size that allows a solution. For comparisons, we compute the optimum solution using a brute-force approach. We create 50 random scenarios of each type and compute the percentage of scenarios solved to optimality by the heuristics, as well as the average error when optimality is not reached, see Table 1.

For the real-life scenarios all heuristics are able to solve the problem to optimality, since the uneven distribution of density makes them considerably more tractable. Yet even for the more complex, artificial scenarios we manage high accuracy, which is probably caused by the geometry of the problem. The center position of an edge is likely to least constrain the rest of the problem and complex constructions that favor other allocations are less likely to occur in practice.

15:00   15:15

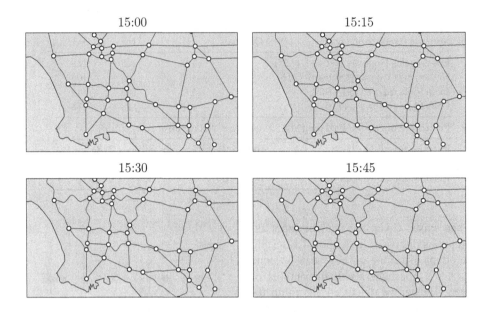

15:30   15:45

Fig. 13. By interpolating the control points of the curve we can generate fluent transitions in both frequency and amplitude. A fraction of the LA dataset is shown.

7 Animating Time-Dependent Data

When the distorted edge length to be displayed changes over time, for example due to traffic conditions, an animation of these changes may give more intuitive information on the underlying processes. Our linear cartograms are very suitable for animations. Here we describe how to use the heuristics from the previous section to create animated data. Further results can be found online[1].

Once more, we first look at the fixed width version and then investigate the fixed frequency version. For animated data, however, the fixed frequency approach appears to be less suitable. There is an increased computational requirement to maintain legal solutions and if the algorithm is run online, we can not guarantee a single consistent frequency across the network.

Fixed Width. Given an input graph we can compute either the maximum edge-width that allows a feasible solution or the maximum edge-width that allows all edges to be centered (see Section 3). As this width is independent of the delay along the edges, it can be computed beforehand and is maintained throughout the animation. All future steps solely alter the frequency along edges. We apply a fluent transition between different states by linearly interpolating the control points of the curves. Additional oscillations are introduced by adding degenerate curves, having all control points at either endpoint of the edge, before interpolating (see Fig. 13).

---

[1] http://www.win.tue.nl/~agoethem/linear_cartograms/, May, 2014

A disadvantage of fixed edge-widths is that the frequency on some edges may become too high. In extreme cases, increasing the frequency can cause problems with visualization.

Fixed Frequency. An alternative to fixed edge-width is using a fixed frequency. We assume, for now, that in each situation a valid configuration exists. Over time the data, and thus the width of edges, will change. If the changes are significant this may cause the current configuration to become invalid. While we can compute a valid configuration, a large change in configuration will create overly complex animations. The number of changes in the selected zones should be minimized. To minimize zone changes we, once more, make use of the heuristics described in Section 6. Instead of the center position, the number of edges that maintain their "current" position is maximized.

If the width of the edges is significantly increased, no valid configuration may exist anymore. To obtain a feasible solution the curve frequency must be increased. Increasing the global frequency, however, creates a large overall change. This risks a visual disconnection between two consecutive states and causes undue attention to be drawn to areas without any significant change. Instead, we increase the frequency of only the invalidating edges. To maintain visual balance, all invalidating edges are set at the same frequency, which is the minimal frequency that generates a feasible solution. As a consequence, at any moment during the animation we display at most two discrete frequencies. A different approach would be to compute the minimum frequency that satisfies all steps beforehand. However, overly increasing the frequency to suit the most restricting time step may reduce visual clarity in less constricted time steps and prevents the algorithm from being run online.

8 Discussion

We introduced a new type of linear cartogram where edges are drawn as sinusoid curves such that their length corresponds to travel time. We assumed that the regions occupied by these curves are triangular or diamond-shaped, and that the curves are C^2-continuous cubic Bézier splines. However, our approach applies to other visual styles as well (e.g., a piecewise linear curve zigzagging in the middle of an edge). The extension to using more than three regions per edge is straightforward, similar to [14]. Interestingly, our improvement under the realistic input assumption can also be used to speed up the line labeling problem in [15].

The optimization problem of maximizing the number of edges that use the centered region is NP-hard. Hence we introduced several heuristics and showed that they work efficiently in practice. Finally, we discussed how to create animations based on time-dependent data. The results give a clear and concise representation of the data without compromising the integrity of the map.

The benefit of our method over traditional edge coloring lies in the comparison of different paths through the network. As length is additional, in contrast to color, it is simpler to compare the required time investment of different paths. This observation, however, hinges on two key aspects. Firstly, the perceived distance between two points in a network should equal the sum of the edge lengths. Research into the distance-similarity metaphor [20] appears to indicate that in a network indeed edge length is used as a measure of distance. Secondly, users should be able to accurately estimate the length of a regular sinusoid curve. When the frequency is fixed across the network, the edge *area*, instead of edge width, is related to the length of an edge. It is unclear if users intuitively will compare the size of the described areas and, therefore, are able to accurately compare the length of different edges. For a fixed edge width the comparison appears to be more intuitive, inherently turning into a symbolization for density on the edges.

While a formal user evaluation was not the main goal of this paper, for future work we recommend further exploration of the effects of this new method. To validate the applicability the perceived length of both types of regular sinusoid curves should be investigated. We note that our method is not restricted to using the exact length. Overemphasizing or underemphasizing edge length to compensate for the perceived length can easily be integrated in the method. Furthermore, a more complete and systematically selected dataset should be evaluated to explore the possible interplay effects with different scenarios.

From an algorithmic perspective for future work it would be interesting to see if a polynomial-time approximation algorithm would be possible maximizing the number of edges in center position. The problem, however, appears to be quite hard as choices can have consequences that reach far across the network. In contrast to many map-labeling problems, we require that all edges select an option, causing choices to propagate along the network. Another direction for future work is to determine if there are restrictions under which we can solve the problem optimally. All real-life scenarios investigated were solved near optimal by the heuristics introduced. If we could show that (most) real-life scenarios adhere to stricter input restrictions, the problem might not be NP-hard under those restrictions.

Acknowledgments. The authors would like to thank all reviewers for their extensive and insightful feedback which helped to improve the paper.

References

1. Robinson, A., Morrison, J., Muehrcke, P., Kimerling, J., Guptill, S.: Elements of cartography. John Wiley & Sons (1995)
2. Bies, S., van Kreveld, M.: Time-space maps from triangulations. In: Didimo, W., Patrignani, M. (eds.) GD 2012. LNCS, vol. 7704, pp. 511–516. Springer, Heidelberg (2013)
3. Cabello, S., Demaine, E., Rote, G.: Planar embeddings of graphs with specified edge lengths. Journal of Graph Algorithms and Applications 11(1), 259–276 (2007)
4. Kaiser, C., Walsh, F., Farmer, C., Pozdnoukhov, A.: User-centric time-distance representation of road networks. In: Fabrikant, S.I., Reichenbacher, T., van Kreveld, M., Schlieder, C. (eds.) GIScience 2010. LNCS, vol. 6292, pp. 85–99. Springer, Heidelberg (2010)
5. Shimizu, E., Inoue, R.: A new algorithm for distance cartogram construction. International Journal of Geographical Information Science 23(11), 1453–1470 (2009)
6. Langlois, P., Denain, J.C.: Cartographie en anamorphose. Cybergeo: European Journal of Geography (1996)
7. Bouts, Q., Dwyer, T., Dykes, J., Speckmann, B., Riche, N., Carpendale, S., Goodwin, S., Liebman, A.: Visual encoding of dissimilarity data via topology preserving map deformation (in preparation, 2014)

8. Barequet, G., Goodrich, M., Riley, C.: Drawing planar graphs with large vertices and thick edges. Journal of Graph Algorithms and Applications 8(1), 3–20 (2004)
9. Goodrich, M., Wagner, C.: A framework for drawing planar graphs with curves and polylines. Journal of Algorithms 37(2), 399–421 (2000)
10. Efrat, A., Erten, C., Kobourov, S.: Fixed-location circular-arc drawing of planar graphs. Journal of Graph Algorithms and Applications 11(1), 145–164 (2007)
11. Duncan, C., Eppstein, D., Goodrich, M., Kobourov, S., Nöllenburg, M.: Lombardi drawings of graphs. Journal of Graph Algorithms and Applications 16(1), 37–83 (2012)
12. Nielsen, C., Jackman, S., Birol, I., Jones, S.: Abyss-explorer: visualizing genome sequence assemblies. IEEE Transactions on Visualization and Computer Graphics 15(6), 881–888 (2009)
13. Wolff, A., Strijk, T.: The map labeling bibliography (2009),
    `http://liinwww.ira.uka.de/bibliography/`
    `Theory/map.labeling.html`
14. Poon, C.K., Zhu, B., Chin, F.: A polynomial time solution for labeling a rectilinear map. Information Processing Letters 65(4), 201–207 (1998)
15. Strijk, T., van Kreveld, M.: Labeling a rectilinear map more efficiently. Information Processing Letters 69(1), 25–30 (1999)
16. Saux, E., Daniel, M.: Data reduction of polygonal curves using B-splines. Computer-Aided Design 31(8), 507–515 (1999)
17. Aspvall, B., Plass, M., Tarjan, R.: A linear-time algorithm for testing the truth of certain quantified boolean formulas. Information Processing Letters 8(3), 121–123 (1979)
18. Halperin, D.: Arrangements. In: Handbook of Discrete and Computational Geometry. Chapman & Hall/CRC (2004)
19. Guibas, L., Hershberger, J., Mitchell, J., Snoeyink, J.: Approximating polygons and subdivisions with minimum-link paths. International Journal of Computational Geometry & Applications 3(4), 383–415 (1993)
20. Fabrikant, S., Montello, D., Ruocco, M., Middleton, R.: The distance–similarity metaphor in network-display spatializations. Cartography and Geographic Information Science 31(4), 237–252 (2004)

3D Network Spatialization: Does It Add Depth to 2D Representations of Semantic Proximity?

Sara Irina Fabrikant[1], Sara Maggi[1], and Daniel R. Montello[2]

[1] Department of Geography, University of Zurich, Zurich, Switzerland
[2] Department of Geography, University of California, Santa Barbara, CA 93106, USA

Abstract. Spatialized views use visuo-spatial metaphors to facilitate sense-making from complex non-spatial databases. Spatialization typically includes the projection of a high-dimensional (non-spatial) data space onto a lower dimensional display space for visual data exploration. In comparison to 2D spatialized displays, 3D displays could potentially convey more information, as they employ all three available spatial display dimensions. In this study, we evaluate if this advantage exists and whether it outweighs the added cognitive, perceptual, and technological costs of 3D displays. In a controlled human-subjects experiment, we investigated how viewers identify document similarity in 3D network spatializations that depict news articles as points connected by links. Our quantitative findings suggest that similarity ratings for 3D network displays are similar to those obtained in a prior 2D study we conducted. With both types of displays, viewers mostly judged document similarity on the basis of metric distances along network links, as opposed to node counts or distance across the network links. However, node counts do affect similarity assessments with 3D displays more than with 2D displays. We also find no significant differences in similarity judgments whether 3D displays are presented monoscopically or stereoscopically. We conclude that any advantage of 3D displays in conveying more information than 2D displays does not necessarily outweigh their additional demands on cognitive, perceptual, and technological resources.

1 Introduction

The exponential growth and availability of online relational text data (e.g., the Web 2.0, online journals, Facebook, Wikipedia, etc.) requires new methods to help people more efficiently select information and construct new knowledge from big text data sources [10]. Ongoing research in GIScience and information visualization has focused on how to effectively depict multivariate, typically non-numeric and non-spatial, data stored in very large databases by means of computational techniques that transform high-dimensional datasets into low-dimensional spatialized data displays [12]. The spatial arrangement of depicted information items in such displays is typically based on the distance-similarity metaphor [8], which states that closer items will be seen as more similar, and more similar items should therefore be placed closer to one another in the spatialized display. The resulting "information spaces" can be visualized in various ways, e.g., as two-dimensional (2D) or three-dimensional (3D) simple point maps, network maps, or continuous terrains [12].

M. Duckham et al. (Eds.): GIScience 2014, LNCS 8728, pp. 34–47, 2014.

Along with the public's increased exposure to low-cost, immersive, stereoscopic 3D display technology (i.e., 3D gaming engines, TVs, and cinemas), there has been growing interest in the information visualization community in designing and using 3D spatializations that depict a corpus of documents in all three available spatial display dimensions [11]. However, it is still unclear how the distance-similarity metaphor will operate in 3D [6]. Few empirical evaluations have examined design guidelines for cognitively inspired and perceptually salient 3D information spatializations [16]. Researchers in information visualization claim that users should be able to extract more information from 3D displays than from 2D displays, as we live in a 3D world [21]. They also argue that less data is lost when high dimensional databases are reduced only to 3D rather than 2D. The 3D displays can supposedly reveal more information, as they contain an additional degree of freedom for display and interactive exploration [6]. For example, as Sedlmair et al. [11] suggest, a common argument for the use of 3D scatterplots is that the intrinsic dimensionality of a dataset is likely to be greater than two dimensions; so 3D displays are able to convey more information. However, researchers have also recognized that this additional supply of information may come with various costs, including perceptual issues (occlusion problems and size-estimation difficulties in perspective views) (e.g., [13]), cognitive demands (the need for direct interactivity and motion parallax to avoid the perceptual issues), and additional technological complexity (the requirement for fast 3D graphics cards and advanced 3D display technology) [6, 11, 14].

In our study, we investigate how users interpret monoscopic and stereoscopic 3D spatialized views and compare what we find to previous work with 2D displays [4]. We are interested in exploring whether the addition of a third display dimension outweighs the potential increased costs of constructing, displaying, interacting with, and interpreting 3D views. We systematically evaluate how different notions of distance might influence the operation of the distance-similarity metaphor in interactive 3D network spatializations.

2   Related Work

2.1   Prospects and Challenges of 3D Visualization

The need to develop 3D design guidelines has gained recognition within the visualization community [11], but we still lack a good understanding of how people perceive and interpret 3D displays [17, p. 259]. We include as 3D displays any graphic or image that appears to extend over three spatial dimensions, even though it is actually a 2D object (e.g., computer screen or piece of paper). We distinguish monocular from stereoscopic (binocular) displays. Monocular displays create the appearance of depth via monocular cues, i.e., features of the display that create depth even when viewed with a single eye [17, pp. 259–260]. These can be further distinguished as static monocular displays (using occlusion, size changes, linear perspective, etc.) or dynamic monocular displays (using movement, including motion parallax, as part of animated displays). Dynamic monocular displays can be further distinguished as interactive or not. In contrast, stereoscopic displays create the appearance of depth via so-called "true 3D," the experience of visual depth that results when the brain combines the offset images from the two eyes during binocular viewing of actual 3D objects or of specially created 2D images (i.e., created to present two offset images separately to each eye).

Of course, 2D information displays have been used for a long time, and respective design guidelines have evolved alongside by long-standing practice, for example, as employed for cartographic maps [1]. The question arises then whether common 2D cartographic design principles can be applied to 3D spatialized data displays, and if so, how? As Bertin [1] writes, a 2D network map is efficiently depicted when the nodes are connected to each other in a manner that minimizes the number of links that intersect or cross each other. Likewise, 3D networks can be depicted in similar fashion, but additional perceptual cues (i.e., graphic variables) are necessary to account for the visually more complex representational structures. Bertin [1] contends that the addition of a third dimension to monoscopic graphs can create a sense of volume, and he also suggested that network links should not cross each other. To depict monoscopic 3D network displays, Bertin suggests changing the thickness of the links according to viewing distance, producing the impression of depth via linear perspective. We are skeptical that this would work unless the network was fairly small and simple in structure. But as is true for Bertin's other design guidelines, those concerning 3D networks have mostly not been examined empirically to this day.

Herman et al. [7] contend that adding an extra dimension to displays can facilitate the depiction of large data structures but might make it difficult for users to find the most appropriate perspective and insightful view on the data space. A good example of this is shown in Figure 1. This monoscopic, static 3D display was created with specialized, state-of-the art 3D network software [20] by Dunne et al. [3] and was voted one of the best scientific visualizations of 2013 in Wired Science[1]. The 3D graph depicts the food web of Estero de Punta Banda trophic species, including its parasites and concomitant links. Green indicates basal taxa, red indicates free-living taxa, and blue indicates parasites. The vertical axis corresponds to short-weighted trophic levels [3]. Unfortunately, relationships between the red, blue, and green balls cannot be identified in this 3D network, due to massive over plotting and extensive crossing of the links. Partial and complete occlusion of the colored balls makes connection properties and distance estimation along the links impossible, due also to depth-perception issues.

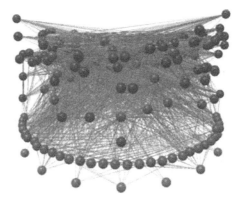

Fig. 1. The food web of Estero de Punta Banda trophic species (extract from Dunne et al., 2013)

---

[1] On the Web at: http://www.wired.com/wiredscience/2013/12/best-scientific-figures-2013/ (accessed Feb. 2014).

This graph might be improved by providing stereoscopic or interactive viewing capabilities. However, Ware [17] points to a list of possible problems with stereoscopic 3D displays, including stereo-blindness (even some people with two functional eyes do not experience stereopsis), diplopia (double vision), the frame cancellation problem, the vergence-focus problem, and stereopsis loss for distant objects; in fact, stereo vision only suggests depth for relatively close distances within about 10 meters. Furthermore, various optical illusions (e.g., the filled-space or Oppel-Kundt illusion, the vertical or vertical-horizontal illusion) have been identified that will modify perceived distances, even in monocular displays [2, 8, 9, 23].

A potential advantage of interactive dynamic 3D graphs is that viewers can find optimal views without intersecting links or occluding features [16]. Indeed, users can manipulate interactive dynamic displays until they find the best view with the least number of occlusions among the depicted features. Ware and Mitchell [19] found that adding 3D depth cues like those available in interactive dynamic displays increased the efficiency and accuracy with which viewers were able to explore very large 3D network displays as compared to non-interactive displays, including static 2D displays. Supplying such interactive 3D would involve additional development time for designers, and might place additional perceptual and cognitive demands on viewers, including those involved in display interaction. Thus, a major 3D spatialization challenge is employing appropriate 3D layout techniques to uncover the essence of buried data relationships, at the same time implementing additional visual cues and human-display interaction mechanisms to support the most effective and efficient human visuo-spatial exploration of the 3D data space.

2.2   Our Prior Spatialization Research

A variety of forms of distance or proximity might work best to convey item similarity in spatialized network displays, including the number of nodes or links between items, metric distance between items along links, or metric distance between items directly across the space within which the network is embedded. We have previously reported on various empirical studies investigating which type of proximity would most likely be intuitively understood by viewers to show the relatedness or semantic similarity between documents in very large databases displayed in 2D and 3D point spatializations, and in 2D network spatializations [4, 6, 8]. These studies also investigated how visual variables besides distance might influence the perception and understanding of the distance-similarity metaphor in spatializations. For example, links that connect nodes in network displays can vary in width, color hue, or color value [4], similar to the visual variables employed for networks such as highways shown on cartographic maps [1]. Another study highlighted how test instructions can influence the use of proximity to judge similarities between items in spatialized displays [5].

In a study on 3D point-display spatializations, we replicated our finding with 2D displays that viewers map judgments of document similarity onto distances between document points, as long as no apparent features such as clusters or lines emerge from sets of points [6]. We also found that variation among participants in their similarity judgments is noticeably larger with 3D than with 2D displays. With 3D displays, we specifically hypothesized that variation in the degree to which participants rotate the displays into the fronto-parallel plane might lead to variation in the apparent distances between assessed points, and thus variation in assessed similarities. "Fronto-parallel" orientation occurs when document points being compared all lie within the same display plane fronto-parallel (normal) to the line of sight. In this orientation, proximal distances (on the retina as well as on the monitor screen) between pairs of points are maximized.

3   Experiment

While many researchers and designers have great enthusiasm for 3D displays, no research we know of has clearly demonstrated their superiority. Based on our previous study with 3D point displays [6], we believe that adding the third dimension will actually detract somewhat from people's ability to see similarity relationships in spatialized displays. This is because people map document similarity onto inter-point distance, as we have shown in earlier work on 2D displays. In order to see distance most clearly, we hypothesize that participants will rotate the 3D displays until all three comparison points are brought into the fronto-parallel plane. This process takes extra time and may not be carried out optimally by all participants or even carried out at all. We thus designed a mixed-factorial experiment to assess the effectiveness of 3D network-spatializations representing documents collected in a very large text document database. Our study design is based on our previous study of 2D network spatializations [4], allowing direct comparison to the results of that study. We investigate how users interpret the distance-similarity metaphor in 3D node-link displays depicting conflicting notions of distance, specifically network metric distance vs. topological proximity (see Fig. 2 for an example stimulus).

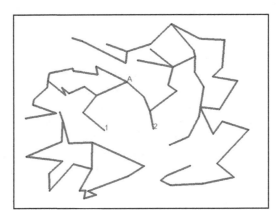

Fig. 2. Example 3D stimulus varying the visual variables of network metric distance and topological node proximity (node count) between assessed entities 1 and 2, with respect to reference entity A.. A is closer to 2 than to 1 in metric distance along the network but equally close in terms of node proximity.

3.1   Methods

Participants. Twenty-eight participants (14 females and 14 males) took part in the experiment, with an average age of 29 yrs. One participant indicated a lack of depth perception, but none claimed to have a color deficiency. Fourteen participants were undergraduate students with geography and other majors, and the other fourteen were not affiliated with a university. We tried to recruit mostly participants with little previous professional experience, training, or college degrees in GIS, cartography, computer graphics or graphic design, to minimize a potential bias due to user background and training. The majority of the recruited participants (19) had less than one year of training in the above-mentioned academic fields, but 7 had between one and five years, and two had more than five years of training in these fields.

Set-up and Materials. The 3D network spatializations were visualized using a Cyviz Geowall, consisting of a Windows PC with a dual-output graphics card, two aligned digital overhead projectors, a back-lit projection screen (2.23 m x 1.80 m), and a pair of polarized glasses for stereoscopic viewing. Participants sat facing the screen at a distance of 2.20 m.

Sixty-five 3D network-displays were created using Vizard 3.0, a Python-based software designed to produce interactive 3D graphics and virtual worlds. The displays consisted of nodes (points) that supposedly represented documents, connected by straight links. Three nodes were distinctly labeled as 'A,' '1,' and '2.' In order to empirically evaluate the way network displays are viewed and interpreted under various conditions, we had participants specifically compare the apparent semantic similarity between documents 'A' and '1' to that between documents 'A' and '2.' The network stimuli were modeled after our earlier study of static 2D spatialized network displays [4]. We thus replicated the 2D configurations and x- and y-coordinates of the nodes from this 2D study but added random z-coordinates to nodes. Compared to the prior 2D study, participants in this 3D study could actively control the rotation of the displays. We randomized the initial orientation of each 3D network upon first being viewed by participants so that it was not in the fronto-parallel (FP) orientation (Figure 3). Participants thus had to rotate the displays if they wanted to get them into FP orientation. When the displays were rotated to FP orientation, the comparison points were maximally distant from each other on the 2D image and at the same viewing distance from the participant. The resulting 2D views matched the displays from our earlier 2D study.

We divided the stimuli into a sequence of four blocks, where the links connecting the nodes were depicted varying a combination of network distance and node proximity, as well as link hue, value, and width; in the present report, we focus only on network distance and node proximity, as we expect those variables to be most sensitive to displaying in 2D vs. 3D. Unlike our previous 2D study, we kept the direct distances between points across the network (i.e., not along network links) constant in this 3D study. In 15 of the trials, we systematically varied metric distance along the network links so that A:2 was equal in length to A:1, twice as far apart, or three times as far apart (in two additional trials we omit below, we varied network distance so that A:2 was either 1.5 or 2.5 times as far apart as A:1, but only for displays that equated node proximity). At the same time, these 15 trials varied node proximity so that A:1 and A:2 were equally far apart (two nodes each), A:1 was three nodes apart while A:2 was two nodes, A:1 was two nodes apart while A:2 was one node, A:1 was three nodes apart while A:2 was one node, or A:1 was four nodes apart while A:2 was one node.

In order to control for a biasing effect due to the horizontal-vertical and filled-interval illusions, we systematically controlled the arrangement of the comparison points A, 1, and 2 along the x and y-axes, but used equal z-coordinates. As the main dependent variable, we recorded participants' similarity ratings of the two pairs of comparison nodes, the viewing angles every 0.016s, as well as the viewing time between each display rotation. We also recorded background questionnaire responses and participants' display preferences for a qualitative analysis.

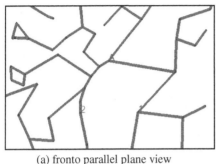

(a) fronto parallel plane view   (b) random starting configuration

Fig. 3. Example 3D stimulus with two different 3D views

Procedure. Participants were randomly divided into two viewing groups: monoscopic mode (only one projector was switched on, without polarized glasses) and stereoscopic mode (two projectors were switched on, with polarized glasses). Participants were individually tested in a session that lasted approximately 45 minutes. After welcoming participants, the Geowall environment was explained and participants signed a consent form. They were then seated in front of the screen and asked to fill out a background questionnaire. They were told that 3D images would appear on the screen and that they would have to interact with the images before answering a test question, using a mouse to input their answer. Subsequent instructions were delivered from slides appearing on the screen. Participants read that they would view 3D-network displays representing information from a database containing documents such as, books, new stories, or journal articles, depicted as black dots. For each display, participants were to judge the similarity of a reference document labeled 'A' with that of two other documents, labeled '1' and '2.' They compared the similarity of these two pairs of documents with the response scale shown in Figure 4, but they were not given any instructions or advice as to how to judge similarity between the documents. Ratings were collected on a 9-point interval scale, with a value of '1' representing "A and 1 much more similar", a value of '9' representing "A and 2 much more similar", and a value of '5' in the middle representing "1 and 2 equally similar to A". Hence, a mean rating less than 5 indicates that participants saw A:1 as more similar, while a mean rating greater than 5 indicates they saw A:2 as more similar.

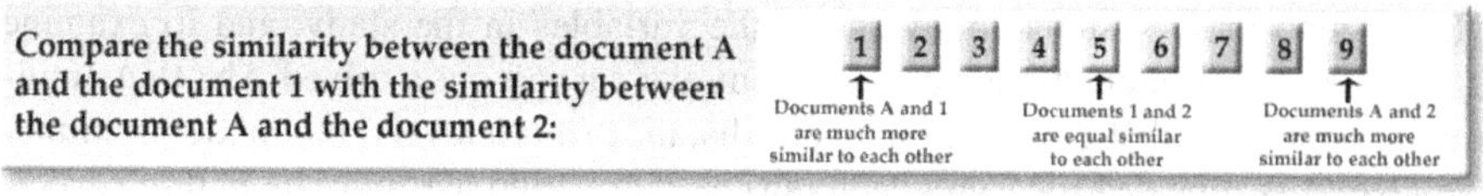

| Compare the similarity between the document A and the document 1 with the similarity between the document A and the document 2: | 1 2 3 4 5 6 7 8 9 ↑ ↑ ↑ ↑ Documents A and 1 are much more similar to each other   Documents 1 and 2 are equal similar to each other   Documents A and 2 are much more similar to each other |

Fig. 4. Response scale used for each test stimulus

Participants viewed thirteen practice trials, where they had the possibility to get acquainted with the test environment, how to respond, and how to rotate the 3D dynamic displays. This was followed by sixty-five trials divided into four blocks for which we recorded participants' similarity judgments and display interactions. Within a block, displays were shown in randomized sequences to avoid potential learning effects. After completing the 3D display portion of the experiment, participants were asked to fill in a post-test questionnaire in order to better understand how they believed they had assessed the similarities of the documents for each kind of display. After completing the test, they were given a meal voucher for the university cafeteria to thank them for participation.

4 Results

4.1 Similarity Ratings

The mean similarity ratings for the 15 trials that contrasted network distance and node proximity are presented in Table 1. The pattern of means suggests that when documents are equal in node proximity and equal in network metric distance, participants rate them equally similar to each other (i.e., average similarity rating does not significantly differ from 5.0), replicating the results from Fabrikant et al.'s [4] 2D study. Also consistent with results from the 2D study, we find that in 3D participants rate documents to be more similar when they are relatively closer in network metric distance. However, unlike our prior 2D study, adding relatively more nodes between documents seems to make them appear less similar, over and above any effects of network distance. When node proximity and network distance conflict, similarity tends to cancel out and earn ratings near 5, similarly to the 2D study results.

Table 1. Mean similarity ratings of the 15 displays varying network metric and node proximity between A:2 and A:1. A mean rating less than 5 indicates that participants saw A:1 as more similar, while a mean rating greater than 5 indicates they saw A:2 as more similar.

		Network metric distance		
		A:1 = A:2	A:1 < A:2	A:1 < A:2
		(1:1)	(1:2)	(1:3)
	A:1 = A:2 (2:2)	5.3	4.3*	3.6**
	A:1 > A:2 (3:2)	4.5	5.4	4.2
Node Proximity	A:1 > A:2 (2:1)	5.3	4.5	5.0
	A:1 > A:2 (3:1)	6.8**	5.4	4.9
	A:1 > A:2 (4:1)	6.1*	6.0*	4.8

* $p<.05$ (significantly different than 5.0), ** $p<.001$ (significantly different than 5.0)

In order to more systematically assess the variables in the study, and to examine possible statistical interactions between them, we carried out a mixed ANOVA (multivariate approach) that examined network distance (three levels) and node proximity (five levels) as repeated-measures factors (i.e., that varied across trials within participants), and viewing mode (two levels) as a between-case factor (monoscopic vs. stereoscopic mode). Network distance was significant as a main effect (F[2, 25] = 9.70, p<.001), as was node proximity (F[4, 23] = 4.19, p<.01). These two effects show that participants found documents less similar if they were further apart in either metric distance or in node count. In contrast, viewing mode had virtually no effect on similarity ratings (F[1, 26] = 0.03, ns); whether displays were viewed monoscopically or stereoscopically did not influence ratings. None of the tests for 2-way or 3-way interactions among the factors were statistically significant either (all p's >.15). Finally, another set of analyses found no significant effects of participants' background and training (i.e., gender, study major, profession, age, map reading abilities, frequency of map use) on similarity ratings.

We further examined the effects of network distance and node proximity by calculating Pearson's correlation coefficients separately for each participant, averaging them after Fisher's r-to-z transform, and then converting them back to r. We correlated the network distance of trials (1, 2, or 3) against the similarity ratings for those trials, arriving at an average correlation of r = -.31, suggesting a modest tendency for participants to rate A:1 as increasingly more similar on trials with greater relative distance between A:2 than A:1. Likewise, we correlated the node proximity of trials (0,1, 2, 3, or 4) against the similarity ratings for those trials, arriving at an average correlation of r = .37, again suggesting a modest tendency for participants to rate A:1 as increasingly more similar on trials with greater relative node count between A:2 than A:1. This latter finding differs from our 2D study in finding that node proximity had an influence on the operation of the distance-similarity metaphor in 3D.

4.2   Rotation Times

We also recorded the times that participants rotated the display along the x and y axes[2] (i.e., pitch and yaw) in the 3D display space during the time they viewed each display, to the nearest 0.016s. The mean rotation times in seconds for the 15 trials that contrasted network distance and node proximity are presented in Table 2. No consistent pattern stands out.

In order to more systematically assess any effects on rotation time, we again carried out a mixed ANOVA (multivariate approach) that looked at network distance and node proximity as repeated-measures factors, and viewing mode as a between-case factor. Confirming the lack of an obvious simple pattern in Table 2, network distance was not significant as a main effect (F[2, 25] = 1.03, ns). Node proximity, however, was significant as a main effect (F[4, 23] = 7.97, p<.001). Viewing mode, whether

---

[2] Although the display could also be rotated along the z axis (roll), we did not record this, as this rotation did not change the apparent distances or nodes between the comparison document points.

displays were viewed monoscopically or stereoscopically, was again nonsignificant as a main effect (F[1, 26] = 3.07, ns). However, viewing mode did interact significantly with network distance (F[2, 25] = 3.23, p<.05). Under monoscopic viewing, distance was significantly related to rotation times, while under stereoscopic viewing, it was not. An examination of the mean rotation times across trials varying in relative network distance between A:1 and A:2 revealed that participants viewing monoscopically rotated the displays 1.5–2 s more when the relative distances were two or three times different than when they were equal. In contrast, when viewed stereoscopically, participants rotated the displays with .6 s no mater the relative distance differences. Neither of the other 2-way interactions nor the 3-way interaction among the factors were statistically significant either (all p's >.4). Finally, another set of analyses found no significant effects of participants' background and training (i.e., gender, study major, profession, age, map reading abilities, frequency of map use) on similarity ratings.

Table 2. Mean time in seconds spent rotating the 15 displays varying network metric and node proximity between A:2 and A:1

		Network metric distance		
		A:1 < A:2 (1:1)	A:1 < A:2 (1:2)	A:1 < A:2 (1:3)
	A:1 = A:2 (2:2)	12.4	11.6	11.6
	A:1 > A:2 (3:2)	11.3	13.4	12.3
Node Proximity	A:1 > A:2 (2:1)	8.9	7.8	7.0
	A:1 > A:2 (3:1)	7.1	11.0	8.1
	A:1 > A:2 (4:1)	9.1	10.6	11.8

We calculated Pearson's correlation coefficients of rotation time with network distance and node proximity separately for each participant, again averaging them after Fisher's r-to-z transform, and then converting them back to r. Consistent with our ANOVA finding of no simple linear relationships between these variables, neither network distances of trials (r = .03) nor node proximity (r = -.19) had substantial linear correlations with rotation time.

5   Discussion

In this 3D network spatialization display study we were able to replicate findings from our prior work on 2D displays [4], with one significant exception. In contrast to the 2D spatialized network displays, where the number of intervening nodes did not have an effect on people's similarity judgments, we do find that node proximity has a significant effect for the employed 3D displays, as suggested by the newly found main effect of node count on similarity ratings and the correlation of node proximity with participants similarity ratings. Participants judged documents connected with fewer intervening nodes as more similar, compared to documents with more nodes in between, irrespective of increasing network metric distance. This is somewhat surprising, but could be explained as follows: Assigning a random z-coordinate to node positions in the 3rd dimension might have added an additional visual cue for document similarity assessment which is not available in 2D. In Figure 3a, the 3D configuration is shown in FP orientation, essentially identical to the view in the prior 2D study. The nodes in this FP/2D view do not seem particularly salient, as not all links change direction at node intersections. This is in contrast to the non-FP views, which are only available in 3D, as shown in Figure 3b. In the z-dimension of the 3D space, document nodes appear much more salient than they do in 2D, as all the links change direction at node intersections, due to randomly assigned z-values. The "ups and downs" of the links might involve increasing attentional costs to assess the distance-similarity metaphor. In fact, this might explain the additional time participants needed for all displays compared to in the 2D study. Perhaps it is easier and faster just to count the nodes for similarity assessment than to visually estimate metric distance along the links in 3D space; estimating distance seems to be much harder in 3D than 2D space. In fact, participants spend significantly more time rotating monoscopic displays to assess network metric distance in the displays, compared to stereoscopic 3D displays, where distance estimation is facilitated by visual depth perception.

Our results provide unique evidence that irrespective of the display mode employed (i.e., mono or stereo viewing), participants rated document similarity in 3D much as they did in 2D. As about 20% of the population cannot see stereoscopically, and considering the extra technological expense to add either motion parallax (3D mono) or distance parallax information for 3D stereo viewing, we argue that these additional costs do not outweigh the potential benefit of facilitating more accurate distance measurements with 3D stereo. In fact, one might argue that the motion parallax provided with interactive 3D displays (i.e., rotation of the 3D network structure) is equally useful or even sufficient for distance judgments (i.e., similar to rotating one's head when trying to judge objects at farther distances in a real world scene), if not more effective and efficient. It is more cost effective when considering hardware needs, display development time, ease of deployment, etc. Motion is also considered one of the strongest visual cues to attract attention [17, 22], irrespective of viewing distance. As Ware and Franck [18] suggest, adding motion to 3D network displays is more important than adding stereo for comprehension of the structure. As participants took significantly less time to rotate the 3D displays in stereo mode on those trials where metric distances differed, one might argue that stereo could be useful if response time (i.e., response efficiency) were an issue, for example, in a decision context of time pressure. However, overall, the difference in the response times suggesting increased cognitive costs with 3D over 2D displays is a compelling reason why 3D spatialized network displays should not be used at all, as similarity ratings are quite similar for 2D and 3D (except, of course, for node proximity).

6   Conclusions

We set out to answer the research question of whether 3D network spatialization would add depth to 2D representations of semantic proximity. Our findings suggest that the interactive 3D viewing mode (i.e., mono vs. stereoscopic displays) did not influence participants' similarity ratings, as compared to static 2D displays evaluated in a prior study [4]. Moreover, similarity ratings for the 3D network displays are very similar to the ones obtained in the 2D network study. That is, viewers mostly map judgments of document similarity onto distances along the network, in 3D as well as 2D space. In contrast to our earlier 2D study, node proximity did have an effect in the present 3D experiment; we believe the nodes became more salient due to direction changes of the links in the 3rd dimension. In other words, viewers find it visually easier to simply count nodes, which is also time efficient, than trying to estimate network metric distance, which is more error prone [13]. Moreover, as in our prior studies on spatialized displays, user-related factors (i.e., group differences), such as gender, age, and previous training, did not significantly affect the similarity ratings.

Similarity ratings of the 3D displays are almost identical to those collected with 2D displays, but it takes participants longer to make decisions in 3D. The potential benefit of adding the third dimension, which allows one to interactively change perspective on the data space and add more information to the representations of abstract data, seems not to benefit participants' decision efficiency. Participants seem not to better understand the distance-similarity metaphor or make faster decisions in 3D, compared to 2D. It might be that an increase in response times in the 3D study corresponds to an increase in users' cognitive load when judging the displays. In fact, response time is further influenced by the viewing mode of the 3D display. If stereo is available, participate take significantly less time to rotate the displays before responding, compared to monoscopic 3D displays, especially when comparing metric network differences in the displays. We have not yet analyzed the qualitative data collected from the post-test questionnaires (e.g., display preferences, how they rated similarities between documents, and which measure they used for each display type), which we aim to do in future work.

These quantitative results thus lead us to conclude that although 3D displays might have the benefit of conveying more information than 2D spatialized views, this advantage is not necessarily enough to overcome the additional demands on cognitive, perceptual and technological resources engendered by interacting with 3D displays.

We recognize that our study had participants judge similarities absent a specific decision-making context, such as document topic or a specific application for the task. We did this to be as general as possible, without limiting our conclusions to a specific context. However, we recognize that real decision-making situations do generally come with a context. Future research should examine how and to what degree context influences the use of 2D and 3D network spatializations.

With this study, we hope to help information visualization designers to create expressive spatialized views that depict document similarity in intuitive ways for effective and efficient decision-making based on increasingly massive (text) databases. Our results to date lead us to conclude that a potential information increase afforded by adding an additional (third) display dimension does not outweigh the increased perceptual and cognitive costs caused by more resource-demanding 3D displays.

References

1. Bertin, J.: Sémiologie Graphique. Les Diagrammes, les Réseaux, les Cartes. Paris, École Des Hautes Études En Sciences Sociales (1967, 1999)
2. Buffardi, L.: Factors Affecting the Filled-Duration Illusion in the Auditory, Tactual, and Visual Modalities. Perception & Psychophysics 10, 292–294 (1971)
3. Dunne, J.A., Lafferty, K.D., Dobson, A.P., Hechinger, R.F., Kuris, A.M., Martinez, N.D., McLaughlin, J.P., Mouritsen, K.N., Poulin, R., Reise, K., Stouffer, D.B., Thieltges, D.W., Williams, R.J., Zander, C.D.: Parasites Affect Food Web Structure Primarily through Increased Diversity and Complexity. PLoS Biol. 11, e1001579 (2013)
4. Fabrikant, S.I., Montello, D.R., Ruocco, M., Middleton, R.S.: The Distance–Similarity Metaphor in Network-Display Spatializations. Cartography and Geographic Information Science 31(4), 237–252 (2004)
5. Fabrikant, S.I., Montello, D.R.: The Effect of Instructions on Distance and Similarity Judgements in Information Spatializations. International Journal of Geographical Information Science 22, 463–478 (2008)
6. Fabrikant, S.I., Montello, D.R., Neun, M.: Evaluating 3D Point-Display Spatializations. In: GIScience 2008 - Fifth International Conference on Geographic Information Science, Park City, Utah, September 23-25, pp. 66–69 (2008)
7. Herman, I., Melançon, G., Marshall, M.S.: Graph Visualization and Navigation in Information Visualization: A Survey. IEEE Transactions on Visualization and Computer Graphics 6, 24–43 (2000)
8. Montello, D.R., Fabrikant, S.I., Ruocco, M., Middleton, R.S.: Testing the First Law of Cognitive Geography on Point-Display Spatializations. In: Kuhn, W., Worboys, M.F., Timpf, S. (eds.) COSIT 2003. LNCS, vol. 2825, pp. 316–331. Springer, Heidelberg (2003)
9. Robinson, J.O.: The Psychology of Visual Illusion. Courier Dover Publications (1998)
10. Salvini, M.M., Gnos, A.U., Fabrikant, S.I.: Cognitively Plausible Visualization of Network Data. In: 25th International Cartographic Conference, Paris, FR, July 3-8, CO-424 (2011)
11. Sedlmair, M., Munzner, T., Tory, M.: Empirical Guidance on Scatterplot and Dimension Reduction Technique Choices. IEEE Transaction on Visualization and Computer Graphics 19(12), 2634–2643 (2013)
12. Skupin, A., Fabrikant, S.I.: Spatialization. In: Wilson, J., Fotheringham, S. (eds.) Handbook of Geographic Information Science, Blackwell Publishers (2007)
13. St. John, M., Cowen, M.B., Smallman, H.S., Oonk, H.M.: The Use of 2D and 3D Displays for Shape-Understanding Versus Relative-Position Tasks. Human Factors 43(1), 79–98 (2001)
14. Tory, M., Sprague, D., Wu, F., So, W.Y., Munzner, T.: Spatialization Design: Comparing Points and Landscapes. IEEE Transactions on Visualization and Computer Graphics 13(6), 1262–1269 (2007)
15. Tory, M., Swindells, C., Dreezer, R.: Comparing Dot and Landscape Spatializations for Visual Memory Differences. IEEE Transactions on Visualization and Computer Graphics 15(6), 1033–1039 (2009a)
16. Tory, M., Arthur, E., Kirkpatrick, M., Atkins, S., Moeller, T.: Visualization Task Performance with 2D, 3D, and Combination Displays. IEEE Transactions on Visualization and Computer Graphics 12(1), 2–13 (2009b)

17. Ware, C.: Information Visualization: Perception for Design, 2nd edn. Morgan Kaufmann Publishers Inc., San Francisco (2004)
18. Ware, C., Franck, G.: Evaluating Stereo and Motion Cues for Visualizing Information Nets in Three Dimensions. ACM Transactions on Graphics 15(2), 121–140 (1996)
19. Ware, C., Mitchell, P.: Visualizing Graphs in Three Dimensions. ACM Transactions on Applied Perception 5(1), 1–15 (2008)
20. Williams, R.J.: Network3D [computer program]. Microsoft Research, Cambridge (2010)
21. Wise, T.A.: The Ecological Approach to Text Visualization. Journal of the American Society of Information Science 53(13), 1224–1233 (1999)
22. Wolfe, J.M., Horowitz, T.S.: What Attributes Guide the Deployment of Visual Attention and How Do They Do It? Nature Rev. Neuroscience 5, 1–7 (2004)
23. Wolfe, U., Maloney, L.T., Tam, M.: Distortions of Perceived Length in the Frontoparallel Plane: Tests of Perspective Theories. Perception & Psychophysics 67, 967–979 (2005)

Uncertainty Analysis of Step-Selection Functions: The Effect of Model Parameters on Inferences about the Relationship between Animal Movement and the Environment

Paul Holloway and Jennifer A. Miller

Department of Geography and the Environment, University of Texas at Austin, USA

Abstract. As spatio-temporal movement data is becoming more widely available for analysis in GIS and related areas, new methods to analyze them have been developed. A step-selection function (SSF) is a recently developed method used to quantify the effect of environmental factors on animal movement. This method is gaining traction as an important conservation tool; however there have been no studies that have investigated the uncertainty associated with subjective model decisions. In this research we used two types of animals – oilbirds and hyenas – to examine how systematically altering user decisions of model parameters influences the main outcome of an SSF, the coefficients that quantify the movement-environment relationship. We found that user decisions strongly influence the results of step-selection functions and any subsequent inferences about animal movement and environmental interactions. Differences were found between categories for every variable used in the analysis and the results presented here can help to clarify the sources of uncertainty in SSF model decisions.

1 Introduction

Movement data is now abundant in GIScience, resulting in applied research to extract patterns and processes from the underlying phenomena. While early applications have involved using data rich travel diaries to explore the relationship between human movement and space [1], there have also been questions related to movement ecology and animal movement [2]. The need to develop techniques to analyze the vast amount of spatio-temporal data was the driving force behind the GIScience sub-field of movement pattern analysis (MPA) [3]. The issue of uncertainty has been a long-standing research focus in GIScience, and uncertainty analysis is now considered a prerequisite for model building [4]. However, the field of MPA has been slower to explore such uncertainties (but see [5]). Uncertainty research within the MPA domain has focused on changes in the temporal scale of movement data and how this affects the calculation of movement parameters [5]. However, there is also considerable uncertainty associated with the statistical methods used to analyze movement data and their outputs that has not been investigate.

M. Duckham et al. (Eds.): GIScience 2014, LNCS 8728, pp. 48–63, 2014.
© Springer International Publishing Switzerland 2014

Step-selection function (SSF) is a powerful new spatial modeling approach that has been developed as an extension of resource-selection function (RSF – a function that is proportional to the probability of the use of a resource unit by an organism –[6]), and is beginning to gain traction as an important tool for studying conservation issues associated with animal movement. SSF was developed by Fortin et al. [7] by combining several methods in order to improve the ability to model resource availability in a home range. These concepts improved upon previous models that did not limit resource availability to an accessible distance of current animal location [8]. Whereas RSF considers the location of an observation, SSF considers the step between two locations. The observed step between two successive locations is compared to a number of alternatively generated steps that the animal could have taken (Fig. 1), and the coefficients from the case-control regression identify which environmental variables characterize the movement steps actually taken. The majority of SSF studies have estimated SSF in the form of:

$$\hat{w}(x) = \exp(\beta_1 x_1 + \beta_2 x_2 + \ldots + \beta_n x_n) \tag{1}$$

where β_n is the coefficient estimated by the conditional logistic regression for the variable x_n. Steps with higher SSF scores $\hat{w}(x)$ have a higher likelihood of being chosen by the animal, meaning that SSF can help to identify the influence of the environment on animal movements by revealing where they are most likely to be at the end of a movement step.

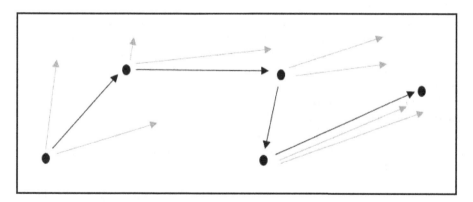

Fig. 1. Conceptual depiction of step-selection functions. Black dots represent successive telemetry locations of an individual, with black arrows representing observed steps. The grey arrows represent the available steps which the individual could have taken.

Since its development, SSF has been used to study a wide range of species, from wrens [9] to wolves [10], and is increasing in its popularity among ecologists due to the power of SSF to identify the influence of environmental variables on the movement of animals. Fitting an SSF involves a number of phases, and we addressed what we considered to be some of the most important and subjective decisions, most of which were described as potentially important by a SSF review paper that was published recently [11]. With SSF currently being used to inform researchers and conservationists about the relationship between animal movement and the environment, the uncertainty associated with the model building phases needs to be explored so that any applications and interpretations can be made with as much confidence as possible.

1.1   Method of Generating Available Steps

As SSFs compare used versus available steps, the method used to generate the available steps could potentially have the most influence on the results [11]. The method introduced by Fortin et al. [7] is still the most commonly applied and involves generating available steps from an empirical distribution of movement parameters (turn angle and step length) from other monitored individuals. Empirical distributions are classified in frequency tables of varying interval size, and draws are then made based on a rejection algorithm, whereby an interval is randomly selected from the table alongside a random number from a uniform distribution. If the random number is less than or equal to the specified probability value, a value from that interval is returned [12]. Fortin et al. [7] used intervals of $20°$ and 50m for the frequency tables for turn angle and step length respectively which equated to 18 bins ($360° / 20°$) for turn angle and 60 bins (maximum length of 3000m / 50m). In theory, the smaller the intervals of the bins, the closer the empirical distribution will match a more continuous real distribution. Other SSF studies have followed, using evenly spaced intervals for turn angle and step length [10], [13]. However, the intervals do not need to be even, and in portions of the distribution where the density is constant, a wider interval may be beneficial compared to when density changes rapidly and narrower bins may be favorable [12]; although, this adaptive method has yet to be employed.

Other studies have generated the available steps using a random distribution [14], a quantile distribution [15], and a parametric distribution [8]. When a random distribution has been used, turn angle was randomly drawn from between $-180°$ to $+180°$, and step length was randomly drawn from any value between 0 and the maximum step length. The same method has been used for quantile distributions, but the values were capped at a specific percentage of the distribution. Bjørneraas et al. [15] used the 99% quantile for step lengths, although this method was not used to select turn angle as the distribution is in degrees and so cannot be capped in the same manner.

A recent study by Forester et al. [8] examined the use of SSFs using three methods of generating available samples (in terms of step length); random (termed uniform), empirical and parametric and compared the results on simulated data and elk data in Yellowstone National Park. They found that empirical and parametric distributions performed better than the uniform distribution which was the method furthest from resembling a continuous real distribution. The level of bias in the estimates differed between empirical and parametric distributions and was dependent on the size of the coefficient. With empirical distributions currently more widely used in SSF studies, we will choose to focus on sampling from this distribution as uncertainty still exists between using even and uneven intervals.

1.2   Number of Steps

Once the available steps have been generated, the next user decision is the number of available steps used in the comparison. In their seminal paper, Fortin et al. [7] used 200 available steps, mainly because their research question was directed at rare habitat selection, although they noted that future SSF studies would not need such a high number. 200 remains the largest number of steps used in SSF studies so far, with the next highest value only 25 steps [10], while the minimum is two [15]. Subsequent studies have not specified the reasoning behind their selection of fewer steps beyond citing the statement of Fortin et al. [7] that they would not need 200. RSFs have been used more extensively than SSFs, and while some studies have suggested using 10,000 available locations (in total) in the comparison [16, 17], using an ad hoc approach still appears to be the norm [18]. Subsequently, it is unknown whether a high number of available locations/steps over-samples the environmental choices available, or whether a low number of available locations/steps under-samples.

1.3   Modeling Approach

The majority of SSF studies have used conditional logistic regression as the statistical method, but there has been variation with respect to how the analysis is conducted. Conditional logistic regression in which regression parameters are fit for all of the individuals together has been the most widely used approach. An alternative to this has recently been employed whereby model parameters are fit for each individual separately and then averaged to attain aggregate information, and has been termed individual modeling [14], [19], [20]. Individual modeling can potentially capture more individual movement traits while still being applicable to the aggregation of individuals and is beginning to be used more regularly.

1.4   Individual Variation

The effect of individual variation on results has only recently begun to be incorporated in SSF studies, but in the wider field of telemetry studies, this is slightly more developed. Lindberg and Walker [21] suggest that more than 20 animals are needed to make reliable statistical inferences about simple population comparisons, and at least 75 animals for realistically complex studies. However, due to the high costs of GPS units, sample sizes are often far from these numbers, with several of the SSF studies containing less than 20 individuals [7], [19], [22]. In studies where sample sizes are lower, individual traits may have more influence on the results, although in a recent study using seven Eleonora's falcons, Gschweng et al. [23] found that removing one from the study did not result in a significant change in habitat use. Few studies have taken into account the idiosyncratic differences among individual animals in SSF models and this is an area that needs further study.

1.5   Uncertainty Analysis

Uncertainty analysis allows researchers to assess the range of outputs associated with the model responses as a result of variations in parameter values used in the model input [4]. This uncertainty analysis will provide SSF practitioners with insight into how to choose appropriate model parameters as well as information on the level of uncertainty associated with the results. Here, we analyze the effect of 1) generating the available steps based on a) turn angle and b) step length, 2) the number of available steps used, 3) the modeling approach used, and 4) the number of individuals used.

2   Methods

2.1   Data Collection

Oilbird (*Steatornis caripensis*) data was obtained from Holland et al. [24] via Movebank [25]. GPS with remote UHF readout was used to collect locations of four individuals with ten minute intervals, resulting in approximately 800 fixes for use, with the number of observations ranging from 133 to 264. Brown hyena (*Hyaena brunnea*) data was collected by Maude [26] in the Makgadikgadi Pans region of northern Botswana between June 2004 and December 2007 with the support of the Makgadikgadi Brown Hyena Project. GPS locations were recorded for each hyena at 1 hour intervals. Ten hyenas were used in this analysis across a time period of a month, which resulted in 4000 fixes for use, ranging from 278-432. The environmental variables used in both of these analyses are briefly described in table 1.

2.2   Uncertainty Analysis: Generating Available Step

Steps were generated from drawing from four distributions for both turn angle and step length; empirical distribution (even bins), empirical distribution (uneven bins), random distribution, and quantile distribution (99%) (n.b. quantile distribution was not used for turn angle). Empirical distributions of even bins formed intervals of 20° for turn angles, and 187.3m and 362m for step length for oilbirds and hyenas respectively. The values of step lengths equates to 2% of the maximum step length. Empirical distributions of uneven bins formed intervals of 5.56% quantiles for turn angle and 2% quantiles for step lengths. These values ensured that the same number of bins would be used for even and uneven empirical distributions.

2.3   Uncertainty Analysis: Number of Available Steps

The number of available steps were chosen to fall between the minimum (2) and the maximum value used (200) in previous SSF studies: 2, 10, 20, 50, 100, and 200.

Table 1. Information about the environmental variables used in the regression models. The IGBP classification scheme was used for the MODIS land cover product [20], which delineates into 16 classes.

Animal	Environmental Variable	Method	Hypothesis	Source
Oilbird	Land Cover – Evergreen Broadleaf Forest	Categorical Variable - Value of land cover at the end of the step	Oilbirds eat the fruit of tropical laurels. The majority of these trees are evergreen broadleaf species. The oilbirds are more likely to move through land covers which contain food sources.	[27]
Oilbird	Cropland – Percentage of cropland	Continuous Variable - Value of cropland at the end of the step	Oilbirds also eat the fruit of oil palms, a commercial crop in South America. The higher the percentage of cropland, the higher chance of oil palms in the area.	[28]
Oilbird	Distance to Roads	Continuous Variable - If step crosses line – distance = 0. Else, distance equals the average distance from start, mid and end of the step	Birds have been found to frequent edge habitats, with roads one of the most common features splitting habitats.	[29]
Hyena	Land Cover – Savanna	Categorical Variable - Value of land cover at the end of the step	Hyenas may visit savanna habitats more often due to a higher potential of prey species.	[27]
Hyena	Land Cover – Open Shrublands	Categorical Variable - Value of land cover at the end of the step	Hyenas may visit shrubland habitats more often due to a higher potential of prey species.	[27]
Hyena	Land Cover – Fragmentation of habitat	Continuous Variable - Number of surrounding habitats of the cover at the end of the step	The more surrounding habitats results in the potential of more prey species. More habitats may mean more species	[27]
Hyena	Distance to Roads	Continuous Variable - If step crosses line – distance = 0. Else, distance equals the average distance from start, mid and end of the step	Hyenas could use roads to travel along or obtain road kill. Alternatively they may avoid them due to the dangers.	[29]

2.4   Uncertainty Analysis: Modeling Approach

Conditional logistic regression (equation 1.0) was done in R using the survival package [30]. This was either conducted with all individuals in one table (herein referred to as aggregate modeling), or for each individual separately with the coefficients averaged to generate an aggregate level model (herein referred to as individual modeling).

2.5   Uncertainty Analysis: Individual Variation

Conditional logistic regression was fit for all the individuals, while systematically dropping one individual from the analysis (n.b. this analysis was undertaken using available steps generated from the empirical distribution of all the individuals). All combinations of turn angle, step length, number of steps and modeling approach was also conducted for each model without certain individuals.

2.6   Data Analysis

In total, 720 oilbird and 1584 hyena regression models were fit. The coefficient values were compared for each variable with Wilcoxon matched pairs signed rank test using a Bonferroni corrected α of 0.05 according to the number of comparisons made. This test converts scores to ranks and compares them across the two conditions. The effect size of the test was calculated by dividing the z value by the square root on N and using the Cohen [31] criteria of 0.1 = small effect, 0.3 = medium effect and 0.5 = large effect.

3   Results

For both hyenas (table 2) and oilbirds (table 3), some combination of coefficients differed for each variable (n.b. tables only contain comparisons where one significant difference exists). The distribution used to generate the turn angle differed between random and empirical (even or uneven), although this effect was very small. Step length distribution caused a variety of significant differences in the coefficients, with Fig. 3 showing the range of coefficient values for savanna selection for hyenas and evergreen broadleaf forest for oilbirds. The number of available samples used in the model resulted in some small significant differences, although these were not consistent between species (tables 2 and 3). Individual modeling resulted in inflated coefficient values when compared to the aggregate model (Fig. 2), with only two hyenas and one oilbird appearing to cause the increased values (Fig. 4). Finally, systematically removing individuals caused significant differences, with the largest differences appearing to match the individuals causing the inflated coefficients in Fig. 4. Removing individuals from the analysis does result in the coefficient of most importance changing in the final regression model in some instances (Fig. 5).

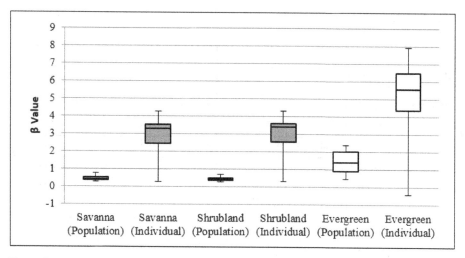

Fig. 2. β values for savanna and shrublands for individual and population modeling of hyena SSF (grey) and evergreen broadleaf forest for oilbirds (white)

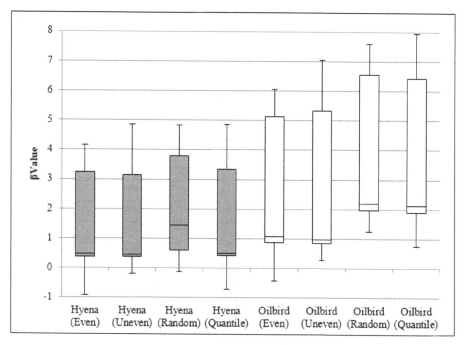

Fig. 3. β values for savanna habitat for hyenas (grey) and evergreen broadleaf forests for oilbirds (white) based on the four methods of generating step lengths of oilbird SSF

Table 2. Wilcoxon Matched Pairs Signed Rank for hyenas, value refers to second group, *medium effect, **large effect

Variable		β Roads	β Savanna	β Shrubland	β Edge
Method	Aggregate - Individ	Lower	Higher**	Higher**	Lower**
Turn Angle	Even - Random	n.s.	Lower	Lower	Higher
	Even - Uneven	n.s.	n.s.	n.s.	Lower
	Uneven - Random	n.s.	Lower	Lower *	Lower *
Step Length	Even - Uneven	n.s.	n.s.	Lower	Higher
	Even - Random	Lower*	Higher**	Higher**	Higher**
	Even - Quantile	n.s.	Higher*	Higher	n.s.
	Uneven - Random	Lower*	Higher**	Higher*	Higher*
	Uneven - Quantile	n.s.	Higher*	Higher	Lower
	Quantile - Random	Lower*	Higher**	Higher**	Higher**
Number of Steps	2 - 50	Lower	n.s.	n.s.	n.s.
	10 - 20	n.s.	Lower*	Lower	n.s.
	10 - 50	n.s.	Lower*	Lower	n.s.
	10 - 100	n.s.	Lower	Lower	n.s.
	10 - 200	n.s.	Lower*	Lower	n.s.
	20 - 50	n.s.	n.s.	Higher	Lower
	20 - 200	n.s.	Lower	n.s.	n.s.
	50 - 200	n.s.	Lower	Lower	n.s.
	100 - 200	n.s.	Lower*	Lower	n.s.
Individuals	All - H1	Higher	Higher**	Higher**	Lower**
	All - H2	n.s.	Lower	Lower**	Lower**
	All - H3	Higher	Lower*	Higher**	Higher**
	All - H4	n.s.	Higher**	Higher**	Higher**
	All - H5	n.s.	Lower	Lower**	Lower**
	All - H6	n.s.	Lower	Lower**	Lower**
	All - H7	Higher	Lower	Lower*	Higher*
	All - H8	n.s.	Higher*	Higher	n.s.
	All - H9	n.s.	Higher**	Higher**	Lower**
	All - H10	Higher	Lower	Higher	n.s.

Table 3. Wilcoxon Matched Pairs Signed Rank for oilbirds, value refers to second group, *medium effect, **large effect

Variable		β Roads	β Evergreen	β Croplands
Method	Aggregate - Individ	Lower	Higher*	Lower**
Turn Angle	Even - Random	n.s.	n.s.	Lower
Step Length	Even - Uneven	n.s.	n.s.	Higher
	Even - Random	n.s.	Higher**	n.s.
	Even - Quantile	n.s.	Higher**	n.s.
	Uneven - Random	n.s.	Higher**	n.s.
	Uneven - Quantile	n.s.	Higher*	Lower
	Quantile - Random	n.s.	Higher*	Higher
Number of Steps	2 - 10	Higher	n.s.	Higher
	2 - 20	Higher	n.s.	Higher
	2 - 50	Higher	n.s.	Higher
	2 - 100	Higher	n.s.	Higher
	2 - 200	Higher	n.s.	Higher
	10 - 20	n.s.	Lower	n.s.
	10 - 200	n.s.	Lower*	n.s.
	20 - 200	n.s.	Lower	n.s.
	50 - 100	n.s.	Lower	n.s.
	50 - 200	n.s.	Lower*	n.s.
	100 - 200	n.s.	Lower*	n.s.
Individuals	All - B1	n.s.	Higher**	Lower**
	All - B2	Lower	Higher**	Higher**
	All - B3	n.s.	n.s.	Lower
	All - B4	Lower	Higher*	Lower**

4   Discussion

The objective of this study was to investigate the uncertainty associated with user decisions and how these decisions influence the results obtained from step-selection functions for two different types of animals (oilbirds and brown hyenas). Oilbirds congregate in caves at night, and forage for food by day, with most of their food source coming from evergreen laurels. Brown hyenas forage solitarily, across large home ranges encompassing a variety of land covers, returning to a clan den at night. Both datasets had a considerable number of observations, as well as differing temporal resolutions (10 minutes for oilbirds, 1 hour for hyenas). Subsequently, any similarities or differences in the results could be attributed to either a user-decision or a species specific trait and should be used by other researchers to help inform their implementation of SSF.

Our results indicate that there is considerable uncertainty associated with the decisions of selecting model parameters and their effects on coefficient values. SSFs associate parameters of movement rules with landscape features, as well as modeling the choices actually presented to the animal as it traverses through the landscape [32]. However, if the user decisions do alter the results observed, then it is difficult to disentangle actual step/habitat preferences from model parameterization decisions. The number of significant differences found in tables 2 and 3, and the size of these differences suggests that researchers are currently unable to distinguish between whether results are representative of step/habitat preferences or whether they are a function of researchers selecting certain methods to generate their SSF model.

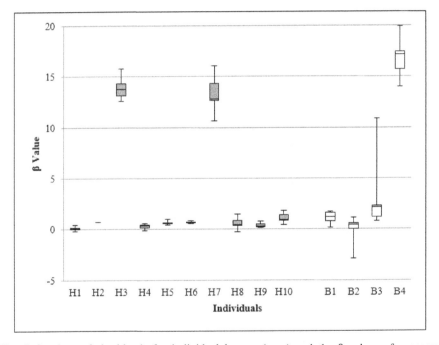

Fig. 4. β values of shrublands for individual hyenas (grey) and the β values of evergreen broadleaf forest for individual oilbirds (white)

The largest differences appeared to be between the modeling approach used and systematically removing one individual from the model. The idea behind an individual modeling approach is that the final model will contain results that highlight environmental interactions that better represent individual to population preferences for habitat use, something that is not possible with aggregate modeling (see Figs. 2 and 4). Fig. 2 shows the range of coefficient values for savanna and shrublands (for hyenas) and evergreen forest (for oilbirds), when calculated using the aggregated and individual modeling approaches. Fig. 4 shows the range of coefficients for each individual hyenas and oilbirds that were averaged in the individual modeling approach. It can be observed that hyena 3 and 7, and oilbird 4 have much higher coefficients than the others (indicating their increased preference for moving into that specific habitat). The aggregate method appears to suppress the idiosyncratic preferences of these individuals, while the individual method incorporates it with a higher average value. Individual modeling therefore incorporates individual information into the coefficient and researchers need to be aware of these differences, as the method used in analysis could over- or under-estimate the importance of an environmental variable based on the idiosyncratic preferences of just one or two individuals (as shown in our research).

The influence of individuals is furthered highlighted by the differences in results when one was dropped from the regression (tables 2 and 3). Removing either hyena 1, 4, 8 and 9 from analysis results in a higher preference for savanna habitats and removing either hyena 1, 3, 4, 8, 9 and 10 from analysis result in a higher preference for shrublands. Similar results are observed for oilbirds (table 3). With only four oilbirds, these differences in results are less surprising, but ten hyenas are not a small sample size, and is larger than the sample size used by Gschweng et al. [23] when they concluded that removing one individual did not result in statistically different results (albeit the results were not based on conditional logistic regression). Fig. 5 identifies the number of times removing an individual results in the coefficient of most importance shifting from the value in the complete model. Overall, these values are relatively small, but it shows that in some instances coefficient importance can change. Therefore, datasets of similar sizes as those used in this research are relatively sensitive to individual preferences. Interestingly, the hyena which had the most impact on a change in coefficient importance is not one of the hyenas already identified as having a substantially higher coefficient value when conditional logistic regression was run for each individual (Figs. 4 and 5). Individuals can therefore influence the value of the coefficients obtained as well as their subsequent importance compared to other environmental variables, although the individual that does so is not necessarily the same for both differences.

A recent review outlining the need for an uncertainty analysis of SSF suggested that the method of generating the available steps could be the most important decision when developing this model [11]. While the method of generating the step length caused significant changes, this was not the case for generating the turn angle (tables 2 and 3). While significant differences did exist between turn angle distributions, they were either of a small or medium effect for hyenas, and only one difference between even and random distributions existed for the coefficient describing the effect of croplands on oilbirds. Step length was a more important variable than turn angle, although it appeared more important for variables where the value was measured at the end of the step (savanna, shrubland, edge habitat for hyenas and broadleaf forest for oilbirds). Distributions which selected randomly up to the maximum (or 99% quantile) step length resulted in higher coefficients than even or uneven distributions (Fig. 3). These results could be compounded by infrequent movement steps that occur over great distances, but are not controlled for through their low probability of occurrence as they would be with an empirical distribution. The higher coefficients obtained using a random distribution are a result of longer steps taking the available steps beyond the evergreen forest for oilbirds and the savanna and shrubland for hyenas. This increases the number of alternative habitats the animal could select, and thus increases the coefficient value of their preferred habitat.

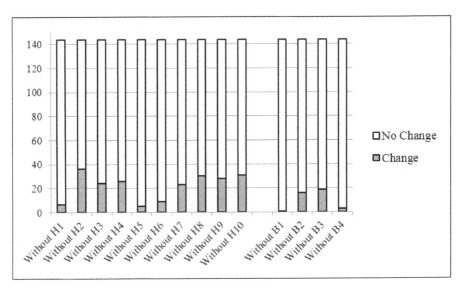

Fig. 5. The number of times removing an individual from the regression (note 144 combinations of variables) results in the coefficient of most importance shifting

Generating steps using a quantile distribution (99%) resulted in higher coefficients for oilbirds, but not hyenas (Fig. 3). The distribution of step lengths for hyenas is relatively normal (note a slight bimodal distribution), while for oilbirds it is L shaped (data not shown). Therefore, using a 99% quantile distribution removes the most extreme values for the hyenas, but fails to do so with the oilbirds, meaning that the quantile distribution continues to produce results similar to a random distribution. We suggest that researcher investigate the type of distribution associated with step lengths first, before making a decision on the type of distribution they use to generate available steps.

The number of available steps was incorporated in the uncertainty analysis as it is unknown whether a high or low number of available steps over- or under-samples the environmental choices available. While results were not consistent for oilbirds and hyenas, differences in the number of available steps used were found for both hyenas and oilbirds (table 2 and 3). Coefficients describing the effects of roads and croplands on oilbird movement differed between all combinations with either 2 steps or 200 steps. This suggests that while values between 10 and 100 do not statistically differ, the smallest and largest values do. The opposite was observed for hyenas, with many of the medium to higher values of steps being significantly different and while 2 steps was not significantly different to a higher number. However, the majority of these differences are small, and it is the category which has the most not significant differences. While it may not be the variable that has the strongest effect on the results, it does have a slight impact (possibly due to rare habitats or homogenous environments) and subsequently the researcher needs to select a number of steps accordingly.

This study used a variety of environmental variables, including discrete, continuous and distance to, with values measured at various points along the step (see table 1). Significant differences in the effect of step length could be attributed to the possibility that this value was measured at the end of the step, while distance to roads (which was often less significant) was averaged across the whole length. More research into how the environmental variables are measured has been suggested by Thurfjell et al. [11] as an area of research that needs further exploration, and certain patterns in the results of this study certainly indicate that this warrants further research.

5   Conclusion

User decisions strongly influence the results of step-selection functions and any subsequent inferences about animal movement and environmental interactions. By assuming that results would be consistent between methods, any conservation management strategies based upon SSF research could be a function of specifying unrealistic movement options. This study found that differences in individual behaviors have the strongest influence on the results observed. Averaging coefficients across individuals results in higher values when studied at an aggregate level, indicating that individual preferences are lost when studied solely at an aggregate level. The influence of removing one individual from the study was surprisingly significant and contradicts recent research [23]. Researchers conducting analysis on such medium sized datasets need to be aware that idiosyncratic preferences could potentially influence the results in terms of coefficient values and importance of variables used in the model, and should check for such occurrences. The method of generating available steps was important, but in this study not as important as some have suggested [11], although variations within the distributions used in this study (i.e. smaller intervals for empirical distributions) could be investigated further. Finally, while the number of steps used in comparison was the least important variable in determining coefficient values, it was still significantly different for a number of combinations. Differences between the extremes (2 and 200) suggest that a medium value may be preferable, but that only 2 steps could potentially mask actual movement preferences. User decisions of SSF practitioners should subsequently be justified based on research objectives where possible and further research into other user decisions of SSF should continue.

Acknowledgments. We would like to thank Richard Holland, Martin Wikelski, Franz Kümmeth and Carlos Bosque for providing the oilbird data, and Glyn Maude for providing the brown hyena data. Research support for this work was provided by the National Science Foundation (#0962198).

References

1. Miller, H.J.: A measurement theory for time geography. Geogr. Anal. 37, 17–45 (2005)
2. Cagnacci, F., Boitani, L., Powell, R.A., Boyce, M.S.: Animal ecology meets GPS-based radiotelemetry: a perfect storm of opportunities and challenges. Philos. T. Roy. Soc. B. 365, 2157–2162 (2010)
3. Gudmundsson, J., Laube, P., Wolle, T.: Computational movement analysis. In: Springer Handbook of Geographic Information, pp. 423–438. Springer, Heidelberg (2012)
4. Crosetto, M., Tarantola, S.: Uncertainty and sensitivity analysis: tools for GIS-based model implementation. Int. J. Geogr. Inf. Sci. 15, 415–437 (2001)
5. Laube, P., Purves, R.S.: How fast is a cow? Cross-Scale Analysis of Movement Data. Trans. GIS. 15, 401–418 (2011)
6. Manly, B.F., McDonald, L.L., Thomas, D.L., McDonald, T.L., Erickson, W.P.: Resource selection by animals: statistical design and analysis for field studies. Springer (2002)
7. Fortin, D., Beyer, H.L., Boyce, M.S., Smith, D.W., Duchesne, T., Mao, J.S.: Wolves influence elk movements: behavior shapes a trophic cascade in Yellowstone National Park. Ecology 86, 1320–1330 (2005)
8. Forester, J.D., Im, H.K., Rathouz, P.J.: Accounting for animal movement in estimation of resource selection functions: sampling and data analysis. Ecology 90, 3554–3565 (2009)
9. Gillies, C.S., Beyer, H.L., St. Clair, C.C.: Fine-scale movement decisions of tropical forest birds in a fragmented landscape. Ecol. Appl. 21, 944–954 (2011)
10. Latham, A.D.M., Latham, M.C., Boyce, M.S., Boutin, S.: Movement responses by wolves to industrial linear features and their effect on woodland caribou in northeastern Alberta. Ecol. Appl. 21, 2854–2865 (2011)
11. Thurfjell, H., Ciuti, S., Boyce, M.S.: Applications of step-selection functions in ecology and conservation. Mov. Ecol. 2, 4 (2014)
12. Beyer, H.L.: Geospatial Modeling Environment, http://www.spatialecology.com/gme/gmehelp.htm (accessed December 12, 2013)
13. Coulon, A., Morellet, N., Goulard, M., Cargnelutti, B., Angibault, J.-M., Hewison, A.M.: Inferring the effects of landscape structure on roe deer (Capreolus capreolus) movements using a step selection function. Landsc. Ecol. 23, 603–614 (2008)
14. Squires, J.R., DeCesare, N.J., Olson, L.E., Kolbe, J.A., Hebblewhite, M., Parks, S.A.: Combining resource selection and movement behavior to predict corridors for Canada lynx at their southern range periphery. Biol. Conserv. 157, 187–195 (2013)
15. Bjørneraas, K., Solberg, E.J., Herfindal, I., Moorter, B.V., Rolandsen, C.M., Tremblay, J.-P., Skarpe, C., Sæther, B.-E., Eriksen, R., Astrup, R.: Moose Alces alces habitat use at multiple temporal scales in a human-altered landscape. Wildl. Biol. 17, 44–54 (2011)
16. Lele, S.R., Keim, J.L.: Weighted distributions and estimation of resource selection probability functions. Ecology 87, 3021–3028 (2006)
17. Barbet-Massin, M., Jiguet, F., Albert, C.H., Thuiller, W.: Selecting pseudo-absences for species distribution models: how, where and how many? Methods in Ecology and Evolution 3, 327–338 (2012)
18. Northrup, J.M., Hooten, M.B., Anderson Jr, C.R., Wittemyer, G.: Practical guidance on characterizing availability in resource selection functions under a use-availability design. Ecology 94, 1456–1463 (2013)
19. Northrup, J.M., Pitt, J., Muhly, T.B., Stenhouse, G.B., Musiani, M., Boyce, M.S.: Vehicle traffic shapes grizzly bear behavior on a multiple-use landscape. J. Appl. Ecol. 49, 1159–1167 (2012)

20. Sawyer, H., Nielson, R.M., Lindzey, F., McDonald, L.L.: Winter habitat selection of mule deer before and during development of a natural gas field. J. Wildlife Manage. 70, 396–403 (2006)
21. Lindberg, M.S., Walker, J.: Satellite telemetry in avian research and management: sample size considerations. J. Wildl. Manag. 71, 1002–1009 (2007)
22. Pereboom, V., Mergey, M., Villerette, N., Helder, R., Gerard, J.-F., Lode, T.: Movement patterns, habitat selection, and corridor use of a typical woodland-dweller species, the European pine marten (Martes martes), in fragmented landscape. Can. J. Zool. 86, 983–991 (2008)
23. Gschweng, M., Kalko, E.K., Berthold, P., Fiedler, W., Fahr, J.: Multi-temporal distribution modeling with satellite tracking data: predicting responses of a long-distance migrant to changing environmental conditions. J. Appl. Ecol. 49, 803–813 (2012)
24. Holland, R.A., Wikelski, M., Kümmeth, F., Bosque, C.: The secret life of oilbirds: new insights into the movement ecology of a unique avian frugivore. PLoS One 4, e8264 (2009)
25. Data from: The secret life of oilbirds: new insights into the movement ecology of a unique avian frugivore, http://hdl.handle.net/10255/move.269
26. Maude, G.: The spatial ecology and foraging behavior of the brown hyaena (Hyaena brunnea) (Doctoral dissertation, University of Bristol) (2010)
27. Land Processes Distributed Active Archive Center (LP DAAC) at the USGS/Earth Resources Observation and Science (EROS) Center, Siox Falls, South Dakota (2013)
28. Ramankutty, N., Evan, A.T., Mondreda, C., Foley, J.A.: Global Agricultural Lands: Croplands (2000); Data distributed by the Socioeconomic Data and Applications Center (SEDAC), http://seda.ciesin.columbia.edu/es/algands.html (2010) (accessed January 3, 2014)
29. Center for International Earth Science Information Network - CIESIN - Columbia University and Information Technology Outreach Services - ITOS - University of Georgia. Global Roads Open Access Data Set, Version 1 (gROADSv1). NASA Socioecomonic Data and Applications Center (SEDAC), Palisades (2013), http://sedac.ciesin.columbia.edu/data/set/groads-global-roads-open-access-v1 (accessed on January 3, 2014)
30. Therneau, T.: A package for survival analysis in S. R package version 2.37.4. (2013)
31. Cohen, J.: Statistical power analysis for the behavioral sciences. Psychology Press (1988)
32. Chekiewicz, C.L.B., St Clair, C.C., Boyce, M.S.: Corridors for conservation: integrating pattern and process. Annu. Rev. Ecol. Evo. Syst. 37, 317–342

Logic Scoring of Preference and Spatial Multicriteria Evaluation for Urban Residential Land Use Analysis

Kris Hatch[1], Suzana Dragićević[1], and Jozo Dujmović[2]

[1] Department of Geography, Simon Fraser University, Canada
[2] Department of Computer Science, San Francisco State University, USA

Abstract. The Logic Scoring of Preference (LSP) is a general multicriteria decision-making method with origins in soft computing and fuzzy reasoning. It allows the nonlinear aggregation of a large number of input criteria without the loss of significance typical for additive GIS-based MCE methods. The objective of this study is to integrate the LSP method with GIS and create LSP suitability maps for the land suitability analysis applied to real geospatial datasets for new urban residential development in the Metro Vancouver Region, Canada. Several factors influencing land use change were selected to construct the aggregation structure for the LSP-GIS method and implemented for decision-making purposes. This method allows simultaneity, replaceability and a range of other aggregators to match various evaluation objectives. The obtained results indicate that the LSP-MCE method provides refined evaluation of urban land suitability and therefore has a high potential for use in urban planning.

1 Introduction

Urban growth and associated urban sprawl has been and will continue to be one of the largest sources of human impact on the terrestrial environment [1]. Continued rise in city populations leads to elevated housing prices that force residents to locate further away from urban centers [2] and further from pre-existing urban transportation network, which in turn leads to increased commuting times and greater reliance on automobiles making humans an ever increasing environmental threat [3]. As a result of the negative consequences associated with urban growth and subsequent sprawl, policies must be formulated to understand and provide improved urban growth forecasting in order to better manage urban sprawl. Spatial decision-making and land-use suitability analysis is one way to identify the most appropriate spatial patterns for future land uses based on a set of preferences and requirements [11]. Multicriteria evaluation (MCE) methods, a specific type of spatial decision support systems (SDSSs), provide complex trade-off analysis between choice alternatives with different environmental and socio-economic factors [12]. In terms of land-use change, MCE has been used to investigate various applications related to land suitability [13]. MCE methods are often used for spatial optimization, and can be linked with GIS and used within a spatial decision support system (SDSS). They are tools designed for decision makers to explore, structure, and solve complex spatial problems [17]. The goal of optimization is to obtain an optimal solution under a set of predefined factors

M. Duckham et al. (Eds.): GIScience 2014, LNCS 8728, pp. 64–80, 2014.
© Springer International Publishing Switzerland 2014

and constraints for analyzing the complex trade-offs between choice alternatives with different environmental and socio-economic impacts [12]. This is accomplished by combining the information from several input parameters into a single suitability index. Spatial optimization models benefit greatly from the addition of MCE analysis due to site-specific evaluation inquiries that need to be undertaken.

Spatial MCE methods are structured using approaches such as the analytic hierarchy process (AHP), simple additive scoring (SAS), multiattribute value technique (MAVT), multiattribute utility technique (MAUT), ordered weighted averaging (OWA), and outranking methods [16]. All MCE methods model human decision-making logic to analyze the complex trade-offs between choice alternatives. However, they have been criticized for producing an oversimplification of reality that is complex [5, 14]. These claims are based on the fact that existing ordinary ranking methods commonly used in GIS-based MCE do not adequately represent human decision making logic and observable properties of human reasoning [5, 16]. The Logic Scoring of Preference (LSP) is a method for analyzing complex trade-offs between choice alternatives, based on precise modeling of human evaluation reasoning [23]. The LSP method provides the flexibility, precision and justifiability of evaluation criteria derived from the structural and logic consistency with observable properties of evaluation reasoning. For each point in a resulting suitability map, LSP criterion functions can use any number of input attributes, and generate an overall suitability score which is defined as a degree of truth of the statement that all requirements are perfectly satisfied. The structure of each LSP criterion function is based on a set of attributes, the corresponding attribute criteria, and a soft computing logic aggregation of attribute suitability scores; that structure is an observable property of human reasoning.

Moreover, the LSP method offers a variety of specific types of elementary attribute criteria. In the area of aggregation of attribute suitability scores, the LSP method offers all aggregators that are observable in human reasoning: hard and soft partial conjunction, hard and soft partial disjunction, pure conjunction and disjunction, conjunctive/disjunctive neutrality, and asymmetric aggregators of conjunctive partial absorption (aggregation of mandatory and optional attributes), disjunctive partial absorption (aggregation of sufficient and optional attributes), as well as complex canonical aggregation structures [6,8,16,21]. In addition, LSP criteria provide separate selection of formal logic parameters of *andness* and *orness* and semantic parameters of relative importance of attributes in aggregation structures. These are unique properties of the LSP method, not available in any of the other techniques such as those based on traditional AHP or OWA [5,16].

Therefore, the main objective of this study is to integrate the LSP method into a GIS-based multicriteria evaluation design in order to evaluate locations for new residential growth given a large set of choice alternatives. The LSP method was developed in a raster GIS environment with a goal to determine suitability potentials for spatial locations for residential land use based upon the combination of factors that are relevant to land-use change.

2   Theoretical Background

The Logic Scoring of Preference method (LSP) follows an aggregation structure where data inputs are represented on a standardized scale and organized into relevant attributes [15]. Inputs are grouped categorically, and arranged on a LSP attribute tree. They are then combined through the use of several LSP aggregators, which represent a spectrum of conditions ranging from *simultaneity* to *replaceability*. The LSP aggregators form the LSP aggregation structure [5]. LSP generates reliable results in relation to the inputs and parameters for chosen aggregation. Features that make the LSP method unique and more effective than other MCE methods include: (i) the use of the step-wise logic aggregation structure which allows for flexibility through its use of continuous logic represented in terms of *simultaneity* and *replaceability* [10], and (ii) the ability to include a large number of inputs into an LSP aggregation structure without loss of significance for any individual input due to the type of logic expressions used in the LSP method. The comparison of LSP, OWA and AHP methods are presented in previous research such as [16], [8], and also in [5], [10], [15].

All additive weighted scoring models (frequently used in MCE) compute an aggregated score as a weighted sum of attribute scores $S=W_1S_1+...W_nS_n$. Such models cannot support mandatory requirements (i.e. $S_i=0$ cannot produce $S=0$) or sufficient requirements (cases where $S_i=1$ must produce $S=1$). The LSP method uses a variety of nonlinear aggregators that are multiplicative in nature (e.g. as a geometric mean $S=S_1^{W_1}S_2^{W_2}...S_n^{W_n}$) and can model mandatory requirements (where $S_i=0$ can produce $S=0$), sufficient requirements, and many other more complex logic structures. It is useful to note that in additive models the weights are normalized ($W_1+W_2+...+W_n=1$) so the average weight is $(W_1+W_{2+...+}W_n)/n=1/n$. The average weight $1/n$ reflects the average significance of an input attribute (by changing an attribute score from 0 to 1 the output score change is limited by the value of corresponding weight). Thus, with an increase of the number of attributes, the significance of each attribute in additive models decreases making some of them completely insignificant. That is not the case with LSP method where the aggregation is nonlinear and each of input attributes can be mandatory, or sufficient or optional. Advantages of the nonlinear LSP models in comparison with traditional additive models are explicitly discussed in reference [8].

The LSP method was initially developed for applications in computer science, such as windowed environment software evaluation [20], evaluation of Java IDEs [21], comparison of search engines [22], as well as other multi-criteria evaluation approaches. Recently, the LSP has been linked with spatial data and GIS, and the multi-criteria evaluation method has been used for solving problems in the field of spatial science. Dujmović et al. [10,15,5,16] proposed the concept of LSP suitability maps which represent a continuous degree of suitability with respect to a particular purpose or objective. The main goal of LSP suitability maps is to provide a suitability degree for a geographic region for purposes such as: suitability for industrial development, agriculture, housing, education, and recreation [15]. In particular, the LSP approach was used to determine suitability for residential land use change [16] where various criteria for residents to move into homes were analyzed, with the LSP method used to aggregate the criteria and determine spatially optimal house locations. The LSP method is implemented in three stages as follows: (1) development of an LSP attribute tree, (2) definition of attribute criteria, and (3) the LSP aggregation structure.

2.1   LSP Attribute Tree

The LSP attribute tree organizes the decision problem using a hierarchical structure of attributes. The decomposition of the overall suitability yields a decomposition tree where elementary attributes are the leaves of the tree. The input attributes are separately evaluated and corresponding suitability degrees are then combined together using the LSP aggregators until one overall output suitability is obtained. It is important that the set of input attributes includes attributes that are both necessary and sufficient for evaluation of suitability and development of LSP suitability maps. Some of the input attributes are denoted as *mandatory*, meaning the input requirements must be satisfied, and others are *optional*, meaning that their satisfaction is desired but not mandatory.

2.2   Attribute Criteria

Each input attribute is separately evaluated using a specific elementary attribute criterion. The attribute criteria reflect stakeholder requirements that input attributes should satisfy. For example, if an attribute represents a distance from a residential area to a park, then distances less than 200 meters can be considered perfect, and distances greater than 2000 meters can be considered unacceptable. Such criteria frequently use linear interpolation between a set of attribute-satisfaction points selected by a decision maker. The choice of attribute criteria for this study is presented in detail in Section 3.2 (Fig. 2).

2.3   LSP Aggregation Structure

In each point of the LSP suitability map the set of elementary attribute criteria generate a set of attribute suitability degrees. The degrees of attribute suitability are then logically aggregated in order to generate an overall suitability degree. The aggregation process based on logical requirements and weighting parameters [8]. LSP aggregators express the combination mandatory, (+), and optional, (-) input criteria. Each LSP aggregator expresses the combination of input parameters on a spectrum of *and-ness* and *orness*, conditions ranging from full conjunction, C, to full disjunction, D. Table 1 depicts different levels of *simultaneity* and *replaceability* and their symbolic notation. It also shows the corresponding values of the weighted power mean exponent r that we use to achieve aggregation that has desired logic properties.

Each LSP aggregator used reflects the degree of *simultaneity* or *replaceability* desired to be expressed between the inputs considered. The further along the spectrum from neutrality (A, the arithmetic mean) to full conjunction (C) (Table 1) the aggregator used is, the stronger and more restrictive the degree of *simultaneity* is. The further in the other direction, from neutrality (A) to full disjunction (D), the stronger is the *replaceability* among inputs. Neutrality (A) is used to express the balance of *simultaneity* and *replaceability*. LSP aggregators can be grouped into one of seven aggregator types [16]. These include: Full Conjunction (LSP aggregator C in Table 1), Hard Partial Conjunction (using aggregators such as C++, C+, C+-, CA, C-+), Soft Partial Conjunction(C-, C--), Neutrality (A), Soft Partial Disjunction, Hard Partial Disjunction, and Full Disjunction (D). Partial disjunction can be realized as De Morgan dual of partial conjunction. If no hard partial disjunction is not needed then we ca directly use soft partial disjunction realized using weighted power means with exponents presented in Table 1. Choosing the LSP aggregator is determined by the desired level of simultaneity or *replaceability* between inputs that the decision maker wants to express. A Hard Partial Conjunction (HPC) operator is used to express the combination of mandatory inputs, whereas a Soft Partial Conjunction operator is less restrictive, and is appropriate for the combination of optional inputs. The analogue is true for Hard Partial Disjunction and Soft Partial Disjunction operators.

Table 1. LSP aggregators representing *simultaneity* and *replaceability*

				Simultaneity				
Symbol C	C++	C+	C+-	CA	C-+	C-	C-	A
r	-∞	-9.06	-3.51	-1.655	-0.72	-0.148	0.261	0.619	1.0

				Replaceability				
Symbol D	D++	D+	D+-	DA	D-+	D-	D-	A
r	∞	20.63	9.521	5.802	3.929	2.792	2.018	1.449	1.0

When combining two or more mandatory inputs, or two or more optional inputs, each of the LSP aggregators combines the inputs using a *generalized conjunction disjunction* (GCD) function described in [6], and realized in the form of weighted power mean. Given a set of input parameters $X_1,...,X_n$, the generalized conjunction disjunction is computed using the following weighted power mean (WPM):

$$GCD(X_1, ..., X_n) = [W_1 X_1^r + \cdots + W_n X_n^r]^{1/r} \qquad (1)$$

where $GCD(X1,...,Xn)$ is the output suitability from the combination of input parameters, $X_1,...,Xn$. The weights $W_1,...,Wn$ are used to express the relative importance of of inputs $X1,...,Xn$, and r is used to express the degree of *simultaneity* and *replaceability* among the inputs $X1,...,Xn$. There are several ways to determine W_i values. These include using the analytical hierarchy process (AHP), preferential neural networks, or using the perceptions of experts (evaluator, stakeholder and domain experts).

The combination of mandatory with optional inputs requires using the conjunctive partial absorption (CPA) function, which uses a different mathematical function described in [9]. Given a mandatory input X, and an optional input Y, the CPA function can be realized as follows:

$$S(X,Y) = [(1 - W_2)[W_1 X^{r_1} + (1 - W_1)Y^{r_1}]^{r_2/r_1} + W_2 X^{r_2}]^{1/r_2} \qquad (2)$$

The CPA aggregation scheme operates so that the optional input penalizes or rewards the overall output value from the combination of mandatory and optional inputs. In other words, given a non-zero mandatory input value (as a zero value will lead to a zero value after the combination of mandatory and optional inputs), the lower the value of the optional input, the greater the penalty applied to the output value after the combination of the mandatory and optional inputs. However if the $y>x$, then a reward is applied. Full information on penalty and rewards with respect to the CPA aggregator can be found in [9].

2.4   Linking LSP and GIS

The LSP method has not been widely used in GIS as it was not developed in geography and it requires understanding of soft computing evaluation logic. Its sophistication brings complexity as using wrong aggregators can create difficulties to first time users. In this study spatially referenced raster data was used as inputs in the LSP method. Fuzzy functions are generated in the raster GIS environment to standardize the inputs as the unit interval. Often this standardized scale is representative of a suitability index, where spatial locations with higher values on the standardized scale have greater suitability with respect to the desired objective. Inputs, intermediate results, and outputs can be represented as GIS raster maps depicting a continuous surface of values of suitability across an entire study site.

3   LSP-GIS Method for Urban Residential Land Suitability

In this study the LSP-GIS method was used to determine spatially optimal locations for urban residential growth across the regional district of Metro Vancouver Canada (Fig. 1). Input criteria consisted of selected factors and data that influence residential growth, with three types of LSP aggregators used for the analysis: neutrality (A), soft partial conjunction (C-- and C-), hard partial conjunction (CA and C+). Three aggregation structures were developed, each of which represent models of strong *simultaneity*, meaning that when combining inputs, all the inputs must have high degrees of satisfaction (indicating high values in their suitability maps) in order to have a high output value. They were developed to also demonstrate that adding more criteria the output suitability maps present more refined results of the suitability of location for new urban developments.

3.1   Study Area and Data

The LSP-GIS method was implemented using the following datasets for year 2006: (i) 20 meter resolution Digital Elevation Model (DEM), (ii) transportation networks (bus, light rail, and roads), (iii) land-use data (Fig. 1), and (iv) Canada Census data. Both the ESRI ArcGIS and the IDRISI Selva GIS software were used to implement the LSP-GIS method.

Three scenarios were designed to best represent the various situations for possible urban residential growth. Scenario 1 is designed with the intended goal of determining suitable locations for suburban development. Density near downtown Vancouver is high. Additionally, housing prices in downtown Vancouver and closer to it are very high and unaffordable, making it important to develop the rural-urban fringe which is the furthest suburbs from downtown Vancouver. Scenario 2 has its focus on family-oriented growth while Scenario 3 was taking in consideration the importance of the transportation network. The geography is a limiting factor in the Metro Vancouver region with the many rivers, bridges and elevation differences making travel time to work and back long. Many residents do not live in close proximity to their workplace especially those living in the suburban areas, far from the central business district and far from efficient public transportation. For these reasons the three scenarios were designed to place greater emphasis on certain categories of importance for residential urban development.

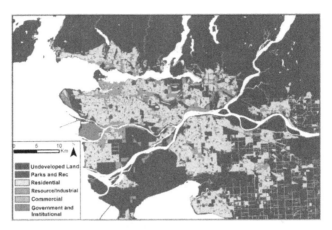

Fig. 1. Land-use map for Metro Vancouver Metropolitan Area for year 2006

3.2　Choosing Elementary Attribute Criteria

For the purpose of illustrating the elementary criteria definition, Scenario 1 is used as an example. A set of elementary criteria is presented in Fig. 2. The corresponding input attributes, classified as mandatory (+) or optional (-), are grouped into four categories (Table 2). Inputs were transformed from their original units into a standard suitability scale using the elementary attribute functions. Suitability values range from 0 to 1 where 0 represent a completely unsuitable area and 1 represent a completely suitable area, and the intermediate values are dependent on the shape of the attributete criterion functions. Figure 2 represents the complete list of fuzzy suitability functions used to transform the input datasets for all three scenarios based on the following categories:

Terrain and Environment: Slopes from 0 to 30 degrees are considered suitable. From 30 to 40 degrees there is decreasing suitability, and a slope beyond 40 degrees is considered completely unsuitable. South facing homes, with aspect values between 135 and 225 degrees are most suitable, due to sunlight exposure. Elevation decreases in suitability from sea level (less than 50 meters) to 1000 meters, beyond which the elevation is completely unsuitable.

Amenities: For each of the amenities, locations in closer proximity are considered more suitable: suitability scores decrease as distance increases, based on driving or walking times. Some amenities penalize (have a lower suitability) if the distance is too small.

Accessibility: Similar to amenities, closer proximity to transportation (roads, bus, light rail, airport) is most suitable, unless the proximity is too close in which case is it considered unsuitable due to noise.

Population: The three distance criteria all favor close proximity locations. Growing communities (larger population growth) are more suitable, and communities with lower median incomes (hence more affordable homes) are more suitable.

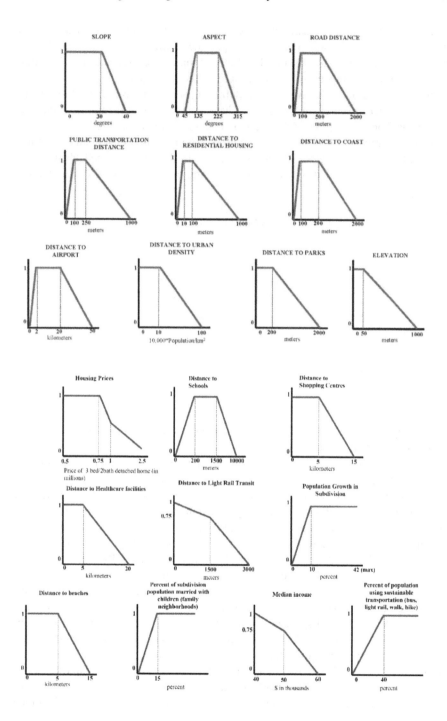

Fig. 2. Elementary criteria transformation functions for each of scenarios 1, 2, and 3

The Scenario 2 uses the same set of elementary criteria and transformation functions as Scenario 1, but with the logical requirements different for some of the input criteria. Scenario 3 uses a subset of 10 of the 19 inputs used for Scenario 1, with the same transformation functions but different logic requirements.

Minor changes in elementary criteria definition (Fig 2) translate to minor changes of input preference scores and cannot produce significant modifications of the suitability maps. For the purpose of this study the choice of the fuzzy functions were done based on the informed knowledge of the users. The proposed elementary criteria are very justifiable and easily acceptable by readers. In the case of urban planning they can be chosen and defined by stakeholders or any experts involved in the decision-making process.

Table 2. Elementary criteria for Scenario 1

Terrain and Environment	Amenities	Accessibility	Population
(+) Slope	(-) Distance to beach	(+) Distance to major roads	(+) Distance to residential housing
(+) Aspect	(-) Distance to coast	(-) Distance to bus lines	(+) Distance to low density areas
(-) Elevation	(-) Distance to parks	(-) Distance to light rail	(+) Distance to family areas
	(-) Distance to shopping	(-) Distance to airport	(-) Population growth
	(-) Distance to care facilities	(-) Amount of sustainable transport	(-) Median income
	(+) Distance to schools		

3.3 LSP Aggregation Structure

In this study, three types of LSP aggregators were used to combine the inputs: neutrality (A), soft partial conjunction (C-- and C-), and hard partial conjunction (CA and C+). The weighted power mean [6] was used when combining mandatory inputs together or optional inputs together and creating compound aggregators.

The suitability maps created from fuzzy linear functions applied to the input datasets form the input criteria for the LSP aggregation structure implemented in the GIS software. The weighted power mean and CPA functions were used to manipulate the suitability maps and create the aggregation for each scenario. Map algebra tools enabled the implementation of WPM and CPA functions in the GIS software.

A framework for the choice of aggregators for combining inputs was determined. The combination of optional and mandatory inputs together necessitates the choice of two aggregators (Fig. 3). The first aggregator with exponent r_1 is usually the neutrality aggregator, A. The second aggregator with exponent r_2 is usually a strong partial conjunction aggregator. The satisfaction of a mandatory input is required, meaning that in its fuzzy suitability map it must have a value greater than 0. While an optional input does not need to be satisfied, the higher value in the fuzzy suitability map (closer to 1) gives a reward to the overall output (higher suitability values in the output map). There are two extreme cases within which a particular penalty or reward is applied when using the conjunction partial absorption function: if the value for the optional input Y is zero, or if the value for Y is one. In the former case, then a penalty is applied when evaluating the overall output of the combination of X and Y. In the latter case, a reward is applied in the combination of X and Y.

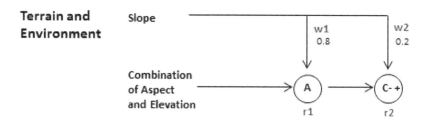

Fig. 3. Example from LSP aggregation structure: the combination of a mandatory (Slope) with optional (Aspect and Elevation) inputs

Both scenarios 1 and 2 operate on two similar aggregation structures (Fig. 4 and 5), with greater influence placed on inputs in the amenities and accessibility categories. Placing greater influence on inputs in the amenities and accessibility categories was performed by using higher weights of preference on inputs (or combinations of inputs) in these categories when combining them with inputs (or combinations of inputs) in other categories. The inclusion of many inputs in Scenarios 1 and 2 demonstrate one of the main benefits of the LSP method to be implemented: the inclusion of many inputs without loss of importance. Scenario 3 (Fig. 6) is focused on the proximity of the transportation network. Within any of the aggregation structures for scenarios 1, 2, or 3 (Figs. 4,5, 6), inputs are combined in a single aggregator (moving left to right in the aggregation structure), the level of *simultanety* increases, and aggregators with stronger partial conjunction (CA, C+-, etc.) are used. This type of aggregation structure is known as a *conjunctive canonical aggregation structure with increasing andness* [16], and is most appropriate for logic aggregation of suitability maps. Therefore, as can be seen in Figs. 4, 5, and 6, within any individual category (ex. amenities), the LSP aggregators used are either *neutrality* (A), or a weak partial conjunction (C--, C-, C-+), and as inputs from multiple categories are combined (moving further to the right in the aggregation structure), the LSP aggregators use stronger forms of partial conjunction (CA, C+-, C+). Penalties and rewards for each scenario, based on [9], are presented in Table 3.

Weights used when combining inputs were determined based on weighting schemes used in previous MCE/LSP studies [8], as well as based on the desired level of influence each input has on the overall suitability for each scenario, and the desired level of influence each input has each time it is combined with other inputs.

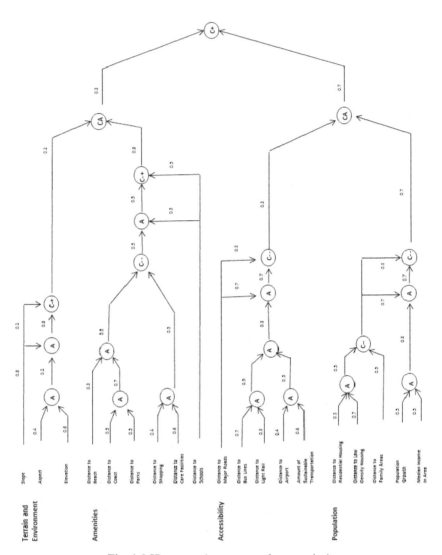

Fig. 4. LSP aggregation structure for scenario 1

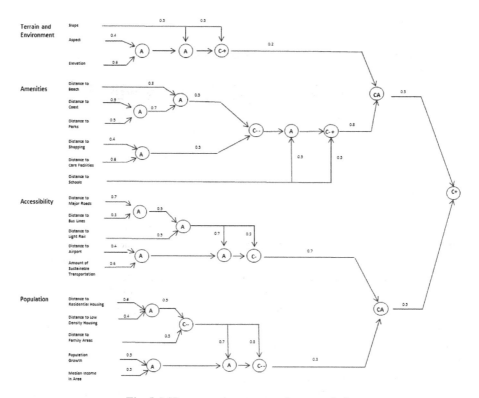

Fig. 5. LSP aggregation structure for scenario 2

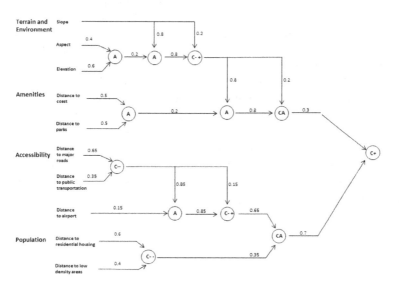

Fig. 6. LSP aggregation structure for scenario 3

Table 3. Penalty/reward table when using conjunctive partial abosportion function in scenarios 1, 2, and 3

Aggregator Used	W1	W2	Penalty (Y=0)	Reward (Y=1)
C-+	0.5	0.5	-0.229	0.15
C-	0.7	0.3	-0.218	0.191
C--	0.7	0.3	-0.214	0.201
C-+	0.8	0.2	-0.164	0.146

4   Results

The residential urban suitability output maps for each scenario are shown in Fig. 7. The open land use class was considered and the LSP output values for the suitability are grouped into five categories with each covering a particular range. Classification of categories was based on natural breaks (Jenks) of the data. The five categories are: high suitability (0.788 and 1.000], medium-high suitability (0.592-0.788], medium suitability (0.404-0.592], medium-low suitability (0.161-0.404], and low suitability [0.000-0.161]. High suitability is an indicator of a compounding satisfaction of several input parameters that follow the logic parameters set out in the attribute tree and aggregation structure.

All three scenarios have high suitability values in the undeveloped areas near the rural-urban fringe. The scenario 1 shows significant growth in the southeast section of the Metro Vancouver region, a much less densely populated part but with much more rural and affordable land for new residential development. The scenario 2 has high suitability values in a significant number of areas across the study site as criteria chosen such as proximity to schools, shopping centers and parks among others are providing suitable site for residential developments that may target new families. The scenario 3 emphasis on transportation is clearly evident: with higher suitability values nearer to freeways, transportation lines, and major roads. Figure 8 represents the municipality of Cloverdale, part of City of Surrey, indicating subsections of output maps depicting the three scenarios. The results illustrate the potential for a fast residential suburban development and the obtained maps could be informative for municipal planners. The LSP map outputs collectively confirm the expected spatial patterns of the scenarios and therefore add significant confidence to the utility of the LSP approach for potential decision making.

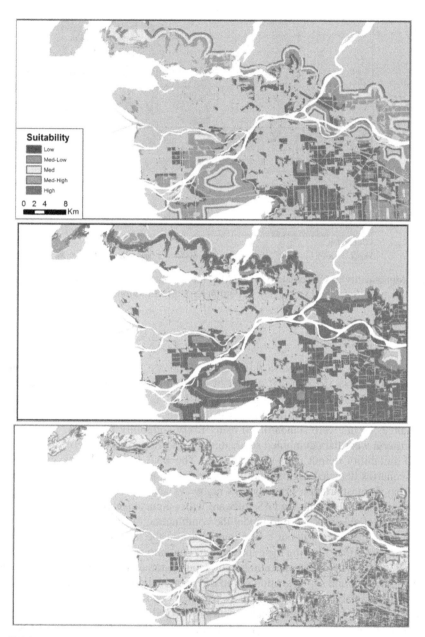

Fig. 7. LSP suitability maps for scenarios 1 (top), 2 (middle), and 3 (bottom). Green is the class with highest, red is with lowest suitability score for housing development.

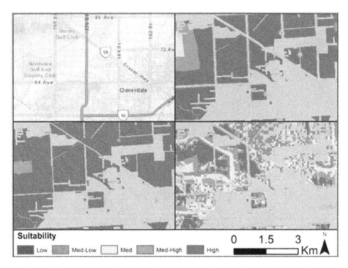

Fig. 8. Municipality of Cloverdale (Google Map top left) and the urban residential land suitability outputs for Scenario 1 (top right), 2 (bottom left), and 3 (bottom right)

5   Conclusions

The LSP method was applied in this study to evaluate the suitability for residential development across the regional district of Metro Vancouver, Canada. Three scenarios were designed with each using a different LSP aggregation structure. The input criteria in the aggregation structures consisted of factors relevant to residential land-use change. Inputs were standardized using fuzzy suitability functions. Standardized input criteria were then combined sequentially by a series of LSP aggregators until an overall suitability output map was obtained. The final LSP suitability maps of the three scenarios depicted high suitability in suburban areas as well as high suitability along major transportation lines such as highways and bus routes. The entire implementation was done within the GIS software with outputs displayed as static maps.

LSP is a reliable procedure to model the combination of several inputs to obtain an output suitability score without losing importance in the individual inputs themselves, and at the same time allowing the expression of continuous logic. The robustness of the model depends heavily on the number of input parameters considered, choices for these parameters, and sophistication of the LSP aggregation structure. Furthermore, the success of the analysis depends highly on parameter choices made at each stage of the model. In this study, the output spatial patterns derived from the LSP maps confirmed the expectations from the scenarios and hence add to the confidence of the LSP approach integrated within GIS-based MCE and applied to real geospatial datasets.

The obtained LSP suitability maps generate justifiable and useful results. The sensitivity analysis and reliability analysis of the LSP method was conducted at a theoretical level [24-26], but this is yet to be tested in a geographical context. The proposed method has a high potential to be used by urban planners as they can incorporate a wider variety of planning perspectives in the form of criteria and aggregators, and obtain many feasible outputs for land parcels suitable for further urban development. For example, an interactive, dynamic web-based LSP suitability map system integrated with Google Maps to evaluate walkability and ten other criteria is provided by SEAS [27]. The LSP-MCE still needs to be fully incorporated into common GIS software as a module that can be used for a wide variety of decision making applications. This method is more refined then the existing MCE modules currently available in GIS software. Within the core of the GIS software and user friendly interface it can become more comprehensible for urban planners and developers, or even in use for participatory decision making process, or by any user in need to solve various geographic suitability problems.

Acknowledgments. This study is fully supported by a Natural Sciences and Engineering Research Council (NSERC) of Canada Discovery Grant awarded to the second author. The authors are thankful to four anonymous reviewers for their valuable comments on an earlier draft of the paper.

References

1. Kaufmann, R.K., Seto, K.C., Schneider, A., Liu, Z., Zhou, L., Wang, W.: Climate Response to Rapid Urban Growth: Evidence of a Human-Induced Precipitation Deficit. J. Clim. 20, 22–99 (2007)
2. Glaeser, E.L., Gyourko, J., Saks, R.E.: Urban growth and housing supply. J. Econ. Geogr. 6, 71–89 (2006)
3. Glaeser, E.L., Kahn, M.E.: Sprawl and Urban Growth. In: Vernon Henderson, J., Thisse, J.-F. (eds.) Handbook of Regional and Urban Economics, vol. 4, pp. 2481–2527 (2004)
4. Malczewski, J.: GIS-based land-use suitability analysis: a critical overview. Prog. Plann. 62(1), 365 (2004)
5. Dujmović, J.J., De Tre, G., Dragićević, S.: Comparison of Multicriteria Methods for Land-use Suitability Assessment. In: European Soc Fuzzy Logic & Technology, Linz (2009)
6. Dujmović, J.J., Larsen, H.L.: Generalized conjunction/disjunction. Int. J. Approx. Reason. 46, 423–446 (2007)
7. Dujmović, J.J.: The LSP Method for Evaluation and Selection of Computer and Communication Equipment. In: Proceedings of MELECON 1987 Mediterranean Electrotechnical Conference and 34th Conference on Electronics, pp. 251–254. IEEE/RIENA, Rome (1987)
8. Dujmović, J.J., Scheer, D.: Logic Aggregation of Suitability Maps. In: International Conference on Fuzzy Systems, pp. 1–8. IEEE, Barcelona (2010)
9. Dujmović, J.J.: Partial Absorption Function. J. Univ. Belgrade, EE Dept., Ser. Math and Phys. 659, 156–163 (1979)
10. Dujmović, J.J., De Tre, G., Van de Weghe, N.: LSP suitability maps. Soft Comput. A Fusion Found. Methodol. Appl. 14, 421–434 (2010)
11. Jankowski, P., Richard, L.: Integration of GIS-based suitability analysis and multicriteria evaluation in a spatial decision support system for site selection. Env. and Plan B 21(6), 323–340 (1994)
12. Carver, S.J.: Integrating multi-criteria evaluation with geographical information systems. Int. J. Geogr. Inf. Syst. 5, 321–339 (1991)
13. Voogd, H.: Multicriteria Methods for Urban and Regional Planning. Taylor & Francis (1983)

14. Malczewski, J.: GISbased multicriteria decision analysis: a survey of the literature. Int. J. Geogr. Inf. Sci. 20, 703–726 (2006)
15. Dujmović, J.J., De Tré, G., Van de Weghe, N.: Suitability Maps Based on the LSP Method. In: Torra, V., Narukawa, Y. (eds.) MDAI 2008. LNCS (LNAI), vol. 5285, pp. 15–25. Springer, Heidelberg (2008)
16. Dujmović, J., De Tre, G.: Multicriteria methods and logic aggregation in suitability maps. Int. J. Intell. Syst. 26, 971–1001 (2011)
17. Densham, P.J.: Spatial Decision Support Systems. Geogr. Inf. Syst. Princ. Appl. 1, 403–412 (1991)
18. Dujmović, J.J., Bayucan, A.R.: A Quantitative Method for Software Evaluation and its Application in Evaluating Windowed Environments. In: Software Engineering Conference. IASTED, San Francisco (1997)
19. Dujmovic, J.J., Nagashima, H.: LSP Method and its Use for Evaluation of Java IDEs. Int. J. Approx. Reas. 41, 3–22 (2006)
20. Dujmović, J.J., Bai, H.: Evaluation and Comparison of Search Engines Using the LSP Method. Comp. Sci. Inf. Sys. 3, 31–56 (2006)
21. Dujmović, J.: Preference Logic for System Evaluation. IEEE Trans. Fuz. Syst. 15(6), 1082–1099 (2007)
22. Dujmović, J.J., Fang, W.Y.: An Empirical Analysis of Assessment Errors for Weights and Andness in LSP Criteria. In: Torra, V., Narukawa, Y. (eds.) MDAI 2004. LNCS (LNAI), vol. 3131, pp. 139–150. Springer, Heidelberg (2004)
23. Dujmović, J.J., Fang, W.Y.: Reliability of LSP Criteria. In: Torra, V., Narukawa, Y. (eds.) MDAI 2004. LNCS (LNAI), vol. 3131, pp. 151–162. Springer, Heidelberg (2004)
24. Dujmović, J.J., Allen III, W.L.: A Family of Soft Computing Decision Models for Selecting Multi-Species Habitat Mitigation Projects. In: Yager, R.R., Reformat, M.Z., Shahbazova, S.N., Ovchinnikov, S. (eds.) Proceedings of the World Conference on Soft Computing, paper 101 (2011)
25. SEAS, LSPmaps (2014), http://www.seas.com

Spatial Weights: Constructing Weight-Compatible Exchange Matrices from Proximity Matrices

François Bavaud

University of Lausanne, Switzerland

Abstract. Exchange matrices represent spatial weights as symmetric probability distributions on pairs of regions, whose margins yield regional weights, generally well-specified and known in most contexts. This contribution proposes a mechanism for constructing exchange matrices, derived from quite general symmetric proximity matrices, in such a way that the margin of the exchange matrix coincides with the regional weights. Exchange matrices generate in turn *diffusive* squared Euclidean dissimilarities, measuring spatial remoteness between pairs of regions. Unweighted and weighted spatial frameworks are reviewed and compared, regarding in particular their impact on permutation and normal tests of spatial autocorrelation. Applications include tests of spatial autocorrelation with diagonal weights, factorial visualization of the network of regions, multivariate generalizations of Moran's I, as well as "landscape clustering," aimed at creating regional aggregates both spatially contiguous and endowed with similar features.

1 Introduction

Weighted unoriented networks are specified by node and edge weights. In spatial statistics, node weights f represent the relative importance of regions, normalized to unity, entering into the definition of weighted averages of the form $\bar{x} = \sum_i f_i x_i$. Also, edge weights e_{ij} constitute spatial weights, entering in the definition of spatially autocorrelated models.

Edge weights are *weight-compatible* if their margins coincide with the set of regional weights f, generally well-defined and known a priori. Symmetric and weight-compatible edge weights define an *exchange matrix E*, whose components can be interpreted as *the probability of selecting a pair of regions*.

On one hand, exchange matrices arguably constitute a style of spatial weights particularly adapted to weighted spatial contexts. On the other hand, exchange matrices E are hardly ever directly known to the fellow worker. Instead, the researcher in general only possesses vague, incomplete spatial information, as expressed in a *spatial proximity matrix G*, whose components provide a spatial measure of proximity between pairs of regions. The proximity or generalized adjacency matrix G may represent adjacencies, the size of the common boundary, the inter-regional accessibility, the inter-regional flow of exchanged units (people, matter, goods, information), as well as many other proxies for neighborhood.

Normalizing G and enforcing symmetry makes the matrix formally equivalent with a distribution on regional pairs, that is with an exchange matrix - see specification U) below. However, the marginal distribution γ resulting from G *does not coincide in general*

M. Duckham et al. (Eds.): GIScience 2014, LNCS 8728, pp. 81–96, 2014.

with the regional spatial weights f: while f measures *regional importance,* γ measures *regional centrality.* Yet, plainly put, a region can be peripheral and important, or central and insignificant, thus establishing the necessity of constructing weight-compatible exchange matrices $E(G, f)$, that is based upon G, but with margin f.

This contribution recalls and reviews a few definitions in spatial autocorrelation (section 2), in unweighted and weighed settings. Particular emphasis is devoted to the comparison of their corresponding canonical measures of spatial autocorrelation, their permutation and normal significance testing, as well as the handling of off-diagonal spatial weights, occurring not that infrequently in applications, such as those involving flows and self-interaction.

The central part (section 3) proposes the construction of a one-parameter family of weight-compatible exchange matrices $E(G, f, t)$ from proximity matrices G. The former, describing a continuous diffusive process generated by G, turns out to be p.s.d., allowing further the definition of a *diffusive squared Euclidean dissimilarity* $\mathcal{D}(G, f, t)$ between regions (section 3.3).

Spatial analysis of French elections illustrate the theory (section 4). Possible applications, briefly outlined in sections (4.1), (4.2) and (4.3), include multivariate generalization of Moran's I, factorial visualization of spatial versus attribute dissimilarities between regions, as well as "landscape clustering," aimed at creating regional aggregates both spatially contiguous and endowed with similar characteristics.

2   (Un)weighted Measures of Spatial Autocorrelation

2.1   Unweighted Setting: Spatial Weights from Spatial Links $V(G)$

In presence of n regions of equal importance (uniform weighting) characterized by the density variable x, Moran's index of spatial autocorrelation is usually defined as (e.g., [1,2,3,4])

$$I \equiv I(V, x) := \frac{n \sum_{ij} v_{ij}(x_i - \bar{x})(x_j - \bar{x})}{(\sum_{ij} v_{ij}) \sum_i (x_i - \bar{x})^2} \qquad \bar{x} := \frac{1}{n} \sum_{i=1}^{n} x_i \qquad (1)$$

where the *spatial weights* matrix $V = (v_{ij})$ is non-negative, symmetric or not.

By construction, $I \equiv I(V, x)$ does depend upon on the spatial field x under investigation, as well as, crucially, upon the specification of spatial weights V: e.g., see [5] for an explicit illustration. The latter authors also propose and investigate various *spatial coding schemes* aimed at extracting a convenient spatial weights matrix $V(G)$ from *spatial link* or *proximity* matrices $G = (g_{ij})$, meant as an immediate, possibly rough but accessible spatial information about proximity relationships between regions i and j. Proximities G between regions may be determined by mutual contiguity, accessibility, inverse distance, flow, etc. In what follows, we assume G to be *symmetric* $g_{ij} = g_{ji}$ as well as *essentially non-negative*, that is such that $g_{ij} \geq 0$ for $i \neq j$. Typical choices are

i) $g_{ij} = a_{ij}$: binary adjacency or contiguity matrix
ii) $g_{ij} = n_{ij}$, where n_{ij} counts the number of units (people, matter, money, information) flowing from i to j

iii) $g_{ij} = F(d_{ij})$ where d_{ij} is a measure of the distance between i and j and $F(d) \geq 0$ is a distance-deterring, decreasing function

iv) $g_{ij} = |\partial A_{ij}|$, the measure of the common boundary between distinct regions i and j.

In particular, [5] together with other workers have considered the following *coding schemes* $V(G)$:

B) the *binary spatial weights* $v_{ij} = 1(g_{ij} > 0)$ taking on value one if $g_{ij} > 0$, and zero otherwise

W) the *row-standardized spatial weights* $v_{ij} := g_{ij}/g_{i\bullet}$ (where $\bullet$ denotes summation over the replaced index, as in $g_{i\bullet} := \sum_j g_{ij}$), prevalent in models of spatial auto-correlation

C) the *globally standardized spatial weights* $v_{ij} := n g_{ij}/g_{\bullet\bullet}$

U) the *standardized spatial weights* $v_{ij} := g_{ij}/g_{\bullet\bullet}$

S) the *variance-stabilizing spatial weights*

$$v_{ij} := \frac{n\, s_{ij}^*}{\sum_{ij} s_{ij}^*} \qquad \text{where} \qquad s_{ij}^* := \frac{g_{ij}}{\sqrt{\sum_j g_{ij}^2}} \ .$$

The above spatial coding schemes respectively constitute the so-called B, W, C, U and S schemes, as referred to and used in the spdep R package [6,7].

2.2 Weighted Setting: E-Coding Scheme $E(G, f, t)$

In all generality, the importance of the n regions differ, as quantified by their relative weights $f_i > 0$ with $\sum_{i=1}^{n} f_i = 1$. Typically, regional weights f_i reflect the relative population (human geography), relative area (physical geography) or relative wealth (economic geography) of region i. Regional spatial weights f_i can be interpreted as the probability $P(i)$ of selecting region i.

Specifying a symmetric probability $P(i, j) = e_{ij}$ to select a pair of neighboring regions (i, j) defines the *exchange matrix* $E = (e_{ij})$ [8]. By construction,

$$e_{ij} = e_{ji} \geq 0 \qquad e_{i\bullet} = f_i > 0 \qquad \sum_{ij} e_{ij} = e_{\bullet\bullet} = 1 \ .$$

In this weighted setup, Moran's index of spatial autocorrelation reads (e.g., [9] and references therein)

$$I \equiv I(E, x) := \frac{\sum_{ij} e_{ij}(x_i - \bar{x})(x_j - \bar{x})}{\sum_i f_i(x_i - \bar{x})^2} \qquad \bar{x} := \sum_{i=1}^{n} f_i x_i \qquad (2)$$

In particular, $-1 \leq I(E, x) \leq 1$ with expected value $E_0(I(E, x)) = -1/(n-1)$ under absence of spatial autocorrelation, *provided E contains no diagonal components* (see section 6 for the general case). Note the equivalent formulation

$$I(E, x) = \frac{\mathrm{var}(x) - \mathrm{var}_{\mathrm{loc}}(x)}{\mathrm{var}(x)} \qquad (3)$$

where

$$\text{var}(x) = \frac{1}{2} \sum_{ij} f_i f_j (x_i - x_j)^2 = \sum_i f_i (x_i - \bar{x})^2$$

is the ordinary (weighted) variance and

$$\text{var}_{\text{loc}}(x) = \frac{1}{2} \sum_{ij} e_{ij} (x_i - x_j)^2$$

is the *local variance*, measuring the average dissimilarity between pairs of spatially associated regions (e.g., [10,11,12,13]). The concept of local variance is related, yet distinct, to the concept of *local indicator of spatial autocorrelation* [2], referring to a weighted decomposition of Moran's I (3) as in $I(E, x) = \sum_i f_i I_i(E, x)$.

The row-standardized matrix of spatial weights $W = (w_{ij})$ obtains from the exchange matrix as $w_{ij} = e_{ij}/f_i$, and constitutes the transition matrix of a reversible Markov chain [14].

3   Obtaining the Exchange Matrix in Two Steps (4) and (5)

Given a symmetric and essentially non-negative, "off-positive" proximity matrix G, i.e., whose off-diagonal components are non-negative, as well as a set of regional weights f, compute the symmetric matrix $\Psi = (\psi_{ij})$

$$\psi_{ij}(G, f) = \frac{1}{\sqrt{f_i f_j}} \frac{\delta_{ij} g_{i\bullet} - g_{ij}}{(g_{\bullet\bullet} - \text{trace}(G))} \ . \tag{4}$$

Then compute the exchange matrix by means of the matrix exponential

$$E(t) := \Pi^{1/2} \exp(-t\,\Psi) \Pi^{1/2} \quad \text{where} \quad \Pi = \text{diag}(f) \quad \text{and} \quad t \geq 0 \ . \tag{5}$$

The free parameter $t > 0$ interprets as the *age of the network*. By construction (proofs below):

1) $E(t)$ in (5) is symmetric and weight compatible: $e_{i\bullet}(t) = f_i$ for any $t \geq 0$
2) $e_{ij}(t) \geq 0$ for all i, j and $t \geq 0$
3) $E(t)$ is p.s.d. (section 3.2)
4) $\lim_{t \to 0} e_{ij}(t) = \delta_{ij} f_i$, expressing complete regional segregation in a "frozen network"
5) $\lim_{t \to \infty} e_{ij}(t) = f_i f_j$, which expresses absence of distance-deterrence effects in a "free" or "complete network."

3.1   Further Formal Considerations

The numerator in (4) contains the so-called *Laplacian* of G (e.g., [15] p.12), defined as $(LG)_{ij} := \delta_{ij} g_{i\bullet} - g_{ij}$, where δ_{ij} are the components of the identity matrix (Kronecker's delta). By construction, LG is *positive semi-definite* (p.s.d.) (see section 3.2) and obeys

$$\text{trace}(LG) = \text{sum}(G) - \text{trace}(G) \ \geq \ 0 \qquad\qquad \text{sum}(LG) = 0 \tag{6}$$

where $\text{sum}(C) := c_{\bullet\bullet}$ and $\text{trace}(C) := \sum_i c_{ii}$. In particular, and $L\,\text{diag}(b) = 0$ *for any diagonal matrix* $\text{diag}(b)$ *with diagonal* b: hence the diagonal elements a_{ii} of the adjacency matrix in (i) (loops), the stayers flow n_{ii} in (ii), or self-boundaries $|\partial A_{ii}|$ in (iv) play *no role* in the construction of LG, Ψ or $E(t)$. $\Psi(G)$ is indeed invariant with respect to transformations of the form $G \to a(G + \text{diag}(b))$, for any scalar $a > 0$ and any vector b (cf. (6) and (4)).

Normalizing $\Psi(G)$ as in (4) amounts in normalizing t in (5), in the hope of making the scale t "intrinsic" or "absolute," that is hopefully comparable among differing data sets. As a matter of fact,

$$\Psi = \Pi^{-1/2} \frac{LG}{\text{trace}(LG)} \Pi^{-1/2} \qquad E(t) = \Pi - \frac{LG}{\text{trace}(LG)} t + O(t^2) \quad (7)$$

The question of the choice of t itself remains fairly open so far. The "self-exchange proportion" $\text{trace}(E(t))$ decreases with t, converging to $\text{trace}(E(\infty)) = \sum_i f_i^2 < 1$, with small time expansion $\text{trace}(E(t)) = 1 - t + 0(t^2)$. This proportion could possibly be estimated by $\text{trace}(N)/\text{sum}(N)$, where N is the inter-regional flows matrix, or some other measure of spatial interaction. For instance, inter-regional migrations between 1985 and 1990 in Switzerland yields $1 - \hat{t} = \text{trace}(N)/\text{sum}(N) = 0.93$ (most people stayed in the same canton during those five years), yielding the possible estimate $\hat{t} = 0.07$.

Equations (4) and (5) constitute a straightforward two-steps procedure generalizing and simplifying the "proposal B" recipe exposed in [9], based upon the construction of a weight-compatible, time-continuous Markov chain generated by a *rate matrix* $R = -\Pi^{-1/2}\Psi\Pi^{1/2} = -\Pi^{-1}LG/\text{trace}(LG)$ reflecting direct spatial transitions, as expressed by the proximity matrix G, whose regional sojourn times are precisely adjusted to make $E(t)$ weight-compatible.

Non-negativity condition 2) above is a direct consequence of the essential non-positivity of Ψ together with the theorem "$\exp(-t\,A)$ is non-negative for all $t > 0$ iff A is essentially non-positive" (e.g., see theorem 8.2 in [16]).

3.2   Spectral Decomposition

Solution (5) can be computed by spectral decomposition of $\Psi = UMU'$, that is $\psi_{ij} = \sum_\alpha \mu_\alpha u_{i\alpha} u_{j\alpha}$. As a matter of fact, $\Psi\sqrt{f} = 0$, thus $\sqrt{f}$ is a trivial eigenvector u (numbered $\alpha = 0$) of Ψ with trivial eigenvalue $\mu_0 = 0$, demonstrating in particular the weight-compatibility

$$E(t)\mathbf{1} = \Pi^{1/2} \sum_{r=0}^{\infty} \frac{(-t)^r}{r!} \Psi^r \sqrt{f} = \Pi^{1/2}\sqrt{f} = f$$

as claimed. The other eigenvalues, *increasingly ordered*, are non-negative, since Ψ turns out to be p.s.d. in view of

$$0 \le \frac{1}{2} \sum_{ij} g_{ij}(h_i - h_j)^2 = \sum_{ij} (\delta_{ij}g_{i\bullet} - g_{ij})h_i h_j$$

for any h. Thus $0 = \mu_0 \leq \mu_1 \leq \mu_2 \leq \ldots \leq \mu_{n-1}$ and

$$E(t) = \Pi^{1/2} \exp(-tUMU') \Pi^{1/2} = \Pi^{1/2} U \exp(-Mt) U' \Pi^{1/2}$$

that is

$$e_{ij}(t) = \sqrt{f_i f_j} \sum_{\alpha=0}^{n-1} u_{i\alpha} u_{j\alpha} \exp(-\mu_\alpha t) = f_i f_j + \sqrt{f_i f_j} \sum_{\alpha=1}^{n-1} u_{i\alpha} u_{j\alpha} \exp(-\mu_\alpha t)$$

thus proving limits 4) and 5) above. Equivalently,

$$e_{ij}^s(t) := \frac{e_{ij} - f_i f_j}{\sqrt{f_i f_j}} = \sum_{\alpha=1}^{n-1} u_{i\alpha} u_{j\alpha} \exp(-\mu_\alpha t) \qquad E^s = U \exp(-Mt) U' \quad (8)$$

where $E^s(t) := \Pi^{-1/2}(E(t) - ff')\Pi^{-1/2}$ is the *standardized exchange matrix*. $E^s(t)$ possesses a trivial eigenvalue $\lambda_0 = 0$ associated with $u_0 = \sqrt{f}$, as well as non-trivial eigenvalues $\lambda_\alpha(t) = \exp(-\mu_\alpha t)$, *decreasingly ordered* for $\alpha \geq 1$, lying in $[-1, 1]$, as required by the Perron-Froebenius theorem on the associated Markov chain W. They even lie in $[0, 1]$, making $E(t)$ p.s.d. or *diffusive*. Note the eigenvectors $U = (u_{i\alpha})$ to be independent of t.

3.3 Diffusive Dissimilarity and Multidimensional Scaling

The p.s.d. nature of $E(t)$ permits to define a "diffusive dissimilarity" between regions, namely

$$\mathcal{D}_{ij}(t) := \frac{e_{ii}(t)}{f_i^2} + \frac{e_{jj}(t)}{f_j^2} - 2\frac{e_{ij}(t)}{f_i f_j} \ . \qquad (9)$$

$\mathcal{D}$ turns out to be *squared Euclidean*, i.e., of the form $\mathcal{D}_{ij} = \|y_i - y_j\|^2$ for some $n \times p$ "diffusive coordinates" $Y = (y_{ia})$, where $p \leq n - 1$. The squared Euclidean nature of $\mathcal{D}$ follows from the *conditional negative-definiteness condition* $\sum_{ij} h_i h_j \mathcal{D}_{ij} \leq 0$ for any h with $\sum_i h_i = 0$ (e.g., see [17]). Determining the diffusive coordinates $Y = (y_{i\alpha})$ constitutes the classical multidimensional scaling (MDS) problem, with solutions in the weighted setting

$$y_{i\alpha}(t) = \frac{\sqrt{\lambda_\alpha(t)}}{\sqrt{f_i}} u_{i\alpha} \qquad \alpha = 1, \ldots, n-1 \qquad (10)$$

where $\lambda_\alpha(t) = \exp(-\mu_\alpha t)$ is the eigenvalue of E^s in (8), and $u_{i\alpha}$ its corresponding eigenvector (e.g., [18,19]). (10) is a member of a family of vertex coordinates on weighted graphs of the form

$$y_{i\alpha}(t) = g(\lambda_\alpha(t)) y_{i\alpha}^s \qquad y_{i\alpha}^s = \frac{u_{i\alpha}}{\sqrt{f_i}} \qquad \alpha = 1, \ldots, n-1 \qquad (11)$$

where $g(\lambda)$ is a non-negative function, and $y_{i\alpha}^s$ is the *standardized* or *raw coordinate* of region i on dimension $\alpha \geq 1$ [19]. Raw coordinates also occur quite naturally in spatial filtering (e.g., [20,21,4,9]).

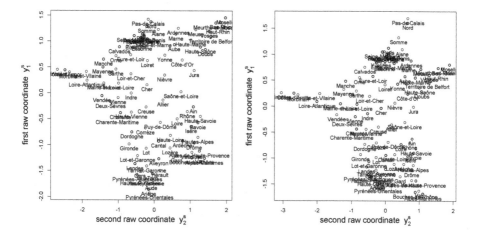

Fig. 1. Raw diffusive coordinates $y_{i\alpha}^s$ (11) for $\alpha = 2$ (abscissa) and $\alpha = 1$ (ordinate), reconstructing the map of French departments from the adjacency matrix A and departmental weights f. Left: uniform weights $f_i = 1/n$. Right: non-uniform "voters weights."

3.4 Summary

Any proximity matrix G between regions, together with any set of regional weights f, yield a one-parameter family of weight-compatible, p.s.d. exchange matrices $E(G, f, t)$. The latter yield in turn a family of squared Euclidean dissimilarities $\mathcal{D}_{ij}(t)$ between regions (9), from which regional coordinates $y_{i\alpha}(t)$ (10) or raw coordinates $y_{i\alpha}^s$ (11) can be extracted by weighted MDS. Hence, any pair (G, f) produces a visualization y or y^s of the spatial configuration between regions, conceivably resembling the true geographical configuration (Figure 1).

4   Illustration: Political Autocorrelation in France

Consider the $n = 94$ departments of "metropolitan France" (Corsica excluded), whose binary adjacency matrix A is chosen as the spatial s matrix. Consider also uniform departmental weights $f_i = 1/n$, but also, in parallel, non-uniform "voters weights" f (section 4.1). Figure 2 depicts the distribution of departmental degrees $a_{i\bullet}$ and non-uniform weights f_i, as well as the non-trivial eigenvalues $\lambda_\alpha(t)$. Figure 1 gives the first two factorial "raw coordinates" (11), in the uniform and non-uniform case. The reconstruction of the geographical map form the adjacency matrix looks fairly adequate.

4.1   Extracting Regional Features by Correspondence Analysis

In general, regions are characterized by uni- or multivariate *features* x, whose variation may or may not be correlated with the diffusive coordinates y in (10); this issue precisely constitutes the topic of *spatial autocorrelation*, as exemplified below. In the

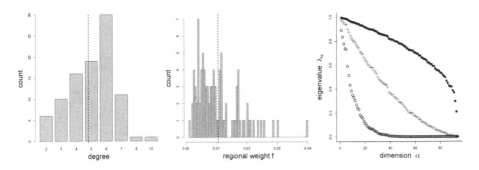

Fig. 2. Left: distribution of the departmental degrees $a_{i\bullet}$ (average degree = 5.06). Middle: distribution of the non-uniform departmental "voters weights" f_i (section 4.1; average weight = 1/94=0.011). Right: scree plot of the eigenvalues $\lambda_\alpha(t)$ of $E^s(t)$ (non-uniform weights) for $t = 0.2$ (solid squares), $t = 1$ (crosses) and $t = 5$ (circles).

sequel, features x will first be computed as regional factor scores, instead of considering directly available regional variables x.

Consider the votes of the first round of the French presidential 2012 election, as recorded by the $n \times p$ contingency table (N_{ik}), fixing the "number of votes for candidate k in department i" where $n = 94$ and $k = 1, \ldots, p = 10$ (Joly, LePen, Sarkozy, Melenchon, Poutou, Arthaud, Cheminade, Bayrou, Dupont-Aignan, Hollande). Figure 3 left yields the scree plot of the proportion of explained chi-square by each of the $\min(n, p - 1) = 9$ factors, whose first and second ones express together 83% of the inertia. Figure 3 right exposes the Correspondence Analysis (CA) biplot depicting the *department* and *candidate coordinates*, showing similarities among departments, among candidates, as well as attraction-repulsion between departments and candidates.

In this context, natural regional weights are provided by $f_i = N_{i\bullet}/N_{\bullet\bullet}$, the *voters weight* of department i, measuring its relative share of voters. By construction, department coordinates $x_{i\beta}$ are centered, standardized and uncorrelated (here $\beta = 1, \ldots p-1$ labels the factors produced in Correspondence Analysis):

$$\sum_i f_i x_{i\beta} = 0 \qquad \sum_i f_i x_{i\beta}^2 = 1 \qquad \sum_i f_i x_{i\beta} x_{i\beta'} = 0 \quad \text{for } \beta \neq \beta' \;.$$

4.2   Spatial Autocorrelation of Voting Pattern

The autocorrelation of each "voting factor" x_β (where $\beta = 1, \ldots, p$), as extracted from the CA of section (4.1), can be tested in turn by computing Moran's indices $I(E, x_\beta)$ in (2). Here $E = E(A, f, t)$ is the weight-compatible, time-dependent exchange matrix constructed in section (3.3), and f stands as before as the non-uniform voters weight.

Associated p-values are computed from B bootstrapped samples associated with the weighted permutation test (section 6). The first factorial political score x_{i1} extracted in section (4.1) turns out to be strongly autocorrelated (Figure 4, left), in contrast to the second score x_{i2} which is not (Figure 4, right).

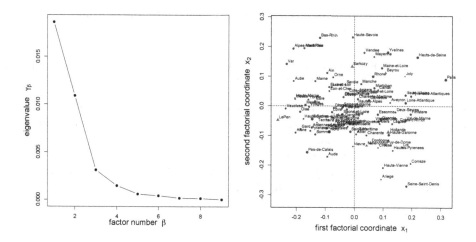

Fig. 3. Correspondence Analysis on the "department × candidate" votes contingency table, in the first round of the French presidential 2012 election. Left: eigenvalues γ_β. Right: biplot. The first axis can be interpreted in terms of right- versus left-wing contrast, but also central-peripheral contrast (53%); the second one seems to oppose poor to rich departments (31%).

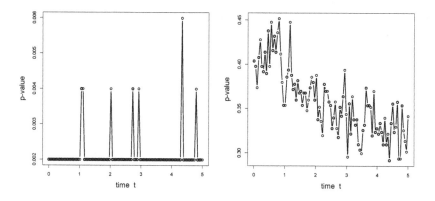

Fig. 4. p-values associated to the significance test of weighted Moran's I (3), for various values of $t \in [0, 5]$, based upon $B = 1000$ permutations of the so-called *spatial modes* (instead of the direct spatial values), in order to take into consideration the heteroscedasticity associated to the weighted setting (see section 6 and [9] for details). The first political component x_{i1} turns out to be strongly autocorrelated (left), in contrast to the second political component x_{i2} (right).

4.3   Relative Inertia

Moran's I can be generalized to *multivariate settings* by considering squared Euclidean dissimilarities D_{ij} between regional profiles, instead of univariate dissimilarities of the form $(x_i - x_j)^2$. In the present analysis, the natural candidate for D is provided by the classical *chi-square dissimilarity* between departments, which can be defined from the contingency table (N_{ik}) or from the "raw" factor scores $x_{i\beta}$, as

$$D_{ij}^{\chi} = \sum_k \frac{N_{\bullet\bullet}}{N_{\bullet k}} \left(\frac{N_{ik}}{N_{i\bullet}} - \frac{N_{jk}}{N_{j\bullet}}\right)^2 = \sum_\beta \gamma_\beta (x_{i\beta} - x_{j\beta})^2 \qquad (12)$$

where γ_β are the eigenvalues (the square of the singular values) of the Correspondence Analysis of section 4.1. Recall (and observe) that weighted multidimensional scaling of D precisely yields the so-called *principal coordinates* $\sqrt{\gamma_\beta}\, x_{i\beta}$, that is *CA is equivalent to weighted MDS on chi-square dissimilarities*.

As claimed, multivariate generalization of Moran's index (3) is provided by the *relative inertia* $\delta \in [-1, 1]$ defined (with $D = D^{\chi}$) as

$$\delta(t) := \frac{\Delta - \Delta_{\text{loc}}(t)}{\Delta} \qquad \Delta = \frac{1}{2} \sum_{ij} f_i f_j D_{ij} \qquad \Delta_{\text{loc}}(t) = \frac{1}{2} \sum_{ij} e_{ij}(t) D_{ij}$$

whose significance can be assessed by usual normal approximation, permutation, or bootstrap tests; see section 6.2. Relative inertia also expresses as (cf. (1))

$$\delta(t) = \frac{\sum_{ij} e_{ij}(t)\, B_{ij}}{\sum_i f_i B_{ii}}$$

where B are the scalar products of MDS, obeying $D_{ij} = B_{ii} + B_{jj} - 2B_{ij}$. In particular, $B_{ii} = D_{if}$ is the squared distance between i and the centroid.

Fig. 5. Soft K-means clustering of French departments, from the initial centroid configuration determined by the $m = 6$ "Hexagon corners": spatial clustering (left), attributes clustering (middle) and landscape clustering (right) defined in (13), with $\nu = 1/2$ and $c(D) = \max_{ij} \mathcal{D}_{ij}$

4.4 "Landscape Clustering"

We are now in possession of *two squared Euclidean dissimilarities*, namely the diffusive dissimilarity $\mathcal{D}_{ij}$ (9), measuring *spatial remoteness* between pairs of regions, and the chi-square dissimilarity D_{ij}^{χ} (12), measuring the voting *attributes contrast* between regional profiles. This circumstance makes it possible to consider various *regional clustering strategies*, namely

a) *spatial clustering*, based upon the diffusive dissimilarity $\mathcal{D}_{ij}$ exhibited in Figure 2 right
b) *attributes clustering*, based upon the attributes dissimilarity D^{χ}_{ij} exhibited in Figure 3 right
c) a presumably new *landscape clustering* based upon minimizing the within-groups inertia associated to the mixed, normalized dissimilarity

$$D^{lan}_{ij} = \nu \, \frac{\mathcal{D}_{ij}}{c(\mathcal{D})} + (1 - \nu) \, \frac{D^{\chi}_{ij}}{c(D^{\chi})} \tag{13}$$

where $\nu \in (0,1)$ controls the spatial versus attribute contributions and $c(D)$ is a normalization factor, such as $\frac{1}{2} \sum_{ij} f_i f_j D_{ij}$ or $\max_{ij} \mathcal{D}_{ij}$. Landscape clustering aims at creating regional aggregates both spatially contiguous and possessing similar features - a natural aim in Quantitative Geography, Spatial Econometrics and Geographic Information Science.

Figures 5 and 6 illustrate spatial clustering (left), attributes clustering (middle) and landscape clustering (right) for the mixed normalized dissimilarities Clusterings result from soft K-means (section 6.3), with initial centroid configuration determined by the $m = 6$ "Hexagon corners" (Bas-Rhin, Nord, Finistère, Pyrénées-Atlantiques, Pyrénées-Orientales, Alpes-Maritimes: Figure 5), or by the $m = 7$ most populated departments (Nord, Bouches-du-Rhône, Paris, Rhône, Pas-de-Calais, Gironde, Loire-Atlantique: Figure 6).

Fig. 6. Soft K-means clustering of French departments, from the initial centroid configuration determined by the $m = 7$ most populated departments: spatial clustering (left), attributes clustering (middle) and landscape clustering (right) defined in (13), with $\nu = 1/2$ and $c(D) = \max_{ij} \mathcal{D}_{ij}$

5   Discussion and Conclusions

Dealing with regions of unequal importance requires a weighted formalism, which arguably helps unifying mathematical enquiries and proposals. For instance, Moran's I and Geary's c appear to be simply related as $c = 1 - I$ in the present "E-scheme."

After briefly reviewing the differences between the unweighted and weighted approaches to spatial autocorrelation, this paper proposes a straight, general prescription aimed at constructing exchange matrices E both compatible with given proximity relations G and regional weights f. The solution contains a freely adjustable parameter t, the age of the network, controlling the importance of direct adjacency, distance deterrence, or inverse bandwidth, when $0 < t < \infty$. At the extremes, the network becomes independent of G, namely with the frozen network $E(0) = \Pi$ and the completely mobile network $E(\infty) = ff'$.

Solution $E(t)$ turns out to be p.s.d., that is modeling a diffusive network. This circumstance permits to define a squared Euclidean dissimilarity on the network, and hence, by MDS, a network visualization. This presumably new *proximity-based dissimilarity* can in turn be compared to some other *features-based* squared Euclidean dissimilarity: this constitutes the very issue of spatial autocorrelation. Both similarities can also be mixed, and fed to partitioning algorithms, yielding "landscape clustering," sensitive to both regional proximities and attributes.

More case studies are most welcome. Further investigations could examine the impact of $E(G, f, t)$ on weighted SAR or CAR models, on the construction of mobility indices, or on the construction of *local indicators* of relative inertia, in the spirit of the well-known proposal of [2].

6 Appendix: Autocorrelation Tests and Soft Clustering

The first part of the appendix derives, under the null hypothesis H_0 of no autocorrelation, the expected value of Moran's I and its variance in the general case of spatial weights, *possibly containing non-zero diagonal components* $v_{ii} \neq 0$, a case little confronted with in the literature. Both unweighted (section 6.1) and weighted settings (section 6.2) are addressed.

6.1 Unweighted Permutation Test

Equation (1) shows that V can be taken as **symmetric** and **normalized**, that is obeying $v_{ij} = v_{ji}$ and $v_{\bullet\bullet} = 1$. Moran index thus expresses as (e.g., [21,4])

$$I(V, x) = \frac{n \sum_{ij} v_{ij}(x_i - \bar{x})(x_j - \bar{x})}{\sum_i (x_i - \bar{x})^2} = n \frac{x' H V H x}{x' H x} \qquad H = I - J/n \ .$$

Here I is the identity matrix, $J = 11'$ is the constant unit matrix and $H = H^2$ is the centering projection matrix, each of order $n \times n$.

The spectral decomposition $H V H = U \Lambda U'$ with U orthogonal and Λ diagonal makes appear a trivial dimension $\alpha = 0$, with constant eigenvector $u_{i0} = 1/\sqrt{n}$ and null eigenvalue $\lambda_0 = 0$. Also, $H u_\alpha = u_\alpha$ for higher-order, non-trivial dimensions $\alpha = 1, \ldots n - 1$.

Define the unweighted *spatial modes* (e.g., see [21,9]) as $y = U'x$, that is $x = Uy$. In particular, $y_0 = \sqrt{n}\,\bar{x}$. Moran's index then reads

$$I(V, x) = n \frac{\sum_{\alpha \geq 1} \lambda_\alpha y_\alpha^2}{\sum_{\alpha \geq 1} y_\alpha^2} = n \sum_{\alpha \geq 1} \lambda_\alpha b_\alpha \qquad b_\alpha := \frac{y_\alpha^2}{\sum_{\beta \geq 1} y_\beta^2} \qquad (14)$$

In the present unweighted setting, the spatial variables X_i are, under H_0, i.i.d. with mean μ and variance σ^2. The resulting spatial modes $Y = U'X$ are uncorrelated with $E(Y_\alpha) = \delta_{\alpha 0} \sqrt{n}\, \mu$ and $\text{Cov}(Y_\alpha, Y_\beta) = \delta_{\alpha\beta}\sigma^2$. In particular, the $n - 1$ quantities b_α $(\alpha \geq 1)$ in (14) are arguably identically distributed under H_0, yet not independently in view of $\sum_{\alpha \geq 1} b_\alpha = 1$. Denoting by $E_\pi(.)$ the expectation under modes permutation, one gets, by symmetry

$$E_\pi(b_\alpha) = \frac{1}{n-1} \quad E_\pi(b_\alpha^2) = \frac{\sum_{\beta \geq 1} b_\beta^2}{(n-1)} =: \frac{t(y)}{(n-1)^2} \quad E_\pi(b_\alpha b_\beta) = \frac{1 - t(y)/(n-1)}{(n-1)(n-2)}$$

for $\alpha \neq \beta$. Taking into account

$$\sum_{\alpha \geq 1} \lambda_\alpha = \sum_{\alpha \geq 0} \lambda_\alpha = \text{trace}(HVH) = \text{trace}(V) - \frac{1}{n} \qquad \text{and}$$

$$\sum_{\alpha \geq 1} \lambda_\alpha^2 = \sum_{\alpha \geq 0} \lambda_\alpha^2 = \text{trace}(HVHHVH) = \text{trace}(VHVH) = \text{trace}(V^2) - \frac{2}{n}\sum_i v_{i\bullet}^2 + \frac{1}{n^2}$$

finally yields

$$E_\pi(I) = \frac{n\,\text{trace}(V) - 1}{n - 1} \qquad \boxed{\text{unweighted setting}}$$

$$\text{Var}_\pi(I) = \frac{t(y) - 1}{(n-1)(n-2)}[n^2\text{trace}(V^2) - 2n\sum_i v_{i\bullet}^2 + 1 - \frac{(n\,\text{trace}(V) - 1)^2}{n-1}] \quad (15)$$

where $t(y) = (n-1)\sum_{\alpha \geq 1} y_\alpha^4 / (\sum_{\alpha \geq 1} y_\alpha^2)^2 \geq 1$ is a measure of modes dispersion, taking on its minimum value $t(y) = 1$ for $y_\alpha = \text{const}$ for $\alpha \geq 1$. In particular, $E_\pi(I) > -1/(n-1)$ whenever spatial weights V contain off-diagonal components.

Under the additional normal assumption $X_i \sim N(\mu, \sigma^2)$, one gets $Y_\alpha \sim N(0, \sigma^2)$ for $\alpha \geq 1$, as well as $E(t(y)) = 3(n-1)/(n+1)$ (e.g., [1], p.43). Then

$$E(\text{Var}_\pi(I)) = \frac{2}{n^2 - 1}[n^2\text{trace}(V^2) - 2n\sum_i v_{i\bullet}^2 + 1 - \frac{(n\,\text{trace}(V) - 1)^2}{n-1}] \quad (16)$$

$$= \frac{1}{(n^2-1)S_0^2}[n^2 S_1 - nS_2 + \frac{2(n-2)S_0^2 + 4nS_0\text{trace}(V) - 2n^2\text{trace}^2(V)}{n-1}]$$

where, for comparison's sake, the normalization condition $v_{\bullet\bullet} = 1$ has been relaxed in the last equation (while retaining the symmetry of V), and the familiar notations

$$S_0 := \sum_{ij} v_{ij} \qquad S_1 := 2\sum_{ij} v_{ij}^2 \qquad S_2 := 4\sum_i v_{i\bullet}^2$$

have been introduced. The last identity in (16) turns out to coincide, up to the terms involving $\text{trace}(V)$, with the formulas proposed in [1], p.44.

6.2   Weighted Permutation Test

In the weighted setup, the spatial field X_i represents a regional aggregate associated to region i. Under H_0, its mean is constant but its variance is inversely proportional to the weight of the region (heteroscedasticity). Hence, $X_i \sim N(\mu, \sigma^2/f_i)$ under normal assumption. The expected value of the weighted Moran coefficient (2) and its variance read [9]

$$E_\pi(I) = \frac{\text{trace}(W) - 1}{n - 1} \qquad \boxed{\text{weighted setting}}$$

$$E(\text{Var}_\pi(I)) = \frac{2}{n^2 - 1}[\text{trace}(W^2) - 1 - \frac{(\text{trace}(W) - 1)^2}{n - 1}] \qquad (17)$$

where $W = (w_{ij})$ with $w_{ij} := e_{ij}/f_i$ is the row-normalized, weight-compatible matrix of spatial weights (section 2.2), and constitutes the transition matrix of a Markov chain, possessing off-diagonal components in general [14].

Unweighted average and variance formulas (15) and (16) should coincide with their weighted analogs (17) whenever $V = E$ constitutes a symmetric, normalized spatial weight matrix with uniform weights $e_{i\bullet} = f_i = 1/n$. Indeed, $W = nV$ with $v_{i\bullet} = 1/n$ in that case, thus demonstrating the expected agreement.

Small Time Limit. In the limit $t \to 0$, (7) and (17) together with $W(t) = I + tR + \frac{t^2}{2}R^2 + 0(t^2)$, where $R = -\Pi^{-1}LG/\text{trace}(LG)$ is the rate matrix, yield

$$I - E_\pi(I) = \frac{t}{\text{trace}(LG)}[\frac{\text{trace}(\Pi^{-1}LG)}{n - 1} - \frac{\sum_{ij} g_{ij}(x_i - x_j)^2}{2 \text{ var}(x)}] + 0(t^2)$$

$$E(\text{Var}_\pi(I)) = \frac{2t^2}{n^2 - 1}[\text{trace}(R^2) - \frac{\text{trace}^2(R)}{n - 1}] + 0(t^3) \boxed{\text{weighted setting, } t \to 0}$$

Interestingly enough, for uniform weights $f_i = 1/n$, the decision variable of the normal test expresses, up to order $0(t)$, as

$$z = \frac{I - E_\pi(I)}{\sqrt{E(\text{Var}_\pi(I))}} = \frac{\tilde{I} - E_\pi(\tilde{I})}{\sqrt{E(\text{Var}_\pi(\tilde{I}))}}$$

where

$$\tilde{I}(x) := 1 - \frac{1}{\text{var}(x)} \frac{\frac{1}{2}\sum_{ij} g_{ij}(x_i - x_j)^2}{g_{\bullet\bullet} - \text{trace}(G)}$$

is *time-independent*, and constitutes an alternative to unweighted Moran's index $I(x)$ (1).

6.3 Soft K-Means

Consider a $n \times m$ *membership matrix* Z with components $z_{ig} \geq 0$ with $z_{i\bullet} = 1$, expressing the probability that region i belongs to group g. The group weight is $\rho_g[Z] = \sum_i f_i z_{ig}$ and the squared Euclidean dissimilarity D_i^g between region i and centroid g is derived form the inter-individual squared dissimilarities D_{ij} as (Huygens principle)

$$D_i^g[Z] = \sum_j f_j^g D_{ij} - \frac{1}{2}\sum_{ij} f_i^g f_j^g D_{ij} \qquad \text{where} \quad f_i^g[Z] = \frac{f_i z_{ig}}{\rho_g} .$$

where D_{ij} is a squared Euclidean dissimilarity. Memberships are iteratively computed (e.g., [22,23,19]) as

$$z_{ig}^{(r+1)} = \frac{\rho_g[Z^{(r)}] \exp(-\beta D_i^g[Z^{(r)}])}{\sum_h \rho_h[Z^{(r)}] \exp(-\beta D_i^h[Z^{(r)}])}$$

where the *inverse temperature* β has been set to 1 in section 4.4, where two variants for the m initial centroids are investigated. After convergence, region i is finally attributed to group $g = \arg \max_h z_{ih}^{(\infty)}$. The alternative iteration

$$f_i^{g(r+1)} = \frac{f_i \exp(-\beta D_i^g[Z^{(r)}])}{\sum_j f_j \exp(-\beta D_j^g[Z^{(r)}])}$$

works as well, as expected.

References

1. Cliff, A.D., Ord, J.K.: Spatial processes: models & applications, vol. 44. Pion, London (1981)
2. Anselin, L.: Local indicators of spatial association — LISA. Geographical Analysis 27(2), 93–115 (1995)
3. Tiefelsdorf, M., Boots, B.: The exact distribution of Moran's I. Environment and Planning A 27, 985–985 (1995)
4. Dray, S., Legendre, P., Peres-Neto, P.R.: Spatial modelling: a comprehensive framework for principal coordinate analysis of neighbour matrices (PCNM). Ecological Modelling 196(3), 483–493 (2006)
5. Tiefelsdorf, M., Griffith, D.A., Boots, B.: A variance-stabilizing coding scheme for spatial link matrices. Environment and Planning A 31, 165–180 (1999)
6. Bivand, R.: Spatial econometrics functions in R: Classes and methods. Journal of Geographical Systems 4(4), 405–421 (2002)
7. Bivand, R., Anselin, L., Berke, O., Bernat, A., Carvalho, M., Chun, Y., Dormann, C., Dray, S., Halbersma, R., Lewin-Koh, N., et al.: spdep: Spatial dependence: Weighting schemes, statistics and models. [R package] (2006)
8. Berger, J., Snell, J.L.: On the concept of equal exchange. Behavioral Science 2(2), 111–118 (1957)
9. Bavaud, F.: Testing spatial autocorrelation in weighted networks: The modes permutation test. Journal of Geographical Systems 15(3), 233–247 (2013)
10. Lebart, L.: Analyse statistique de la contiguïté. Publications de l'Institut de Statistique des Universités de Paris XVIII, pp. 81–112 (1969)
11. Le Foll, Y.: Pondération des distances en analyse factorielle. Statistique et Analyse des Données 7(1), 13–31 (1982)
12. Meot, A., Chessel, D., Sabatier, R.: In: Lebreton, D., Asselain, B. (eds.) Opérateurs de voisinage et analyse des données spatio-temporelles, pp. 45–71 (1993)
13. Thioulouse, J., Chessel, D., Champely, S.: Multivariate analysis of spatial patterns: A unified approach to local and global structures. Environmental and Ecological Statistics 2(1), 1–14 (1995)
14. Bavaud, F.: Models for spatial weights: A systematic look. Geographical Analysis 30(2), 153–171 (1998)
15. Chung, F.R.: Spectral Graph Theory, vol. 92. American Mathematical Society (1997)
16. Varga, R.S.: Matrix iterative analysis, vol. 27. Springer Science & Business (2000)

17. Cressie, N.: Statistics for Spatial Data. Wiley Series in Probability and Statistics. Wiley-Interscience, New York (1993)
18. Cuadras, C.M., Fortiana, J.: Weighted continuous metric scaling. In: Gupta, A.K., Girko, V.L. (eds.) Multidimensional Statistical Analysis and Theory of Random Matrices, Netherlands, VSP, pp. 27–40 (1996)
19. Bavaud, F.: Euclidean distances, soft and spectral clustering on weighted graphs. In: Balcázar, J.L., Bonchi, F., Gionis, A., Sebag, M. (eds.) ECML PKDD 2010, Part I. LNCS, vol. 6321, pp. 103–118. Springer, Heidelberg (2010)
20. Griffith, D.A.: Eigenfunction properties and approximations of selected incidence matrices employed in spatial analyses. Linear Algebra and its Applications 321(1), 95–112 (2000)
21. Griffith, D.A.: Spatial autocorrelation and spatial filtering: Gaining understanding through theory and scientific visualization. Springer, Berlin (2003)
22. Celeux, G., Govaert, G.: A classification EM algorithm for clustering and two stochastic versions. Computational Statistics & Data Analysis 14(3), 315–332 (1992)
23. Rose, K.: Deterministic annealing for clustering, compression, classification, regression, and related optimization problems. Proceedings of the IEEE 86(11), 2210–2239 (1998)

Spatial Graphs Cost and Efficiency: Exploring Edges Competition by MCMC

Guillaume Guex

Department of Geography, University of Lausanne, Switzerland

Abstract. Recent models for spatial networks have been built by determining graphs minimizing some functional F composed by two antagonist quantities. Although these quantities might differ from a model to another, methods used to solve these problems generally make use of simulated annealing or operations research methods, limiting themselves to the study of a single minimum and ignoring other close-to-optimal alternatives. This contribution considers the arguably promising framework where the functional F is composed by a graph *cost* and a graph *efficiency*, and the space of all possible graphs on n spatially fixed nodes is explored by MCMC. Covariance between edges occupancy can be derived from this exploration, revealing the presence of cooperative and competition regimes, further enlightening the nature of the alternatives to the locally optimal solution.

1 Introduction

Spatial networks constitute a particular case in networks studies, where nodes and edges are embedded in a metric space. The study of these networks received a special attention in the recent years, as they model a large quantity of complex geographic systems, such as transportation networks (road, railroad and airlines networks), power grids networks and internet [1–12]. The particularity of these networks is that the underlying space directly controls the cost of edges, thus impacting their topology. Previous empirical studies have examined different spatial network structures and demonstrated that the effect of space greatly differs, depending on the nature of networks (reviewed extensively in [3]). Nevertheless, their designs typically attempt to maximize some utility function while minimizing some kind of cost function, making abstraction of other geographical or economical constraints encountered in real-world situations.

This article attempts to study a particular class of models of optimal networks defined as networks minimizing some functional F specified below. These models exhibit a great variety of interesting results, depending on the ingredients entering the composition of F, and are aimed at modelling numerous different geographic systems of interest. Here, we will consider the case where $F = C - I \cdot E$, where C is the *cost* of the network (the sum of all edges length) and E the *efficiency* (the mean length of shortest-paths between all pairs of nodes), while the parameter I, the *investment*, acts as a balance between those quantities. This simple and intuitive model, already studied in [1–3, 9], gives results similar to our railroads, highways or power grid networks. Previous researches concentrated on finding a single graph minimizing F, discarding the study of the nature of the space of all possible graphs on n fixed nodes, controlled

M. Duckham et al. (Eds.): GIScience 2014, LNCS 8728, pp. 97–108, 2014.

by F. By contrast, we attempt here to explore this space with a *Monte Carlo Markov Chain* (MCMC) algorithm [13–18] or more precisely, a variant of simulated annealing model, implying heating as well as cooling schedules (see section 2.4). By examining the history of the algorithm, edge competition and synergies can be revealed, enabling the design of close-to-optimal graphs.

This article is divided in two parts. The first one sets the formalism and the mathematical tools needed to perform the algorithm and the second one examines a few case studies in more detail.

2   Formalism

2.1   Generalities and Notations

A *graph* is a couple $\mathcal{G} = (\mathcal{V}, \mathcal{E})$ where $\mathcal{V}$ are the vertices (or nodes) set of size n and $\mathcal{E}$ the edges set of size m. A graph is said to be *spatial* when all vertices are embedded in a Euclidean space. A spatial graph is entirely defined by two matrices: X the matrix of vertex coordinates in space and the $n \times n$ symmetric adjacency matrix $A = (a_{ij})$, where $a_{ij} = 1$ if $\{i, j\} \in \mathcal{E}$, $a_{ij} = 0$ otherwise.

This article considers simple unoriented spatial graphs in $\mathbb{R}^2$ equipped with the Euclidean distance d^E. In this context, every edge $e = \{i, j\}$ possesses a *length* l corresponding to the Euclidean distance between nodes composing it, i.e. $l(\{i, j\}) = d^E(i, j)$. Edge lengths permit to define an alternative version of the well-known shortest-path distance, referred to, in the literature, as the *weighted shortest-paths distance* d^{wsp} (or route distance in [1–3, 9]):

$$d^{wsp}(i, j) = \min_{\xi \in \mathcal{P}(i,j)} \sum_{e \in \xi} l(e)$$

where $\mathcal{P}(i, j)$ is the set of all paths between i and j.

2.2   Functional Minimization

Some other quantities can be defined on spatial graphs. Define the *cost* C of a graph $\mathcal{G}$ as the sum of all edge lengths:

$$C(\mathcal{G}) = \sum_{e \in \mathcal{E}} l(e)$$

Furthermore, define the *efficiency* E of graph $\mathcal{G}$ as the mean, along all pairs of vertices, of the inverse of the weighted shortest-path distance:

$$E(\mathcal{G}) = \frac{1}{n(n-1)} \sum_{i \neq j \in \mathcal{V}} \frac{1}{d^{wsp}(i, j)}$$

Obviously, for any set of vertices, the empty graph yields a null cost and efficiency, while the complete graph gives their maximum. From a concrete point of view, the efficiency represents the ability of the network to effectively transport agents from any node to another, while the cost is self speaking. Therefore, an optimal network planning may seek to maximize the efficiency while minimizing the cost, leading to the minimization of the function F defined by:

$$F(\mathcal{G}) = C(\mathcal{G}) - I \cdot E(\mathcal{G})$$

where the parameter $I \geq 0$ is the *investment*, acting as an arbiter between the conflicting objectives. When $I \to 0$ the graph minimizing F is the empty graph, while $I \to \infty$ generates the complete graph. For carefully chosen intermediate values, depending in turn on several parameters such as the real cost of the edges and the insistence on the efficiency of the network, the solutions are similar to some real spatial networks, like railroad, highways or power grid networks [2, 3, 9]. Note that, unless $I \to \infty$, the resulting graph may not be connected (see e.g. left plot in Fig. 1). If we replace the weighted shortest-path distance by the standard shortest-path distance in the formula for efficiency, optimal graphs will possess a structure of "Hub-and-spoke," similar to an airline network [3, 9].

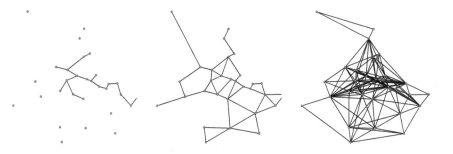

Fig. 1. Local minima for different investment values on 30 fixed points in $\mathbb{R}^2$ with abscissas and ordinates generated as $\mathcal{N}(0, 1)$. On the left $I = 1$, in the middle $I = 100$ and on the right $I = 10^6$.

2.3   MCMC Exploration of the Space of all Graphs

For a fixed set $\mathcal{V}$ of n vertices in space, the *space of all graphs* on these vertices is noted $\Gamma_{\mathcal{V}}$. This space is similar to the atomic spins space in the Ising model, where every possible edges can be in two states { presence, absence } [15, 19]. Thus the size of $\Gamma_{\mathcal{V}}$ is $2^{n(n-1)/2}$. Let $W = (w_{kl}) = (w_{\mathcal{G}_k \mathcal{G}_l})$ be the transition matrix of a *Markov chain* on $\Gamma_{\mathcal{V}}$ defined by:

$$w_{kl} = w_{lk} = \begin{cases} \frac{2}{n(n-1)} & \text{if } \mathcal{G}_k \text{ and } \mathcal{G}_l \text{ differ exactly by one edge} \\ 0 & \text{otherwise} \end{cases}$$

With this simple transition matrix, the Markov chain will jump from the graph $\mathcal{G}_k$ to any graph $\mathcal{G}_l$ having exactly one more or one less edge with equal probability. It is obvious that this Markov chain can reach any nodes from any starting point, and therefore the chain is *irreducible*. The MCMC *Metropolis-Hasting* (MH) algorithm [13, 16, 17], is designed to create a new Markov chain having a desirable stationary distribution p_k on the states from any irreducible Markov chain:

1. From the state k, generate a new state l with probability w_{kl}
2. Jump to l with probability $\tilde{w}_{kl}$ defined by:

$$\tilde{w}_{kl} = \min\left(1, \frac{p_l w_{lk}}{p_k w_{kl}}\right)$$

      otherwise stay in k
3. Iterate

Since $w_{kl} = w_{lk}$, one has that $\tilde{w}_{kl} = \min(1, \frac{p_l}{p_k})$.

For any initial configuration, the algorithm will converge to the invariant distribution p_k, itself determined so as to favor near-optimal graphs, as in the *Gibbs sampling* of p_k [13, 14]:

$$p_k = \frac{1}{Z}\exp(-\beta F(\mathcal{G}_k))$$

where $Z = \sum_k \exp(-\beta F(\mathcal{G}_k))$ is the standardization constant and $\beta = \frac{1}{T}$ is the *inverse temperature* parameter ($T \geq 0$ is the *temperature*). In fact, the value of β controls the randomness of the MH algorithm jumps, as seen by:

$$\tilde{w}_{kl} = \min\left(1, \frac{\exp(-\beta F(\mathcal{G}_l))}{\exp(-\beta F(\mathcal{G}_k))}\right) = \min\left(1, e^{\beta(F(\mathcal{G}_k) - F(\mathcal{G}_l))}\right)$$

If $\beta \to 0$, then $\tilde{w}_{kl} = 1$, i.e. the MH algorithm will jump to any candidate state l, while $\beta \to \infty$ implies that $\tilde{w}_{kl} = 1$ iff $F(\mathcal{G}_l) \leq F(\mathcal{G}_k)$ and $\tilde{w}_{kl} = 0$ otherwise.

2.4   Cooling Schedule and Exploration History

While a local minimum can easily be obtained with the MH algorithm and a simulated annealing cooling schedule (left plot in Fig. 2, as seen in [20, 21]), we are more interested here by the history of the exploration of space $\Gamma_\mathcal{V}$. Indeed, local minima are arguably often not really compatible with some real life constraints and we would be interested in finding alternative, but still efficient, ways to build the network. That is why we need our cooling schedule to be reheated periodically (right plot of Fig. 2) in order to avoid to be stuck in the same local minimum and to explore different parts of the space.

Recording the graph history $\{\mathcal{G}_1, \mathcal{G}_2, \ldots, \mathcal{G}_t\}$ of a MH run by keeping track of the states of every edges modified at least once $\{e_1, e_2, \ldots, e_p\}$, permits to obtain statistics on the behavior of the MH algorithm. Let the *history* matrix $H = (h_{rs})$ defined as followed:

$$h_{rs} = \begin{cases} 1 & \text{if the edge } e_s \text{ was present in the graph } G_r \\ 0 & \text{otherwise} \end{cases}$$

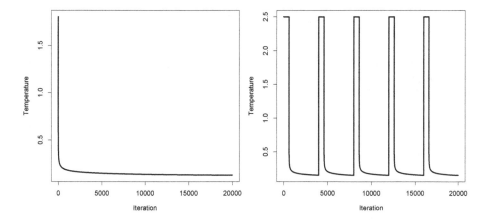

Fig. 2. Two types of cooling schedule: the first case represents a classical simulated annealing cooling schedule designed to find only one local minima: $T(t) = c/\log(1 + t)$ with $c = 1.25$. In the second case, we periodically set a high temperature during 400 iterations followed by a similar cool-down, in the hope of finding another minimum.

This matrix can be viewed as an usual "individuals × variables" matrix, enabling the computation of various indices. For instance, we can calculate the probability of the appearance of an edge as $p(e_s) = h_{\bullet s}/t$, its variance $\text{var}(e_s) = p(e_s)(1 - p(e_s))$ and the variance-covariance matrix between edges as $\Sigma = \frac{1}{t}H^{c\prime}H^c$ (where H^c is the matrix H after column centration). This variance-covariance matrix permits in turn to apply a *principal component analysis*, where the factor scores of all observed graphs in the history will underline recurrent configurations in Γ_V and the saturations between edges will highlight the competition or the cooperation existing between them.

3   Case Studies

3.1   Randomly Located Nodes

Let us first analyze the behavior of the MH algorithm on small sized graphs. 30 nodes in $\mathbb{R}^2$ with coordinates following a $\mathcal{N}(0, 1)$ are drawn, I is set to 50 and the temperature follows the cooling schedule exhibited on the right in Fig. 2 during $t = 20,000$ iterations.

As apparent on Fig. 3, the algorithm does not explore very efficiently the graph space. Each time we reheated the system it escaped from a local minimum before converging again on each cool down. The different minima seem to be close to each other, at least according to what appears in the first two factor scores (explaining only 13% and 9% of the variance). Fig. 4 confirms the closeness among the different minima, since some critical edges appear more frequently than others. The saturation plot shows that edges appearing frequently are correlated positively between themselves.

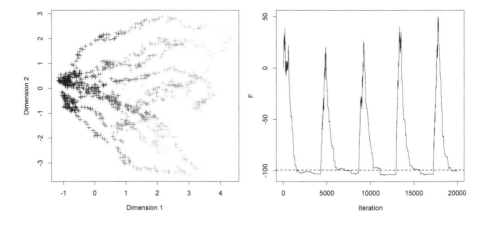

Fig. 3. Results for the graph history in the MH run. Left: first two dimensions of the factor scores, each point represents an iteration (proportion of variance explained: 13%, respectively 9%). The lower the value of F is for each iteration, the darker the point. Right: value of the functional F versus iteration.

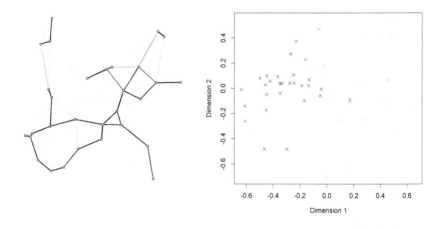

Fig. 4. Results for the edges in the MH run. Left: all edges created at least once during the process. Right: the saturation plot, where each points representing an edge and proximity capture correlation, i.e. two edges appearing frequently together will be close to each other. On both graphics, the darkness of an element is proportional to its apparition frequency during the run.

These results, while interesting, are a bit tarnished by the presence of high temperature states. While the presence of these states is essential to escape local minima, they bear very little information on optimal and alternative solutions to the efficient network

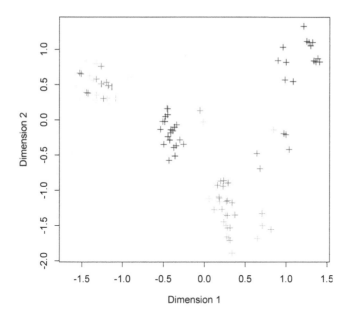

Fig. 5. The first two factorial coordinates of the states of the MH runs where states with high values of F have been removed. The five different cold temperature phases appear clearly, illustrating five different local minima.

building. Therefore, a second analysis is performed after removing all states having a functional value higher than -100 (corresponding to the dotted line on the Fig. 3, functional plot). By construction, the selected states constitute near-optimal solutions.

The graphic in Fig. 5 illustrates the emergence of five different "cold" temperature regimes during the MH run differ more than what it appear at first glance, showing that they indeed correspond to different local minima of F. Points in the middle yield the lowest value of the functional and correspond to the third cool-down. Graphics in Fig. 6 emphasize the edges created during the process. Here, we can observe some competition between edges. For example, edges numbered 7 and 36, 42 and 49, 20 and 39, 22 and 46, are placed on the opposite side one to another in the saturation plot. In the graph, we can see that both pairs represent building alternative to a close-to-optimal graph. On the other hand, edges 11, 10 and 26 are very centered, meaning that they have a very low variance and represent a kind of "backbone" appearing in any close-to-optimal graph. Iterations 8,643 and 12,903 in Fig. 7 exhibit some built variations. Note that state 12,903 to have a lower functional value than state 8,643 (-103.8 versus -100.4).

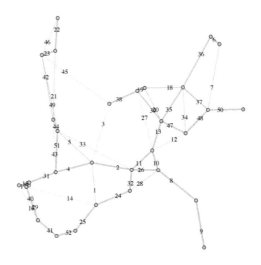

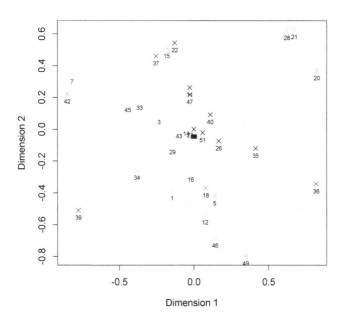

Fig. 6. Top: edges occupation frequencies. Bottom: saturations, with the same labeling

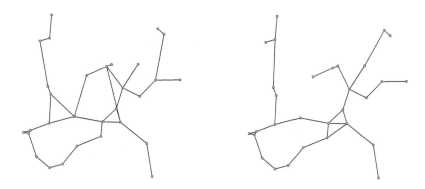

Fig. 7. Graph states at iteration number 8,643 (on the left) and 12,903 (on the right). Their F value are respectively -100.4 and -103.8.

3.2   US Cities

To study a real life case, the algorithm will be run on nodes representing US cities with more than 500,000 inhabitants (Fig. 8). Latitudes θ_i and longitudes α_i have been extracted from the R data `world.cities{maps}}` and we consider the *geodesic dissimilarity* between those cities: $D_{ij} = \arccos^2(\kappa_{ij})$, where $\kappa_{ij} = \sin\theta_i \sin\theta_j + \cos\theta_i \cos\theta_j \cos(\alpha_i - \alpha_j)$. Again, 20,000 iterations of MH are run with an investment

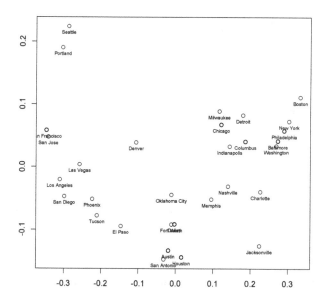

Fig. 8. Representation of the US cities with more than 500,000 inhabitants created by multidimensional scaling from their geodesic dissimilarities

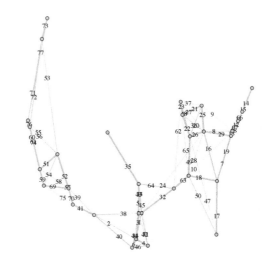

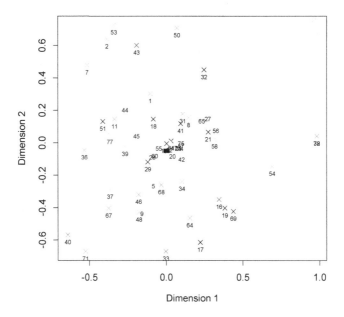

Fig. 9. Top: edges occupation frequencies. Bottom: saturations resulting from a complete MH run with 20,000 iterations.

of $I = 50$ (distances have been multiplied by 30 to match distances of the previous example) and higher temperature states have been removed from analysis.

Here again, edges frequencies and saturations in Fig. 9 reveal the occurrence of competing edges in the construction of the network together with some robust edges. Note the possibility of weighting each node relatively to its population resulting in a weighted efficiency functional, currently under investigation. Nevertheless, the present result can constitute a good start to explore ways of building real networks, where particular edges can be discarded in a second time, due to some morphological or economical constraints.

4 Conclusion

Exploring the possible graphs states on n nodes by MCMC not only reveals alternatives to the optimal network, but also gives insights on the structure of this space as controlled by the functional F. In the present case, the functional makes the shortest-paths requirement conflicts with the length of the edges, and permits to preliminary explore how the shortest-paths distance is linked to the Euclidean distance in this context. The investment, the cooling schedule, the starting state and the number of iterations are shown to greatly affect this exploration, and a careful design should be made depending on what is searched. The question of how to precisely set the parameters according to spatial configuration in hand remains largely open, and a deeper study should be performed before implementing this algorithm in real life applications. The numerical complexity and computational demands of the algorithm are also quite heavy, requiring a way of optimizing the parameters before applying the algorithm to a larger set of nodes. Nevertheless, case studies already show promising results and, provided the procedure can be efficiently refined, its flexibility should permit numerous applications to a large variety of situations.

References

1. Aldous, D.: Optimal spatial transportation networks where link-costs are sub linear in link-capacity. arXiv (0803.2037v2) (2008)
2. Aldous, D., Shunn, J.: Connected spatial networks over random points and a route-length statistic. arXiv (1003.3700) (2010)
3. Barthélemy, M.: Spatial networks. arXiv (1010.0302v2) (2010)
4. Berg, J., Lassig, M.: Correlated random networks. Physics Review Letters 89, 228701 (2002)
5. Brede, M.: Coordinated and uncoordinated optimization of networks. Physical Review E 81, 066104 (2010)
6. Courtat, T., Gloaguen, C., Douady, S.: Mathematics and morphogenesis of the city: A geometrical approach. Physical Review E (2010)
7. Crucitti, P., Latora, V., Marchiori, M.: A topological analysis of the italian electric power grid. Physica A 228, 92–97 (2004)
8. Gendron, B., Crainic, T., Frangioni, A.: Multicommodity capacited network design. Springer, Berlin (1999)
9. Gastner, M., Newman, M.: The spatial structure of networks. European Physical Journal B 49, 247–252 (2006)

10. Mathias, N., Gopal, V.: Small-worlds: How and why. Physical Review E 63, 021117 (2001)
11. Valverde, S., Cancho, R.F., Sol, R.V.: Scale-free networks from optimal design. Europhysics Letters 60, 512–517 (2002)
12. Yang, H., Bell, M.: Models and algorithms for road network design: A review and some new developments. Transport Reviews 18(3), 256–278 (1998)
13. Andrieu, C., de Freitas, N., Doucet, A., Jordan, M.I.: An introduction to MCMC for machine learning. Machine Learning 50, 5–43 (2003)
14. Carter, C., Kohn, R.: On Gibbs sampling for state space models. Biometrika 81(3), 541–553 (1994)
15. Ferenberg, A., Swendsen, R.: New monte carlo technique for studying phase transitions. Physical Review Letters 61, 2635 (1988)
16. Hastings, W.: Monte carlo sampling methods using Markov chains and their applications. Biometrika 57, 97–109 (1970)
17. Metropolis, N., Rosenbluth, A., Rosenbluth, M., Teller, A., Teller, R.: Equations of state calculations by fast computing machines. Journal of Chemical Physics 21, 1087–1092 (1953)
18. Newman, M., Barkema, G.: Monte Carlo methods in statistical physics. Oxford University Press (1999)
19. Ising, E.: Beitrag zur theorie des ferromagnetismus. Zeitschrift für Physik A Hadrons and Nuclei (1925)
20. Hajek, B.: Cooling schedules for optimal annealing. Mathematics of Operations Research 13(2), 311–329 (1988)
21. van Laarhoven, P.J.M., Simulated, E.A.: annealing: Theory and applications. Mathematics and Its Application 37, 7–15 (1987)

Geosemantic Network-of-Interest Construction Using Social Media Data

Sophia Karagiorgou[1], Dieter Pfoser[2], and Dimitrios Skoutas[3]

[1] School of Rural and Surveying Engineering, National Technical University of Athens
[2] Department of Geography and GeoInformation Science, George Mason University
[3] Institute for the Management of Information Systems, R.C. ATHENA

Abstract. An ever increasing amount of geospatial data generated by mobile devices and social media applications becomes available and presents us with applications and also research challenges. The scope of this work is to discover persistent and meaningful knowledge from user-generated location-based "stories" as reported by Twitter data. We propose a novel methodology that converts geocoded tweets into a mixed geosemantic network-of-interest (NOI). It does so by introducing a novel network construction algorithm on segmented input data based on discovered mobility types. The generated network layers are then combined into a single network. This segmentation addresses also the challenges imposed by noisy, low-sampling rate "social media" trajectories. An experimental evaluation assesses the quality of the algorithms by constructing networks for London and New York. The results show that this method is robust and provides accurate and interesting results that allow us to discover transportation hubs and critical transportation infrastructure.

1 Introduction

An important resource in today's mapping efforts, especially for use in mobile navigation devices, is an accurate collection of point-of-interest (POI) data. However, by only considering isolated locations in current datasets, the essential aspect of how these POIs are connected is overlooked. The objective of this work is to take the concept of POIs to the next level by computing *Networks of Interest* (NOIs) that encode different types of connectivity between POIs and capture peoples type of movement and behavior while visiting these POIs. This new concept of NOIs has a wide array of application potential, including traffic planning, geomarketing, urban planning, and the creation of sophisticated location-based services, including personalized travel guides and recommendation systems. Currently, the only datasets that consider connectivity of locations are road networks, which connect intersection nodes by means of road links purely on a geometric basis. POIs however, encode both geometric and semantic information and it is not obvious how to create meaningful links and networks between them. We propose to capture, both, *geometric* and *semantic* information in one NOI by analyzing social media in the form of spatial check-in data. We use the concept of check-in as a generic term for users actively volunteering their presence at a specific location. Existing road maps and POIs encode mostly geometric information and consist of street

M. Duckham et al. (Eds.): GIScience 2014, LNCS 8728, pp. 109–125, 2014.
© Springer International Publishing Switzerland 2014

maps, but may also include subway maps, bus maps, and hiking trail maps. To complement this dataset, *geometric trajectories* consist of geo-referenced trajectory data, such as GPS tracking data obtained from people moving on a road network. This type of data is assumed to have a relatively high sampling rate. Typical examples include vehicle tracking data sampled every 10 or 30 seconds. Such datasets are constructed using *map construction* (cf. [1], [2] for surveys).

In this work, we will use *behavioral trajectories* as a data source. They are obtained from social media in the form of spatial check-in data, such as geocoded tweets from Twitter. Similar to GPS tracking, the user contributes a *position sample* by checking in at a specific location. Compared to geometric trajectories, such check-in data result in very low-sampling rate trajectories that when collected for many users provide for a less dense, but semantically richer "movement network" layer. The main challenge arises from the fact that trajectories composed from geocoded tweets differ technically and semantically from raw GPS-based type of trajectories. Unlike trajectories obtained from GPS devices in typical tracking applications, such data are typically quite sparse since individuals tend to publish their positions only at specific occasions. However, we advocate that by combining and analyzing time and location of such data, it is possible to construct event-based trajectories, which can then be used to analyze user mobility and to extract visiting patterns of places. The expectation towards behavioral trajectories is that by integrating them into a Network of Interest, the resulting dataset will go beyond a homogeneous transportation network and will provide us with a means *to construct an actual depiction of human interest and motion dependent on user context and independent of transportation means.* As early maps were traces of people's movements in the world, i.e., view representations of people's experiences, NOIs try to fuse different qualities of such trace datasets obtained through intentional (e.g., social media, Web logs) or unintentional efforts (e.g., routes from their daily commutes, check-in data) to provide for a *consequent modern map equivalent.*

Specifically, in this paper we address the challenge of extracting a geosemantic NOI from noisy, low-sampled geocoded tweets. To do so, we introduce a new NOI construction algorithm that segments the input dataset based on sampling rate and movement characteristics and then infers the respective network layers. To fuse the semantic and geometric network layer into a NOI, we introduce a semantics-based algorithm that takes position samples (check-ins) to create network hubs. A detailed experimental evaluation uses two real-world datasets of geocoded tweets and discusses the NOI construction results in terms of quality and significance.

The remainder of this paper is organized as follows. Section 2 reviews related work on spatiotemporal inference techniques. Sections 3 and 4 present our algorithms for trajectory segmentation and re-association to build the NOI in a layered fashion. In Section 5, we evaluate the quality of the NOI construction method. Finally, Section 6 concludes the paper and outlines future research directions.

2   Related Work

Various approaches have been proposed for using user-generated geospatial content to extract useful knowledge, such as identifying travel sequences, interesting routes or socio-economic patterns. In the following, we present a review of the literature using a categorization of the approaches according to the type of problem solved.

Several methods focus on *sub-sequence extraction (routes) from moving object trajectories* by mining spatiotemporal movement patterns in tracking data. Kisilevich et al. [3] present an automatic approach for mining semantically annotated travel sequences using geo-tagged photos by searching for sequence patterns of any length. In [4], Chen et al. extract important routes between two locations by observing the traveling behaviors of many users. Although, they mine a transfer network of important routes, they accept that the distance between any two consecutive points in a trajectory does not exceed 100m, which becomes unrealistic. Zheng et al. [5] use online photos from Flickr and Panoramio to analyze people's travel patterns at a tour destination. They extract important routes, but no transportation network. Asakura et al. [6] investigate the topological characteristics of travel data, but they focus on identifying a simple index of clustering tourist's behavior. Mckercher and Lau [7] identify styles of tourists and movement patterns within an urban destination. Our approach analyzes, both, traffic patterns and topological characteristics of travel routes, while most existing work focuses on traffic patterns only. Choudhury et. al [8] explore the construction of travel itineraries from geo-tagged photos. In contrast, in this work an itinerary is defined as a spatiotemporal movement trajectory of much finer granularity.

There also exist various methods based on *trajectory clustering*. The majority of the proposed algorithms such as k-means [9], BIRCH [10] and DBSCAN [11] work strictly with point data and do not take the temporal aspect into consideration. Several approaches match some sequences by allowing some elements to be unmatched as in the Longest Common Sub Sequence (LCSS) similarity measure [12]. However, our goal in this work is rather to apply a trajectory clustering approach and also take into consideration the temporal aspect of the data. Similarity measures for trajectories that take the time and derived parameters, such as speed and direction, into account have been proposed in [13]. This approach is close to ours with respect to the examined aspects of temporal dimension, however, our method applies clustering techniques in order to infer the connectivity of a NOI. In a previous work [14], we derived a connected road network embedded in vehicle trajectories, while in [15] we inferred a hierarchical road network based on different movement types. The current approach differs in that it deals with uncertain social media check-in data by taking into account the spatial as well as the temporal dimension to derive a NOI.

Characterized by its spatial and temporal dimension, geocoded tweets can be regarded as one kind of spatiotemporal data, which also connects this study to the knowledge extraction-based techniques of the spatiotemporal data mining domain. Crandall et. al [16] investigate ways to organize a large collection ($\sim$ 35 million) of geo-tagged photos and determine important locations of photos, such as cities, landmarks or sites, from visual, textual and temporal features. Kalogerakis et. al [17] estimate the geolocations of a sequence of photos. Similarly, Rattenbury et. al [18] and Yanai et. al [19] analyzed the spatiotemporal distribution of photo tags to reveal the inter-relation between word concepts (photo tags), geographical locations and events. Girardin et al. [20] extract the presence and movements of tourists from cell phone network data and the geo-referenced photos they generate. Similarly, [21] proposes a clustering algorithm of places and events using collections of geo-tagged photos. These approaches efficiently deliver focal spatial data extractions from diverse data sources, while the aim of this work is to also extract *how this data is connected (links)*. In [22], Kling studies urban dynamics based on user generated data from Twitter and Foursquare using a probabilistic model. However, these dynamics have not been translated to a (transportation) graph structure.

All these works target the extraction of some kind of knowledge and patterns from photos or geo-referenced sources with textual and spatiotemporal metadata, while we focus on mining transportation and mobility patterns from check-in data, such as geocoded tweets from Twitter.

Overall, what sets this work aside is that *social media data is used as a tracking data source*. We use it not only to extract features or knowledge patterns of human activities, but a complete Network of Interest.

3   NOI Layer Construction

As explained in Section 1, our goal is to extract a Network of Interestthat captures interesting information about user movement behaviors based on social media tracking data. User check-in data are tuples of the form $U = \langle u, x, y, t \rangle$, denoting that the user u was at location (x, y) at time t. These data are organized into trajectories, which represent the sequence of locations a user has visited. Typically, multiple trajectories are produced for each user by splitting the whole sequence of check-ins, e.g., on a daily basis. Hence, each resulting trajectory is an ordered list of spatiotemporal points $T = \{p_0, \ldots, p_n\}$ with $p_i = \langle x_i, y_i, t_i \rangle$ and $x_i, y_i \in R, t_i \in R^+$ for $i = 0, 1, \ldots, n$ and $t_0 < t_1 < t_2 < \ldots < t_n$.

The goal is to construct a Network of Interest that reveals the *movement behavior* of users. This Network of Interest is a directed graph $G = (V, E)$, where the vertices V indicate important locations and the edges E important links between them according to observed user movements. In particular, we are interested in two aspects of the Network of Interest. A *geometric NOI aspect* provides a representation of how users actually move across various locations, thus preserving the actual geometry of the movement, while a *semantic NOI aspect* represents the qualitative aspect of the network by identifying significant locations and links between them. In our approach, we treat these two aspects as different layers of the same Network of Interest. In the following, we describe the steps for constructing these layers and fusing them to produce the final Network of Interest.

3.1   Segmentation of Trajectories

Behavioral trajectories, as in our case derived from geocoded tweets, contain data to construct both the geometric and the semantic layer of a Network of Interest. Conceptually, users tweet when they stroll around in the city as well as when they commute in the morning. While all these tweets will result in behavioral trajectories, *some of them depict actual movement paths*, while others simply are tweets sent throughout the day. In what follows, we try to separate our input data into two subsets and to extract the trajectories corresponding to the respective layer.

A main challenge when inferring a movement network from check-in data is that this data is very heterogeneous in terms of their sampling rate, i.e., often being very sparse. However, even the sparse subsets of the data are helpful in identifying significant locations, whereas the denser subsets can be used to capture more fine grained patterns of user movement.

For this purpose, we analyze the trajectories and group them into subsets with different temporal characteristics. In our approach, we treat these two aspects by applying a (i) *mean speed* threshold to capture the user movement under an urban transportation mode and by applying (ii) a *sampling rate* threshold to identify "abstract" and "concrete" movement. This allows us to treat each subset separately later on in the network construction phase. The "abstract" type of movement corresponds to the *semantic NOI aspect* and the "concrete" corresponds to the *geometric NOI aspect*.

Users with frequent check-ins, i.e., a high sampling rate, provide us with the means to derive a geometric NOI layer, while low sampling rates only allow us to reason about abstract movement, i.e., derive a semantic NOI layer.

Notice that typically the same individual, within one daily trajectory may have recorded their data using different sampling rates. In this case, the trajectory needs to be segmented according to the frequency of user position samples. A simple process for achieving this separation is the following. First, a duration and a speed (length divided by duration) is recorded for each segment of a trajectory. Each segment is assigned a corresponding duration type of movement. Focusing on urban transportation, we use a mean speed to filter out trajectories and then the duration between samples to determine "abstract" and "concrete" movement. Figure 1a shows the trajectories classified to different sampling rates using the example of geocoded tweets for London. Using a heatmap coloring schema, concrete and abstract movements are shown in blue and red, respectively.

The process is outlined in Algorithm 1. For each line segment L_j of each trajectory T, we compute a duration and a mean speed value (Algorithm 1, Lines 6-7), and the segment is then assigned to the corresponding segmented set of trajectories T_G, T_S according to the min and max time interval (Lines 9-13). The algorithm produces segmented sets of trajectories (Lines 10 and 13) based on the corresponding time interval attributes.

3.2   Geometric Layer Construction

To construct the geometric NOI layer we use frequently sampled trajectories. The sampling rate threshold was established through experimentation. In the examples of Section 5, the sampling rate threshold was set to $5min$. I.e., for the construction of the geometric layer the duration in between position samples of trajectory dataset is less than $5min$ (cf. Table 1), approximately covering 57% of the original tweets collection.

The geometric NOI layer construction approach follows a modified map construction approach (e.g., [14,15]) by (i) initially clustering position samples to derive network nodes, (ii) linking nodes by using the trajectory data and (iii) refining the link geometry.

To derive network nodes we employ the DBSCAN clustering algorithm [11] using a distance threshold and a minimum number of samples threshold parameter. We revisit the segmented trajectories to identify how the network nodes are connected by

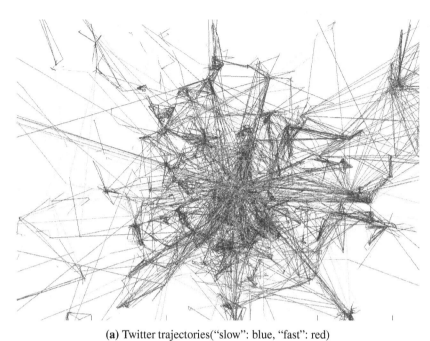

(a) Twitter trajectories("slow": blue, "fast": red)

(b) Respective OSM network

Fig. 1. Twitter Trajectories and OSM Network London (bounding box: [51.18N, 0.85W],[51.80N, 0.86E])

Algorithm 1. Segmentation of Trajectories

Input: A set of trajectories T
Output: Two sets of segmented trajectories T_G, T_S

1 **begin**
2 | /*__Trajectories segmentation according to time intervals__*/
3 | V_{max} ▷ maximum mean speed
4 | **foreach** $(T_i \in T)$ **do**
5 | | **foreach** $(L_j \in T_i)$ **do**
6 | | | $\bar{t}(L_j) \leftarrow \delta t(P[i-1], P[i])$ ▷ Time interval
7 | | | $\bar{v}(L_j) \leftarrow \frac{\delta x(P[i-1], P[i])}{\delta t(P[i-1], P[i])}$ ▷ Mean speed
8 | | | **if** $\bar{v}(L_j) \leq V_{max}$ **then**
9 | | | | **if** $\bar{t}(L_j) \leq T_{min}$ **then**
10 | | | | | $T_G \leftarrow L_j$
11 | | | | **end**
12 | | | | **else if** $\bar{t}(L_j) \geq T_{min}$ *and* $\bar{t}(L_j) \leq T_{max}$ **then**
13 | | | | | $T_S \leftarrow L_j$
14 | | | | **end**
15 | | | **end**
16 | | **end**
17 | **end**
18 **end**

---

creating links. The links represent clustered trajectories as two nodes can be connected by different trajectories. For each link (i) a *weight* is derived representing the number of the trajectories comprising the link and also (ii) a *length* representing the Euclidean distance between the nodes that constitute the link. In addition to this, we apply a reduction step to simplify the constructed network. The intuition is that due to varying sampling rates, links between nodes might exhibit redundancy. This reduction step eliminates redundant links by substituting longer links with links of more detailed geometries. We reconstruct links of longer duration by using links of shorter duration if their geometries are similar. We achieve this by using the degree of constructed nodes. Starting with nodes of a higher degree of incoming links, i.e., significant nodes, for such a node, we sort all incident links based on descending duration order. We then reconstruct those, which temporally and spatially cover other links that can be reached in less time. Figure 2a gives an example by showing in dark gray links before reduction, and in light gray a portion of the underlying OSM transportation network. Figure 2b shows then in dark gray the resulting links after applying the reduction step. Part of the larger geometry has been substituted with a more detailed geometry.

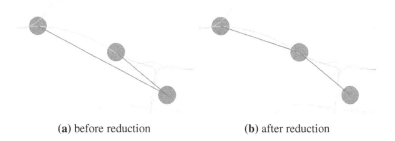

(a) before reduction   (b) after reduction

Fig. 2. Network Reduction Example - constructed network is shown in dark gray and the road network in light gray

3.3 Semantic Layer Construction

To construct the Semantic NOI layer, we rely on trajectories exhibiting low sampling rates (using approximately 19% of the original tweets collection), i.e., potentially cover large distances in between position samples making it difficult to reconstruct the actual movement (cf. Table 1). By initially applying the DBSCAN clustering algorithm (see Table 1 for parameter details), we extract a set of nodes that correspond to the *hubs* of the semantic layer. Performing a linear scan of the trajectories reveals the respective portions that connect the sets of nodes. For each link sample (i) a *weight* is derived representing the number of the trajectories comprising a link. At this step, we do not apply any reduction method as the geometries of the semantic layer are less accurate. Overall, this layer allows us to extract a network with less spatial accuracy but of greater semantic value.

4   NOI Construction and Layer Fusion

The final part of the Network of Interest construction process consists of (i) the extraction of hubs, i.e., significant locations that user frequently visits, and (ii) the fusion of the layers, i.e., the geometric and the semantic layer to produce the integrated network.

4.1   Network Hubs

Hubs are POIs that users frequently depart from and arrive at. In particular, specific indicators for hubs are (i) number of constituting position samples, (ii) stemming from many different users, (iii) over extended periods of time.

The Network Hubs Inference algorithm takes as input the *entire trajectory dataset* used in geometric and semantic layer construction (Algorithm 2, Line 9) and determines the k-NNs of each position sample (Line 12), which are subsequently filtered according to the number of users and the period of time covered (Lines 13-15). On these filtered position samples, we apply the DBSCAN clustering algorithm using a distance threshold and a minimum number of samples (Line 16). The centroids of the resulting clusters are the candidate hubs (Line 17). A final filtering step is applied as follows. For each candidate hub, we also record two properties. A *weight* for the hub is derived as the total number of nodes the hub was derived from, i.e., the size of the corresponding cluster. In addition, we record the *degree* of each hub, i.e., the number of incoming and outgoing edges of the cluster. A candidate hub is included in the output if both the following two conditions hold: (a) both the in-degree and out-degree are above a specified threshold and (b) the in-degree and out-degree do not differ significantly (threshold determined by experimentation). These conditions are used to ensure that the identified hubs correspond to places where a sufficiently large number of users frequently depart from and arrive at (Lines 23-24).

Algorithm 2. Hub Inference

> **Input**: A set of segmented trajectories T_G, T_S
> **Output**: Network Hubs

1 **begin**
2 | /*Clustering position samples of segmented trajectories to compute network hubs*/
3 | $H^* \leftarrow \emptyset \triangleright$ Candidate Hubs
4 | $H \leftarrow \emptyset \triangleright$ Hubs
5 | $d_{max} \triangleright$ proximity threshold
6 | $u_{min} \triangleright$ min. number of users
7 | $h_{min} \triangleright$ min. number of time periods
8 | $deg_{in}, deg_{out}, deg_{min}, \epsilon$
9 | $\triangleright$ position samples from combined trajectories
10 | $P \leftarrow \text{UNION}(T_G, T_S)$
11 | $\triangleright$ Samples $\rightarrow$ Hubs
12 | **foreach** $(P[i])$ **do**
13 | | $\nu_i \leftarrow \text{FINDNN}(P[i], d_{max})$
14 | | $u_p \leftarrow \text{COUNTUSERS}(\nu_i)$
15 | | $h_p \leftarrow \text{COUNTHOURS}(\nu_i)$
16 | | **if** $(u_p \geq u_{min})$ *and* $(h_p \geq h_{min})$ **then**
17 | | | $C \leftarrow \text{DBSCAN}(\nu_i, d_{max}) \triangleright$ Clusters
18 | | | $H^* \leftarrow \text{CENTROID}(C) \triangleright$ Hub candidates
19 | | **end**
20 | **end**
21 | **foreach** $H^*[i]$ **do**
22 | | $deg_{in} \leftarrow \text{GETINDEG}(H^*[i])$
23 | | $deg_{out} \leftarrow \text{GETOUTDEG}(H^*[i])$
24 | | **if** $deg_{in} \geq deg_{min}$ *and* $deg_{out} \geq deg_{min}$ *and* $\left| \frac{deg_{in}}{deg_{out}} - 1 \right| \leq \epsilon$ **then**
25 | | | $H \leftarrow H^*[i]$
26 | | **end**
27 | **end**
28 **end**

4.2   Layer Fusion

The final part of the process comprises the fusion of the geometric and semantic NOI layers. We construct the NOI by starting with the semantic layer and merging the geometric layer onto it. The intuition for this is that the semantic layer corresponds to a geometrically abstract but semantically richer user movement that contains relevant transportation hubs. The geometric layer corresponds to a less semantic but more accurate depiction of movement, i.e., fills in the gaps of the semantic layer. The fusion of these layers should result in a comprehensive movement network.

The fusion task involves (i) finding hub correspondences among the different network layers and (ii) introducing new links to the semantic layer for the uncommon portions of the NOI.

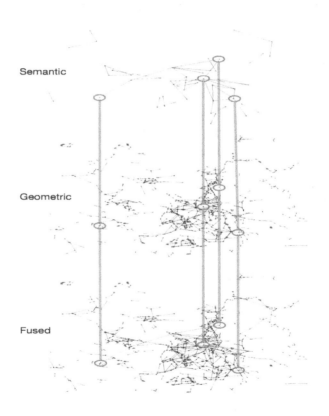

Fig. 3. London - Fused Network

Using, both, layers and the hubs, we try to identify common nodes by spatial proximity (Algorithm 3, Lines 11-13). Any node from the geometric layer that has not been introduced yet since it is not connected to the semantic layer will be added (Lines 22-23). The next step involves introducing new links for uncommon portions of the layered

---

Algorithm 3. NOI Fusion

Input: Networks to be conflated S, G
Output: Network of Interest

1 **begin**
2 | /*Network layers fusion to extract the final map*/
3 | ▷ edges and nodes of **S**emantic and **G**eometric layers
4 | $E_S \leftarrow \text{EDGES}(S)$, $N_S \leftarrow \text{NODES}(S)$
5 | $E_G \leftarrow \text{EDGES}(G)$, $N_G \leftarrow \text{NODES}(G)$
6 | H ▷ Hubs
7 | H_G ▷ hubs $\cap$ geometric nodes
8 | H_S ▷ hubs $\cap$ semantic nodes
9 | H_O ▷ $H - H_G - H_S$
10 | ▷ Node alignment
11 | **foreach** $H[i]$ **do**
12 | | ▷ finding Nearest Neighbors $H_G \leftarrow (H[i], \text{NN}(H[i], N_G))$
13 | | $H_S \leftarrow (H[i], \text{NN}(H[i], N_S))$
14 | **end**
15 | ▷ Node alignment
16 | **foreach** $H_G[i]$ **do**
17 | | $H_O \leftarrow (H_G[i], 1\text{-NN}(H_G[i], H_S))$
18 | | ▷ Node insertion to semantic layer
19 | | **foreach** $(H_G[i] \notin H_O)$ **do**
20 | | | $E_i = \text{ON}(E_S, H_G[i])$
21 | | | **if** $E_i \neq \text{NULL}$ **then**
22 | | | | $H_S.add(H_G[i])$
23 | | | | $E_S.delete(E_i)$
24 | | | **end**
25 | | **end**
26 | | ▷ Link insertion
27 | | **foreach** $(H_G[i] \notin H_S)$ **do**
28 | | | $H_S.add(H_G[i])$ ▷ remaining nodes
29 | | | **foreach** $(E_G[i] \notin E_S)$ **do**
30 | | | | $E_S.add(E_G[i])$ ▷ remaining links
31 | | | **end**
32 | | **end**
33 | **end**
34 **end**

---

network. Here links of the geometric layer are introduced by adding them to the semantic layer (Lines 28-30). Typically this accounts for the cases of adding complete (local) network portions.

A result of applying this conflation algorithm to network layers is shown in Figure 3. Indicated are the circled hub correspondences between the semantic, the geometric layer, and the resulting fused Network of Interest

5   Experimental Evaluation

An assessment of the quality of a Network of Interest is a challenging task as there is no ground-truth data. In the case of map-construction algorithms, an existing road network can be used. However, a NOI represents a geosemantic construct containing aspects of both, regular transportation networks (roads, public transport, etc), but also the overall movement sentiment of users in a city. For the following evaluation, we will use a combination of existing POI datasets and (public) transportation networks to assess the constructed NOIs. Before giving details of the experimental results and constructed NOIs, we first describe the characteristics of the datasets used and our overall evaluation methodology.

5.1   Experimental Setup

We conduct experiments on two real-world datasets comprising geocoded tweets retrieved for London and New York City over a period of 60 days using the Twitter Public Stream API. Data from London covers the period of December 2012 to January 2013. The New York as collected from November 2013 to December 2013. To focus on trajectories of active users, we kept only the trajectories of the top 200 most active users with respect to geotweets for each city. Moreover, we only consider trajectories that consist of at least 5 geotweets. Figure 1a visualizes the movements of 200 Twitter users during the course of a single day in London. Notice that some very prominent areas, such as highways, can be distinguished visually even before any processing of the data takes place.

Through experimentation, we established the parameters for the various steps of the algorithm as summarized in Table 1. To compare the generated network, we consider as ground-truth data the corresponding public transportation network obtained from OSM [23]. What follows is a brief description of the trajectories collected from the geocoded tweets, as well as the networks obtained from OSM.

In London, the actual public transportation network consists of 27,021 links (edges) and 47,575 nodes (vertices) and has a length of 21,287km. It covers an area of 420km × 118km including the metropolitan area of London. The geocoded tweets cover a great portion of this network, specifically an area of 365km × 104km, and have a total combined length of 256,400km (Figure 1a). The dataset consists of 463 trajectories with a median length of 7.4km. The median sampling rate, i.e., rate at which a user geotweets, is 12min, while the median speed is 37km/h.

For New York the actual public transportation network consists of 84,367 links and 75,070 nodes and has a length of 9,846km. It covers an area of 105km × 85km. The geocoded tweets consist of 37,962 trajectories, with a median length of 2.9km and total length of 214,090km, covering an area of 92km × 74km largely overlapping with the public transportation network. The median sampling rate is 8min, while the median speed is 22km/h.

Table 1. Parameter Summary

Algorithm	Value
Segmentation of Trajectories	
Mean Speed	10km/h
Time Interval	5, 60min
Geometric NOI	
Distance Threshold	100m
Minimum Number of Samples	2
Semantic NOI	
Distance Threshold	300m
Minimum Number of Samples	2
Extraction of Hubs	
Minimum Number of Samples	10
Minimum Number of Users	2
Minimum Number of Time Periods	10
Distance Threshold	300m
Layer Fusion	
Distance Threshold	50m

5.2   Visual Comparison

A first and quick overview of the quality of the inferred Network of Interest can be obtained by *visual inspection*, i.e., by comparing it to the ground-truth public transportation network and looking for similarities and differences.

Figure 4 visualizes the NOIs of the cities of London (Figure 4a) and New York (Figure 4b). In each case, the constructed network is visualized using black lines, while the ground-truth network is shown using light gray lines. As evident, especially for the case of New York, the constructed NOI lines up with the transportation network and identifies major hubs.

5.3   Quantitative Evaluation

For a more systematic and quantitative assessment of NOIs, we devise two means, (i) comparing the constructed NOI to the geometry of a respective transportation network and (ii) comparing the nodes of our NOI with a POI dataset to discover semantics in terms of their type. This approach allows us to assess the similarity with respect to the ground-truth network and to draw conclusions not only with respect to the spatial accuracy of the result, but also the semantics of the nodes.

To *compare networks* we select all the nodes of the constructed network and identify corresponding nodes in the ground-truth network by means of nearest-neighbor queries. Using the OSM public transport data, we select for every hub of the Network of Interest the nearest node in the OSM data. If the inferred nodes are close to the actual transportation network nodes, then the constructed NOI closely relates to the transportation network.

To discover the *type of transportation* a hub represents, e.g., bus, metro, tram and railway, we again use OSM data. We apply reverse geocoding (identify POIs based on

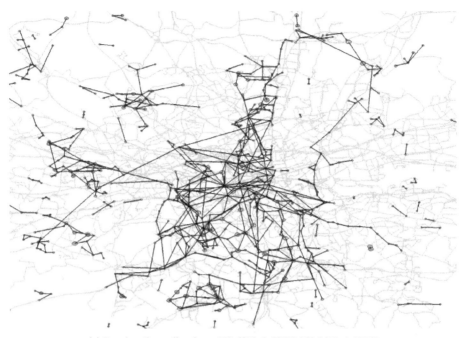

(a) London (bounding box: [50.60N, 0.50W],[52.00N, 1.25E])

(b) New York (bounding box: [40.54N, 74.10W],[40.92N, 73.70W])

Fig. 4. Networks of Interest

coordinates) to relate OSM POIs to NOI locations. This then allows us to identify public transportation nodes in our generated Network of Interest. The results are summarized in Figure 5, which shows the degree of a node, i.e., the number of incoming and outgoing links. In this case, we use the degree as an indicator for the importance of the node and the fact that high-degree nodes were identified as transportation nodes allows us to reason about the type of network we constructed. Identified transportation nodes (i.e. bus, metro, etc) have higher degrees (> 20) when compared to *other* nodes with lower degree (< 5).

In this experimentation, (i) nearest-neighbor queries evaluate the spatial accuracy of the NOI, while (ii) the reverse geocoding assesses the semantics of the hubs. The higher

Table 2. Evaluation Summary

	Nearest Neighbor Statistics		Reverse Geocoding Statistics	
	Found Total	Ratio %	Found Total	Ratio %
London	1389 1562	89	964 1562	62
New York	1423 1649	86	873 1649	53

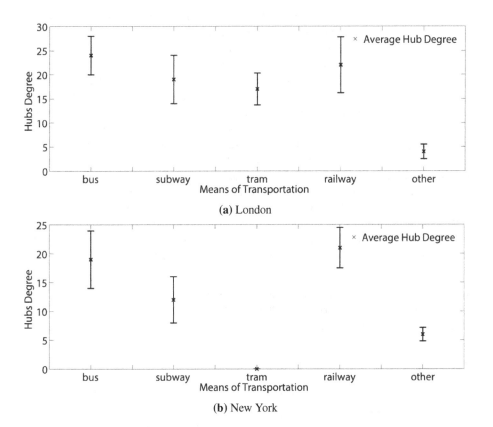

(a) London

(b) New York

Fig. 5. Hubs Statistics

the number of correctly constructed nodes, the higher also the quality of the network. As shown in Table 2, transportation nodes are inferred with high accuracy. 89% of the extracted hubs in London and 86% in New York are identified as transportation nodes in the OSM ground-truth network. In the case of the reverse geocoding test, the ratios are a bit lower due to the fact that the reverse geocoding service returns only POIs that are located exactly or very closely to specific coordinates.

An overall sentiment of our experimentation could be that the network construction process results in a Network of Interest that *captures certain aspects of a public transportation network*. A core problem in such experimentation is that using social media as a tracking data source to construct a network has the inherent challenge that no actual ground-truth data is available to assess the quality of the result. Using in our case a public transportation network allows us to show some similarities, however, the constructed NOI could not be completely mapped (explained) by it as it represents a more complex network whose characteristics cannot be captured by a single existing network dataset. These concerns are also issues we want to address in future work.

6   Conclusions

Social media applications and their data have been used in a wide range of data mining applications. However, to the best of our knowledge this work is the first to construct a geosemantic Network of Interest using social media as a tracking data source. The NOI construction algorithm is based on segmenting geocoded tweets and constructing two separate network layers. A geometric and a semantic layer of a NOI are derived and using network hubs, these layers are then fused to generate a Network of Interest. Performing an experimental evaluation using two large-scale datasets, the algorithm produces NOIs of considerable accuracy, which identify important transportation hubs and capture portions of the respective public transport networks.

The directions for future work are to refine the NOI construction process and scaling the algorithms and to use it for larger datasets and more complex NOIs. Here, we will also have the opportunity to identify temporal aspects of the NOIs, e.g., transportation routes to and from the city, temporal variations, as well as characteristics of the NOI graph itself (connected components). We are also in the process of applying the proposed methods to mobile phone tracking data, a dataset that is "in between" GPS tracking data and check-in data in terms of positional accuracy and sampling rate.

Acknowledgments.   This work was supported by the EU FP7 Marie Curie Initial Training Network GEOCROWD (FP7-PEOPLE-2010-ITN-264994) http://www.geocrowd.eu.

References

1. Ahmed, M., Karagiorgou, S., Pfoser, D., Wenk, C.: A comparison and evaluation of map construction algorithms. Under submission (2013)
2. Biagioni, J., Eriksson, J.: Map inference in the face of noise and disparity. In: Proc. 20th ACM SIGSPATIAL GIS Conference, pp. 79–88 (2012)

3. Kisilevich, S., Keim, D.A., Rokach, L.: A novel approach to mining travel sequences using collections of geo-tagged photos. In: Proc. 13th AGILE Conference, pp. 163–182 (2010)
4. Chen, Z., Shen, H.T., Zhou, X.: Discovering popular routes from trajectories. In: Proc. 27th International Conference on Data Engineering, pp. 900–911 (2011)
5. Zheng, Y.T., Zha, Z.J., Chua, T.S.: Mining travel patterns from geotagged photos. ACM Transactions on Intelligent Systems and Technology 3(3), 56:1–56:18 (2012)
6. Asakura, Y., Iryo, T.: Analysis of tourist behaviour based on the tracking data collected using a mobile communication instrument. Transportation Research Part A: Policy and Practice 41(7), 684–690 (2007)
7. Mckercher, B., Lau, G.: Movement patterns of tourists within a destination. Tourism Geographies 10(3), 355–374 (2008)
8. De Choudhury, M., Feldman, M., Amer-Yahia, S., Golbandi, N., Lempel, R., Yu, C.: Constructing travel itineraries from tagged geo-temporal breadcrumbs. In: Proc. 19th World Wide Web Conf., pp. 1083–1084 (2010)
9. Lloyd, S.: Least squares quantization in pcm. IEEE Transactions on Information Theory 28(2), 129–137 (2006)
10. Zhang, T., Ramakrishnan, R., Livny, M.: Birch: An efficient data clustering method for very large databases. In: Proc. 1996 SIGMOD Conference, pp. 103–114 (1996)
11. Ester, M., Kriegel, H.P., Sander, J., Xu, X.: A density-based algorithm for discovering clusters in large spatial databases with noise. In: Proc. 2nd SIGKDD Conference, pp. 226–231 (1996)
12. Bollobás, B., Das, G., Gunopulos, D., Mannila, H.: Time-series similarity problems and well-separated geometric sets. In: Proc. 13th Annual Symposium on Computational Geometry, pp. 454–456 (1997)
13. Pelekis, N., Kopanakis, I., Marketos, G., Ntoutsi, I., Andrienko, G., Theodoridis, Y.: Similarity search in trajectory databases. In: Proc. 14 International Symposium on Temporal Representation and Reasoning, pp. 129–140 (2007)
14. Karagiorgou, S., Pfoser, D.: On vehicle tracking data-based road network generation. In: Proc. 20th ACM SIGSPATIAL GIS Conference, pp. 89–98 (2012)
15. Karagiorgou, S., Pfoser, D., Skoutas, D.: Segmentation-based road network construction. In: Proc. 21th ACM SIGSPATIAL GIS Conference, pp. 470–473 (2013)
16. Crandall, D.J., Backstrom, L., Huttenlocher, D., Kleinberg, J.: Mapping the world's photos. In: Proc. 18th World Wide Web Conf., pp. 761–770 (2009)
17. Kalogerakis, E., Vesselova, O., Hays, J., Efros, A.A., Hertzmann, A.: Image sequence geolocation with human travel priors. In: Proc. 11th International Conference on Computer Vision, pp. 253–260 (2009)
18. Rattenbury, T., Good, N., Naaman, M.: Towards automatic extraction of event and place semantics from flickr tags. In: Proc. 30th ACM SIGIR Conference, pp. 103–110 (2007)
19. Yanai, K., Kawakubo, H., Qiu, B.: A visual analysis of the relationship between word concepts and geographical locations. In: Proc. ACM International Conference on Image and Video Retrieval, pp. 13:1–13:8 (2009)
20. Girardin, F., Calabrese, F., Fiore, F.D., Ratti, C., Blat, J.: Digital footprinting: Uncovering tourists with user-generated content. IEEE Pervasive Computing Magazine 7, 36–43 (2008)
21. Kisilevich, S., Mansmann, F., Keim, D.: P-dbscan: A density based clustering algorithm for exploration and analysis of attractive areas using collections of geo-tagged photos. In: Proc. 1st International Conference and Exhibition on Computing for Geospatial Research and Application, pp. 38:1–38:4 (2010)
22. Kling, F., Pozdnoukhov, A.: When a city tells a story: Urban topic analysis. In: Proc. 20th ACM SIGSPATIAL GIS Conference, pp. 482–485 (2012)
23. OpenStreetMap Foundation: Openstreetmap: User-generated street maps (2013), http://www.openstreetmap.org

Data Quality Assurance for Volunteered Geographic Information

Ahmed Loai Ali[1,3] and Falko Schmid[1,2]

[1] Cognitive Systems Group, University of Bremen, Germany
[2] SFB/TR 8 Spatial Cognition, University of Bremen, Germany
[3] Faculty of Computers and Information, Assiut University, Egypt

Abstract. The availability of technology and tools enables the public to participate in the collection, contribution, editing, and usage of geographic information, a domain previously reserved for mapping agencies or companies. The data
of Volunteered Geographic Information (VGI) systems, such as OpenStreetMap
(OSM), is based on the availability of technology and participation of individuals.
However, this combination also implies quality issues related to the data: some
of the contributed entities can be assigned to wrong or implausible classes, due to
individual interpretation of the submitted data, or due to misunderstanding about
available classes. In this paper we propose two methods to check the integrity of
VGI data with respect to hierarchical consistency and classification plausibility.
These methods are based on constraint checking and machine learning methods.
They can be used to check the validity of data during contribution or at a later
stage for collaborative manual or automatic data correction.

1 Introduction

During the last decade, low-cost sensing devices like handheld GPS receivers or smartphones became available and accessible for many consumers. In the same period powerful open GIS software and web technologies have been developed. The availability
of technology and tools enables the public to participate in the collection, contribution,
editing, and usage of geographic information, a domain previously reserved for mapping agencies or large organizations. Volunteered Geographic Information (VGI) [1],
the voluntary collection and contribution of geo-spatial data by interested individuals
became a large and vital movement. VGI projects like OpenStreetMap[1] (OSM) result
in large scale data sets of geographic data covering many parts of the world. This new
way of geographic data production changed not only the way of data processing but
also applications and services built on it [2–4].

There exist a huge number of services based on e.g., OSM data, such as map providers,
trip advisers, navigation applications. Depending on the service, reliable data is necessary. However, without coordinated action, the experience and training of experts, and
industrial grade sensing devices it is hard to guarantee data of homogeneous quality.

The absence of a clear classification system in, e.g., OSM, the ambiguous nature of
spatial entities, and the large number of users with diverse motivations and backgrounds

---

[1] http://www.OpenStreetMap.org

M. Duckham et al. (Eds.): GIScience 2014, LNCS 8728, pp. 126–141, 2014.

foster the generation of data of mixed quality. Whatever a body of water is a pond or a lake, whatever a grassland is a meadow, natural reserve, a park, or a garden is not just a question of a proper, crisp definition, but also a question of perception, conceptualization, and cultural background. What is a pond somewhere, might be a lake in a different environment, a river might be a creek or a stream. In addition to rather conceptual issues, many contributed entities are incompletely classified or wrongly attributed due to the open and loose attributation mechanism in OSM. As a result, a significant amount of data is not correctly classified and can cause errors whenever they are addressed by algorithms, such as rendering, analysis, or routing. This situation triggers questions about the quality of VGI data, suitable mechanisms for guaranteeing and fostering high quality contributions, and correcting problematic data.

Hence, it becomes increasingly important to analyze the heterogeneous quality of VGI data. Several studies investigate the quality of VGI by applying geographic data quality measures, such as feature completeness, positional accuracy, and attribute consistency [5–7]. These approaches usually require using reference data sets to evaluate the VGI data. However, these data sets are in many cases not available.

In this paper we present two approaches for analyzing the quality of VGI data: one by constraint checking and one by machine learning, i.e., we are analyzing the available data only with respect to consistency and plausibility based on contributions themselves. The results can be used to re-classify existing data and to provide guidance and recommendations for contributors during the contribution process. Recommendations can be directly generated from the data source itself by analyzing the distribution of the contributed feature in the surrounding area, thus the locality of entitles is preserved and no global rules are applied to locally generated data.

2   Related Work

In VGI, contributors produce geographic information without necessarily being educated surveyors or cartographers. In open platforms such as OSM, the motivation for contribution can be highly diverse, and the quality of contributions also depends on the used equipments and methods. Thus, the combination of diverse educational backgrounds, different views on required data and its quality, as well as technical constraints lead to data of mixed quality. Hence, the assessment of VGI data quality became a focus in VGI related research.

Quality of VGI data has various perspectives and notions: completeness, positional accuracy, attribute consistency, logical consistency, and lineage [8]. The quality can be assessed by basically three different methods: comparison with respect to reference data, semantic analysis, and intrinsic data analysis.

One approach to assess the quality of VGI data is by means of a direct comparison with reference data collected with a certain quality standards. The challenge of this approach is to identify a robust mutual mapping function between the entities of both data sets. In [6, 9] the authors are able to show a high overall positional accuracy of OSM data in comparison with authoritative data. In terms of completeness, some studies conclude that some areas are well mapped and complete relative to others. They also show a tight relation between completeness and urbanization [9, 10].

Different aspects have influence on the quality of VGI data, e.g., the combination of loose contribution mechanisms, and the lack of strict mechanisms for checking the integrity of new and existing data are major sources of the heterogeneous quality of VGI data [11]. Amongst others, semantic inconsistency is one of the essential problems of VGI data quality [12]. In [13] and [14] the authors present methods for improving the semantic consistency of VGI. The analysis of semantic similarity is applied to enhance the quality of VGI by suggesting tags and detecting outliers in existing data [13, 14], as well as by ontological reasoning about the contributed information (e.g., [15]). Another approach for tackling quality issues is the development of appropriate interfaces for the data generation and submission. In [16, 17] the authors demonstrate that task-specific interfaces support the generation of high quality data even under difficult conditions.

An alternative approach is evaluating the available data along three intrinsic dimensions [8]:

- *Crowdsourcing evaluation*: the quality of data can be evaluated manually by means of cooperative crowdsourcing techniques. In such an approach, the quality is ensured through checking and editing of objects by multiple contributors, e.g., by joint data cleaning with gamification methods [18].
- *Social measures*: this approach focuses on the assessment of the contributors themselves as a proxy measure to the quality of their contributions. [6, 9] use the number of contributors as a measure for data quality, [19] analyzes the individual activity, [11] investigates positive and negative edits, [20] is researching fitness-for-purpose of the contributed data.
- *Geographic context*: this approach is based on analyzing the geographic context of contributed entities. This approach relates to Tobler's first law of geography which states that "all things are related, but nearby things are more related than distant things" [21].

3   Managing Quality of VGI Data

A big challenge for VGI is the quality management of the contributed data because of its multidimensional heterogeneity (knowledge and education, motivation for contribution, and technical equipment). The problem requires the development of tools advising contributors during the entity creation process, but also to correct already existing data of questionable quality. Amongst others, quality problems can be general accuracy issues, geometric or topological constraint violations, hierarchical inconsistencies, and wrong or incomplete classification. In this work we focus on hierarchical inconsistencies and wrong or incomplete classification. Whenever we use the term "wrong" in our study we mean the assignment of a *potentially* wrong class or *tag* to the respective entity due to labeling ambiguity. "Wrong" entities will be detected by our classification and consistency checking algorithms. This is only an indicator for a potential conflict.

In the case of OSM, it is known that the data set contains large amounts of problematic data (e.g., see Section 2). On the other hand, we can assume that a significantly larger part of the data is of sufficient quality: the large amount of volunteers constantly improving the data set and the large number of commercial applications built on top of the data set are good indicators for it. Given that this rather unprovable statement is true, we can use the data itself for quality assessment by learning its properties and using the results as an input for the processes described in our approach.

Figure 1 describes the two phase approach: in the *Classification* phase, we can either apply machine learning algorithms to learn classifiers of the so far contributed data, or we can define classification constraints the data has to satisfy. Some of the before mentioned quality issues could be solved if at the point of data generation or contribution the integrity with existing data is checked. Depending on the potential problem to be addressed, different automatic approaches for satisfying inherent constraints are available, e.g., [22].

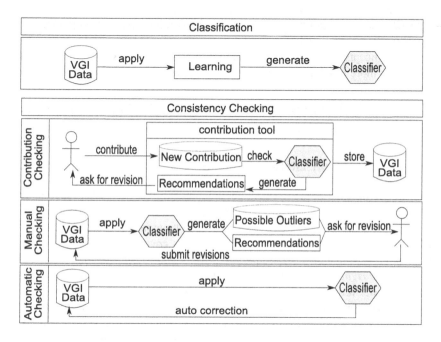

Fig. 1. Proposed approaches to ensure VGI quality, see Section 3 for a detailed description

Hence, in the *Consistency Checking* phase we propose three approaches for checking the consistency of the data: during *Contribution Checking* the contribution tool should inform users during the contribution process about potentially problematic data based on the generated classifier. Contributors can now consider the hints generated by the system about an object and can take actions to correct it if necessary. After contribution, the new data can be used to train the classifier again (if checking is based on an learning approach). *Manual Checking* should provide tools allowing the identification of problematic entities in the existing data set. They can be presented to volunteers for checking and correcting, ideally based on plausible suggestions. And finally, *Automatic Checking* can correct obviously wrong data automatically, if the correction can be computed without human assistance.

4 Tackling Areal Consistency and Classification Plausibility

The majority of data quality studies focus on point-like or linear geographic entities, such as points of interest or road networks (see Section 2). In this work we focus on quality issues related to areal entities, that is extended geometric entities. Our methods can be applied to entities of all possible scales, from very large administrative or natural entities to rather small ones like buildings or park benches.

The focus of our work is the quality of the *classification* of the contributed data. We are particularly interested in:

- *Hierarchical consistency* of administrative data: we check if administrative elements are used according to intrinsic, logical rules.
- *Classification plausibility* of areal entities: the correct classification of entities can be difficult, especially when contributors are not aware of potential conflicts due to similar concepts. Here we focus on ambiguity issues resulting from the availability of two or more possible classification options of entities (e.g., park vs. garden vs. grass).

Our study is build on OSM data. We will use notions typically used in the OSM tagging scheme, such as: *keys* and *values*.

5 Hierarchical Consistency Analysis

Administrative boundaries are political geographic entities with a strict inherent structure, such as *continents* consist of *countries*, *countries* consisting of *states* and *states* consisting of *districts*, etc. In OSM[2] administrative boundaries are defined as *subdivisions of areas/territories/jurisdictions recognized by governments or other organizations for administrative purposes. Administrative boundaries range from large groups of nation states right down to small administrative districts and suburbs, with an indication of this size/level of importance, given by tag "admin_level" which takes a value from 1 to 10*. However, as countries can have different administrative partitioning, some levels might not be applicable or the classification schema may not be sufficient. In this case it can be extended to 11 levels (e.g., in Germany and Netherlands).

Typically, administrative boundaries around administrative Units U are structured such that every administrative unit typically belongs to *one* administrative level of 1 to 11 (exceptions are, e.g., city states):

$$\forall u \in U_i \text{ where } 1 \leq i \leq 11 \tag{1}$$

Each administrative unit where $i > 1$ is contained in an administrative unit of a higher level; all together the contained units *exhaustively* cover the territory of the containing unit:

$$\forall u_a \in U_{i>1}, \exists u_b \in U_{j>i} : u_a \subset u_b \tag{2}$$

Administrative units on one level can share borders but do *not intersect* each other:

$$\forall U_j, U_k \subset U_i : U_j \cap U_k = \emptyset \tag{3}$$

---

[2] http://wiki.openstreetmap.org/wiki/Key:admin_level#admin_level

However, there are exceptions from this strict hierarchy, such as exclaves, enclaves, city states, or embassies. Still, the vast majority of administrative units follow a clear and exhaustive hierarchical ordering. This allows checking the integrity of the available administrative data in OSM by checking the following type of outliers:

- *Duplication*: in the case of duplication, entities belong to two or more different administrative units. See Figure 2(a).
- *Inconsistency*: hierarchical inconsistency occurs when entities of higher administrative units are contained in units of lower levels or the same level. See Figure 2(b)
- *Incorrect Values*: incorrect values occur throughout the OSM data set, probably due to the import from different classification schemes. Typically the value of *admin_level* tag is not a numerical value between 1–11.

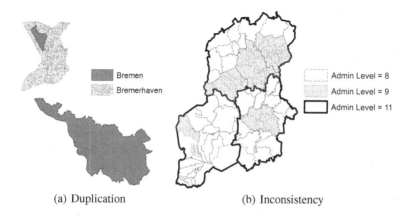

(a) Duplication   (b) Inconsistency

Fig. 2. Incorrect classification plausibility (Duplication & Inconsistency). In a) a part of Bremen city is within Bremerhaven, in b) units on level 11 contain elements of level 8 and 9.

5.1   Consistency Analysis Results and Discussion

We applied the consistency rules on the complete OSM data set downloaded at January 20th, 2014. At the time of analysis, the OSM data contained 259,667 geographic entities classified as administrative units (admin_level = *value*). 24,410 entities, thus about 10% of all administrative units contained problematic assignments, see Figure 3. We identified 14,842 duplications, 9,305 inconsistencies and 263 incorrect values.

Figure 2(a) illustrates an example for *duplication*: a part of the administrative unit representing Bremen city, is part of another unit representing Bremerhaven city. Figure 2(b) shows an instance of inconsistency: some administrative units of level 8 and 9 are contained by administrative units of level 11.

Of course, not all of the 24,410 detections represent wrong data, some cases already represent the mentioned special cases, some inconsistencies might be detected due to incomplete presence of administrative hierarchies. However, a plausibility check as sketched in Section 3 would draw the attention of the contributor towards potential errors.

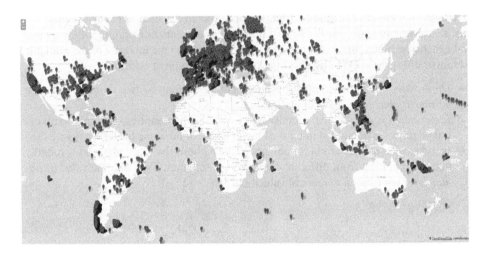

Fig. 3. Distribution of potentially incorrect hierarchical classification of administrative units

6   Classification Plausibility Analysis

When users contribute data to OSM, they have a large range of possibilities to classify the data. In some cases classifying entities is not straightforward; depending on the perspective of the contributor different possible classes may be applicable. A water body can still be a pond or already be a lake, the grass covered area can be a park, a garden, meadow or grassland. In many cases there is no definite answer, especially as in OSM there is no explicit classification system, just recommendations. However, utilizing spatial data requires homogeneous handling of data of identical concepts. Only if the same type of entities are identically classified, algorithms can access them properly for analysis, rendering, or reasoning. However, in many cases users contribute data with wrong classifications either due to conceptual ambiguity or due to a different understanding of the available concepts.

In this work we exemplify our approach on analyzing classification plausibility of entities, which are classified either as *park* or *garden*. We chose these classes as they are good examples for classification ambiguity: within OSM, parks and gardens lack a clear definition distinguishing them. Thus, contributions of these features mainly depend on individual conceptualizations. Many entities are obviously not correctly classified when we inspected them with a commonsense understanding of parks and gardens. Typically parks are public, accessible areas of a cultivated nature. Gardens, in contrast are typically private areas also featured with cultivated nature. However, one large difference of both entities is not only their infrastructural containments, but also their size: parks are usually significantly larger than gardens. As usual when it comes to geospatial reality, we can observe everything such as large public gardens or small parks. However, the vast majority of gardens and parks follow this vague classification (see Figure 6 for a support of this statement), especially relative to entities in their surrounding (parks and gardens can have significantly different dimensions in different areas of the world, usually correlated to the available territory in relation to the population). In the following we analyzed entities classified with the tags leisure=*park* and leisure=*garden*.

6.1 Classification Learning to Ensure VGI Quality

Due to the large amount of data in OSM, it is possible to apply machine learning techniques to tackle data quality issues. Machine learning algorithms can learn from existing data and extract implicit knowledge to build a classifier. Then such a classifier can be used for ensuring the quality as sketched in Figure 1, either during contribution or by applying on already existing data. In our approach learning the classifier on the contributed data is used to predict the correct class of an entity (i.e. park or garden in our example). This is done in two steps: a learning or training step, and a validation step.

In the first step our system learns a classifier based on the properties of pre-classified entities of a *training set* [23, 24]. In this work, the training set consists of entities representing parks and gardens, $D_{train} = (E_1, E_2, ..., E_n)$, where each Entity E is represented by a set of features (such as: size, location ...etc.) and is assigned to a class C (i.e. park or garden), $E = (F_1, F_2, ..., C)$. This step tries to identify a function, $f(E) = C$ to predict the class C of a given entity E.

In the second step the generated classifier is used for classification: we apply it on a test set to measure the accuracy of the classifier. The test set only contains entities not used for training. The classifier performance is evaluated according to classification accuracy on the test entities [23, 24].

6.2 Experiments and Setup

As described previously, we focus on classification plausibility in case of similarly applicable classes, in our case parks (leisure = *park*) and gardens (leisure = *garden*). We use data from Germany, the United Kingdom (UK), and Austria. According to [6, 9], OSM data is of acceptable quality in Germany and the UK. In our study we use data downloaded on December 20th, 2013.

We selected data from the ten densest (population/area) cities of each country. Figure 4 shows the selected cities and the present number of parks and gardens within each city. We decided to use cities as spatial units, as they define graspable spatial regions. In our experiments we follow the locality assumption of Tobler's first law of geography: different cities in the same country might have a closer understanding of parks and gardens than cities of different countries. Thus, it will be more likely to produce meaningful results if we apply a learned classifier from one city on the data of another city in the same country. Learning areal properties in Hong Kong and applying them on data of Perth/Australia might not be valid due to the size of the available territory. The same holds for the idea of learning *global* parameters for parks and gardens — spatial entities have a strong grounding in local culture and history of a particular country, applying global rules on local data will lead in many cases to wrong classifications due to different local concepts.

In the following we learned the classifiers of 10 cities per country, and applied them mutually to every other city. By assessing the classification accuracy, this method

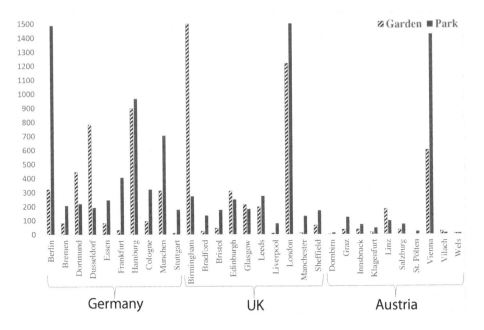

Fig. 4. Number of Parks and Gardens within the selected data set

allows identifying the most accurate classifiers for a city, and the identification of biased classifiers due to biased or ambiguous classification practices within specific cities.

In our study we applied a straightforward approach to distinguish between parks and gardens: we compared their size. Size is not probably enough to reliably distinguish between gardens and parks, especially if we consider other related classes such as meadows or grassland. When we have a closer look into how the classes are populated, we can see that the distribution can be rather clear, as it is, e.g., the case in Birmingham (see Figure 5(a)). There are also places with a less clear separation, e.g., the case of London (see Figure 5(b)), where parks and gardens seem to have a large conceptual overlap. However, our intention behind choosing the area is to detect incorrect classification at a very early point of contribution, when no other features are yet provided. Confronted with an "early-warning," users can reconsider the class they selected and modify it if required. However, especially a review of the existing data, as suggested in Section 3, can be fed by such a classifier. Figure 6 shows the mean areas of parks and gardens. It clearly shows that the areas per class are generally distinct and can be used to distinguish between entities of the two classes.

Feature Selection. The areas of each class have a specific distribution in each city. Figure 6 shows that parks are more likely to be large (i.e., tens of thousands to millions sqm), while gardens are more likely to cover rather smaller areas (i.e., a few sqm to a few thousands sqm). Although there are rare cases (i.e. Royal Botanic Gardens in the UK about one million sqm, however, they can be considered to be parks) corrupting

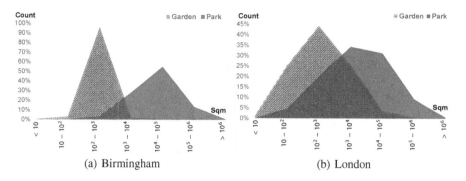

(a) Birmingham   (b) London

Fig. 5. Distribution of parks and gardens areas in London and Birmingham

the distribution; the majority of entities follow a common distribution. This distribution might also be similar in other cities, even if the data does not reflect it. By learning these distributions, we can distinguish between parks and gardens, and apply the learned classifiers to other cities and check the existing data or to guide contributors during the contribution process.

Classifier Training. Building a classifier basically can be done using *Eager Learning* (EL) or *Lazy Learning* (LL). In EL a training set is used to build a complete classifier before receiving any test entities. Bayesian classification, support vector machines (SVM), neural network (NN), and decision trees are examples for EL algorithms. In LL, generalization beyond the training data is delayed until a query is made to the system. K-nearest neighbors (KNN) and case based reasoning (CBR) are examples of lazy learning [23, 24]. In OSM a set of pre-classified entities is already stored, and the classification process is performed on new entities at contribution time. The new entity is classified based on similarity to existing entities. Hence, it is a good idea to follow the lazy learning paradigm to develop a classifier.

We decided to use KNN [25, 26] for building a classifier. KNN classifies entities based on closest training examples. It works as follows: the unclassified entity is classified by checking the K nearest classified neighbors. The similarity between the unclassified entity and the training set is calculated by a similarity measure, such as Euclidean distance.

Classifier Validation. During the validation process we use independent data sets for training and testing or we use the same data set for mutually applied classifiers (with this method, we evaluate if a classifier from a different city can be applied to another city). In the latter case, we use *K-fold cross validation* (CV) [27] to show the validity of our classification. In CV a training set is divided into K disjointed equal sets, where each set has roughly the same class distribution. Then the classifier is trained K times[3], and each time a different set is used as a test set. Afterwards the performance of the

---

[3] 5 and 10 are recommended values for K.

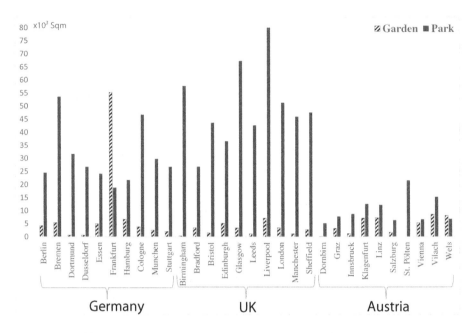

Fig. 6. Mean area size of parks and gardens for the selected data set

classifier is measured as the average of developed classifiers [27]. We build classifiers
for each city in a country. The results can be inspected in Tables 1, 2 and 3. The rows of
the tables represent the accuracies of different classifiers for the data of each city as a
test set. These classifiers were generated based on the data of other cities as training sets
and are represented in the columns. The last column "Class. Acc." shows the average
classification accuracy of parks and gardens within each city based on the top three
classifiers (italic red values).

Classifier Assessment. Depending on just one training and test set might result in biased classifiers. Furthermore, we aim to detect possible incorrect classifications based
on the similarity between cities within the same country. Thus, we build mutual classifiers between cities at the same country. One challenge is to assess the classifier performance. The accuracy of a classifier applied on a given test set is expressed by the
percentage of correctly classified entities (please see the next section for a deeper discussion on the measurability of the results). However, in some cases accuracies are
biased due to overfitting or underfitting [23, 24]. A reason can be unbalanced population of the training or the test set. This happens for instance when the classifiers created
from Liverpool or Manchester are applied on the Birmingham data (see Table 2). The
Receiver Operation Characteristics (ROC) curve is a useful measure to asses the performance of classifiers. The ROC curve represents the relative trade-off between benefits
and costs of the classifier. In particular the Area Under the ROC Curve (AUC) is a useful measure to asses a classifier. The closer the value of a AUC is to 1, the higher its

Table 1. Classification accuracy for parks and gardens of cities in Germany

Test Set	Berlin	Bremen	Dortmund	Dusseldorf	Essen	Frankfurt	Hamburg	Cologne	Munchen	Stuttgart	Class. Acc.
Berlin	80.43	76.78	76.23	72.25	74.07	82.03	56.44	79.38	78.94	82.2	75.23
Bremen	71.93	72.28	70.18	70.18	69.12	72.28	59.30	72.98	71.23	71.93	71.70
Dortmund	54.14	55.79	83.31	82.26	82.41	32.93	76.84	81.05	76.84	32.93	82.26
Dusseldorf	43.59	59.08	85.74	91.38	91.18	19.69	86.36	87.28	78.26	19.69	89.95
Essen	77.44	71.95	79.27	79.88	82.32	75.00	66.16	80.49	78.35	75.00	80.69
Frankfurt	89.68	79.13	75.00	62.39	65.37	92.66	47.94	78.67	78.21	92.89	88.07
Hamburg	54.15	55.87	59.03	61.27	61.76	51.69	61.06	58.97	57.90	51.79	61.36
Cologne	78.13	79.09	81.49	80.05	80.05	77.16	66.35	80.53	80.29	77.16	80.13
Munchen	72.50	71.02	79.37	77.90	79.17	69.16	62.48	78.49	78.88	69.25	78.65
Stuttgart	93.58	74.33	80.75	65.24	67.38	94.65	54.01	74.33	78.61	94.65	76.11

Table 2. Classification accuracy for parks and gardens of cities in the UK

Test Set	Birmingham	Bradford	Bristol	Edinburgh	Glasgow	Leeds	Liverpool	London	Manchester	Sheffield	Class. Acc.
Birmingham	99.73	0.99	70.03	92.65	90.79	92.67	0.94	69.27	1.29	94.73	92.73
Bradford	59.49	84.81	73.42	54.43	67.09	70.25	84.81	74.68	81.65	68.99	72.78
Bristol	72.73	79.55	78.64	67.27	75.91	79.09	79.55	76.82	79.55	81.82	78.03
Edinburgh	65.23	44.44	59.14	59.32	63.26	63.26	44.62	59.50	51.61	60.75	60.63
Glasgow	74.30	45.55	67.18	70.23	69.72	73.03	45.80	67.94	61.07	69.97	71.76
Leeds	75.96	57.87	72.34	70.43	77.45	75.96	58.09	73.40	58.94	77.66	77.02
Liverpool	86.05	89.53	88.37	80.23	87.21	89.53	89.53	87.21	89.53	90.70	87.60
London	68.26	64.88	72.51	66.77	72.02	72.22	65.05	73.03	68.12	72.83	72.63
Manchester	67.38	92.20	80.85	63.83	73.05	78.01	92.20	79.43	91.49	79.43	73.29
Sheffield	71.55	72.41	78.88	70.26	74.14	77.59	72.41	73.71	73.71	78.02	75.72

performance. Good classifiers should have AUC value between 0.5 and 1 [28]. Tables 1, 2, and 3 represent the accuracies of the generated classifiers, while AUC measures are dropped due to space restrictions. A combination of accuracy and AUC is used to determine the classification accuracy of parks and gardens for each city. We select the three top classifiers with the highest AUC measures (italic red values), and neglect biased classifiers with AUC less than or equal 0.5 (blue values). The classification accuracy is measured on the basis of the average accuracy.

Table 3. Classification accuracy for parks and gardens of cities in Austria

Test Set	Training Set										Class. Acc.
	Dornbirn	Graz	Innsbruck	Klagenfurt	Linz	Salzburg	St. Pölten	Vienna	Vilach	Wels	
Dornbirn	100	84.62	84.62	84.62	23.08	53.85	84.62	76.92	15.38	76.92	82.05
Graz	63.06	77.71	64.33	77.71	31.85	68.15	77.71	74.52	35.03	60.51	51.59
Innsbruck	80.19	66.04	83.02	66.04	52.83	50.94	66.04	66.98	47.17	47.17	67.30
Klagenfurt	72.13	73.77	70.49	70.49	31.15	62.30	73.77	75.41	47.54	49.18	65.57
Linz	41.52	34.66	43.32	34.66	62.09	37.91	34.66	38.63	61.01	40.07	48.01
Salzburg	56.60	67.92	59.43	67.92	39.62	70.75	67.92	64.15	42.45	58.49	60.38
St. Pölten	100	100	100	100	25.00	80.00	100	95.00	30.00	55.00	X
Vienna	59.39	70.36	58.45	70.36	38.93	62.10	70.36	68.28	37.50	61.86	65.69
Vilach	34.29	31.43	34.29	31.43	68.57	48.57	31.43	31.43	77.14	22.86	59.02
Wels	56.25	56.25	56.25	56.25	31.25	56.25	56.25	50.00	50.00	37.50	56.25

Results Discussion. Our results show that the cities in Germany and the UK have a classification accuracy from 70% to 90% for parks and gardens (see Tables 1 and 2). This means, according to our generated classifiers and their mutual application in other cities, about 10% to 30% of all analyzed entities within each city might be incorrectly classified. In Austria (see Table 3) we achieve poorer results. This might be due to the relative low number of entities in the available data set, or to already existing classification problems. In some of the cities, e.g., St. Pölten only one class of entities is available or predominant and causes the classifier to be highly biased and practically unusable (see Figure 4 and Table 3).

Of course, the classification results have to be interpreted with care. In none of the selected data sets, we had a qualified reference data set of known good quality. We selected the data sets as they were, and tried to identify two size classes within them: one for gardens and one for parks. In most cities we could identify good classifiers, however, their accuracies are not verifiable to full extend. As we have no clear ground truth, we cannot claim the correctness of the classifiers. With our approach we were able to identify a large set of entities worth looking at again. All samples we inspected showed clear evidence for entities that have been classified in an inappropriate way: "parks" around residential buildings in residential areas, as well as "gardens" with typical park facilities such as ways, playgrounds, or larger water bodies.

Although these samples were randomly chosen, they showed indicators for the validity of our approach. There are other evidences about that our results point in the right direction. In April 2014 we reviewed all entities that were detected as outliers in this paper. Of the originally 24,410 detected conflicts of the hierarchy consistency analysis (see Section 5) 10,635 entities had been already corrected or removed by the OSM community. Thus, in about 40% our approach pointed to entities identified as incorrect by crowdsourcing reviewers. The classification plausibility analysis resulted in 2,023 problematic entities in Germany, 2,516 in the UK, and 1,062 in Austria. About 8% of the German entities, 8% of the UK entities, and 11% of the Austrian entities have been revised since then. It is necessary to state that they have been revised without explicitly pointing to them. An appropriate infrastructure, e.g., a website or a gamified entity checker, can help to point users to the detected entities and revise them if necessary.

Also, the developed a very simple classifiers. If we want to successfully distinguish more than two classes, we need to consider more features than just size, thus we have to learn, e.g., typically contained or surrounding features of entities. By applying the approach as discussed in Section 3, we can select the detected entities and present them in a crowdsourcing manner to volunteers for inspection. The potentially re-classified entities could be used for rebuilding the classifier with clearer evidence.

7   Conclusion and Future Work

In this work we propose a new approach to manage the quality of VGI data during contribution, and on the existing data set manually or automatically. We presented two approaches to tackle VGI quality. We mainly focused on the problem of potentially wrong classifications that might lead to heterogeneous data quality. We developed two methods to tackle hierarchical consistency and classification issues based on ambiguity of potential entity classes.

With our first method, constraint based checking of hierarchical elements, we are able to detect all inconsistencies in the existing OpenStreetMap data set. With our second method, we can identify potentially wrong areal classifications in the OpenStreetMap data set by learning classifiers of different entity classes. The results show that we can identify a large number of existing problems in OSM data with both approaches. These detected conflicts could be presented to voluntary users to validate the entities' class, potentially based on suggestions generated along with it. For more complex classifiers being able to detect multiple possible classes, like, e.g., the "green areas" on a map (parks, gardens, meadow, grassland, scrub, etc.) we need to develop meaningful classifiers considering sets of features to be learned. We also need to think about appropriate ways to implement the proposed quality assurance methods, e.g., by means of gamification of user-based validation of the detect problematic data.

Acknowledgments. We gratefully acknowledge support provided by the German Academic Exchange Service (DAAD), as well as the German Research Foundation (DFG) via the Transregional Collaborative Research Center on Spatial Cognition SFB/TR8. Furthermore, we would like to thank the anonymous reviewers for their valuable comments.

References

1. Goodchild, M.F.: Citizens as sensors: the world of volunteered geography. GeoJournal 69(4), 211–221 (2007)
2. Coleman, D.J., Georgiadou, Y., Labonte, J., et al.: Volunteered geographic information: the nature and motivation of produsers. International Journal of Spatial Data Infrastructures Research 4(1), 332–358 (2009)
3. Feick, R., Roche, S.: Valuing volunteered geographic information (VGI): Opportunities and challenges arising from a new mode of GI use and production. In: Proceedings of the 2nd GEOValue Workshop, HafenCity University Hamburg, Germany, pp. 75–79 (2010)
4. Zook, M., Graham, M., Shelton, T., Gorman, S.: Volunteered geographic information and crowdsourcing disaster relief: a case study of the Haitian earthquake. World Medical & Health Policy 2(2), 7–33 (2010)
5. Girres, J.F., Touya, G.: Quality assessment of the french OpenStreetMap dataset. Transactions in GIS 14(4), 435–459 (2010)
6. Ludwig, I., Voss, A., Krause-Traudes, M.: A comparison of the street networks of Navteq and OSM in Germany. In: Advancing Geoinformation Science for a Changing World, pp. 65–84. Springer (2011)
7. Neis, P., Zielstra, D., Zipf, A.: The street network evolution of crowdsourced maps: OpenStreetMap in Germany 2007–2011. Future Internet 4(1), 1–21 (2011)
8. Goodchild, M.F., Li, L.: Assuring the quality of volunteered geographic information. Spatial Statistics 1, 110–120 (2012)
9. Haklay, M.: How good is volunteered geographical information? a comparative study of OpenStreetMap and Ordnance Survey datasets. Environment and Planning. B, Planning & Design 37(4), 682 (2010)
10. Neis, P., Zielstra, D., Zipf, A.: Comparison of volunteered geographic information data contributions and community development for selected world regions. Future Internet 5(2), 282–300 (2013)
11. Mooney, P., Corcoran, P.: The annotation process in OpenStreetMap. Transactions in GIS 16(4), 561–579 (2012)
12. Elwood, S., Goodchild, M.F., Sui, D.Z.: Researching volunteered geographic information: Spatial data, geographic research, and new social practice. Annals of the Association of American Geographers 102(3), 571–590 (2012)
13. Mülligann, C., Janowicz, K., Ye, M., Lee, W.C.: Analyzing the spatial-semantic interaction of points of interest in volunteered geographic information. In: Egenhofer, M., Giudice, N., Moratz, R., Worboys, M. (eds.) COSIT 2011. LNCS, vol. 6899, pp. 350–370. Springer, Heidelberg (2011)
14. Vandecasteele, A., Devillers, R.: Improving volunteered geographic data quality using semantic similarity measurements. ISPRS-International Archives of the Photogrammetry, Remote Sensing and Spatial Information Sciences 1(1), 143–148 (2013)
15. Schmid, F., Kutz, O., Frommberger, L., Kauppinen, T., Cai, C.: Intuitive and natural interfaces for geospatial data classification. In: Workshop on Place-Related Knowledge Acquisition Research (P-KAR), Kloster Seeon, Germany (2012)
16. Schmid, F., Frommberger, L., Cai, C., Dylla, F.: Lowering the barrier: How the what-you-see-is-what-you-map paradigm enables people to contribute volunteered geographic information. In: Proceedings of the 4th Annual Symposium on Computing for Development, p. 8. ACM (2013)
17. Schmid, F., Frommberger, L., Cai, C., Freksa, C.: What you see is what you map: Geometry-preserving micro-mapping for smaller geographic objects with mapit. In: Geographic Information Science at the Heart of Europe, pp. 3–19. Springer International Publishing (2013)

18. Arteaga, M.G.: Historical map polygon and feature extractor. In: Schmid, F., Kray, C. (eds.) Proceedings of ACM MapInteract, 1st International Workshop on Map Interaction. ACM (2013)
19. Neis, P., Zipf, A.: Analyzing the contributor activity of a volunteered geographic information project: The case of OpenStreetMap. ISPRS International Journal of Geo-Information 1(2), 146–165 (2012)
20. Barron, C., Neis, P., Zipf, A.: A comprehensive framework for intrinsic OpenStreetMap quality analysis. Transactions in GIS 18 (2014)
21. Tobler, W.R.: A computer movie simulating urban growth in the detroit region. Economic Geography 46, 234–240 (1970)
22. Devogele, T., Parent, C., Spaccapietra, S.: On spatial database integration. International Journal of Geographical Information Science 12(4), 335–352 (1998)
23. Bishop, C.M.: Pattern Recognition and Machine Learning (Information Science and Statistics). Springer-Verlag New York, Inc., Secaucus (2006)
24. Han, J., Kamber, M., Pei, J.: Data Mining: Concepts and Techniques, 3rd edn. Morgan Kaufmann Publishers Inc., San Francisco (2011)
25. Cover, T., Hart, P.: Nearest Neighbor pattern classification. IEEE Transactions on Information Theory 13(1), 21–27 (1967)
26. Witten, I.H., Frank, E.: Data Mining: Practical Machine Learning Tools and Techniques, 2nd edn. Morgan Kaufmann, San Francisco (2005)
27. Kohavi, R., et al.: A study of Cross-Validation and Bootstrap for accuracy estimation and model selection. In: Proc. International Joint Conference on Artificial Intelligence (IJCAI), pp. 1137–1145 (1995)
28. Fawcett, T.: An introduction to ROC analysis. Pattern Recognition Letters 27(8), 861–874 (2006)

Re-Envisioning Data Description
Using Peirce's Pragmatics

Mark Gahegan and Benjamin Adams

Centre for eResearch, The University of Auckland, New Zealand

Abstract. Given the growth in geographical data production, and the various mandates to make sharing of data a priority, there is a pressing need to facilitate the appropriate uptake and reuse of geographical data. However, describing the meaning and quality of data and thus finding data to fit a specific need remain as open problems, despite much research on these themes over many years. We have strong metadata standards for describing facts about data, and ontologies to describe semantic relationships among data, but these do not yet provide a viable basis on which to describe and share data reliably. We contend that one reason for this is the highly contextual and situated nature of geographic data, something that current models do not capture well — and yet they could. We show in this paper that a reconceptualization of geographical information in terms of Peirce's Pragmatics (specifically firstness, secondness and thirdness) can provide the necessary modeling power for representing situations of data use and data production, and for recognizing that we do not all see and understand in the same way. This in turn provides additional dimensions by which intentions and purpose can be brought into the representation of geographical data. Doing so does not solve all problems related to sharing meaning, but it gives us more to work with. Practically speaking, enlarging the focus from data model descriptions to descriptions of the pragmatics of the data — community, task, and domain semantics — allows us to describe the **how**, **who**, and **why** of data. These pragmatics offer a mechanism to differentiate between the perceived meanings of data as seen by different users, specifically in our examples herein between producers and consumers. Formally, we propose a generative graphical model for geographic data production through pragmatic description spaces and a pragmatic data description relation. As a simple demonstration of viability, we also show how this model can be used to **learn** knowledge about the community, the tasks undertaken, and even domain categories, from text descriptions of data and use-cases that are currently available. We show that the knowledge we gain can be used to improve our ability to find fit-for-purpose data.

1 Rethinking the Way we Describe Geographic Data

Our efforts to create better geographic data models and communicate richer data descriptions have led to very fruitful avenues of research, such as the representation of semantics, the visualization of uncertainty, the propagation of error, and others [43,18,41,26,21,28]. The era of volunteered geographic information (VGI) further complicates the picture with new challenges for understanding spatial data meaning, accuracy, and quality [19,1]. Research to date may allow us to describe the quality

M. Duckham et al. (Eds.): GIScience 2014, LNCS 8728, pp. 142–158, 2014.

(or perhaps even the semantics) of a single dataset, with effort, but we cannot propagate — with suitable modification — this information into derived products. Thus the onus remains firmly on the data producer to document quality and meaning of every new dataset. This has never been sustainable; most datasets do not have comprehensive quality information at the level of sophistication that consumers need. It is even less sustainable in the era of VGI and mash-ups, where more data is combined in hitherto unanticipated ways than ever before.

Furthermore, there is a real danger that all these different research strands have moved us further away from the actual problem, of describing these important aspects of data in an integrated and combinable manner, for example so that they can be used together in a query to find useful data. Without a way to bring these threads back together, our fruitful research avenues are in danger of becoming *cul-de-sacs*. Our approach to modeling geographic data is drastically in need of an overhaul.

Finally, as a community, we have been guilty of concentrating too heavily on the perspective of the data producer: describing "facts" about data, but not acknowledging the tacit world-view that can render these "facts" true and useful (or not) within a given context. Knowing which "facts" remain true when the context is changed, and also which facts remain relevant are both key to describing geographical data better. We term this idea the *pragmatics* of data, after Peirce [32].

1.1   An Alternative Approach

We suggest the following five propositions offer an alternative way forward:

1. We do not know what the eventual user will need to know about the data they wish to use, and we cannot know, in advance, the likely utility of any of the descriptions we may strive to add as data producers (such as ontologies, workflows, and accuracy assessments). And despite the huge volume of work published on conceptual geographic data models[1], we are no closer to knowing which ones have lasting value. We need empirical evidence, not more rhetoric, to produce a better model.

2. Consequently, we deliberately move away from the search for a single, definitive conceptual model of geographical data, and propose instead a meta-model where we can evaluate the actual utility of various forms of descriptions, from the perspective of specific tasks and research needs, using evidence gathered from actual use-cases.

3. We propose this simple meta-model as a set of description spaces, each comprised as "facets," that represent themes that we believe may have utility — but we do not claim that these are either necessary or sufficient — they are rather a place to begin. Within these facets we measure compliance to some kind of desired "optimal" state — as simply as we can (see section 2). Again, we make no claim that these facets are right, rather that they may prove to be useful under evaluation and (hopefully) that they are simple enough to be assigned and read with ease.

4. We broaden the scope of data description to consider the perspective of the data consumer. So we begin by asking: "What kinds of things might a consumer of the data want to know?" Rather than: "What kinds of things might a producer of the data be

---

[1] Including work published by the authors of this paper!

persuaded to say?" Furthermore, current approaches emphasize the **where, when,** and **what** aspects of data, with various degrees of success and completeness, but often leave aside the deeper questions of **who, how,** and **why.** These questions carry much meaning for a potential consumer of the information (they speak to reputation, quality, and motivation). We believe there are aspects of these deeper questions that can be captured that allow us to start framing more practical (and answerable) questions that often substitute for deeper ontological and epistemological questions: e.g., "for what task did you make this data?" can act as a surrogate for: "what does this data mean to you?" or "which organization produced the data?' may in some circumstances substitute for "what is the likely quality of the data?" These substitutions are certainly not perfect, but in a Bayesian sense they are better than nothing; and what's more, we can readily compute the degree to which they help elucidate the pragmatics we seek, as we show in Section 4.

5. The benefits of such an approach are many: (i) descriptive facets can be added or retracted according to need; (ii) the system could learn over time which kinds of data descriptions are most useful, so that data producers can focus their efforts when creating time-consuming data descriptions; (iii) multiple perspectives onto the meaning and use of data can be supported concurrently — allowing for the natural fact that we do not all see the world in the same way; (iv) shifting the emphasis from producing more metadata to learning from use-cases lifts an unmanageable burden from the data producers; (v) the conceptual model is not now a fixed thing, but can grow or change as new needs arise, as we learn more about which facets offer the most useful descriptions of data, or as new computational technologies provide us with additional descriptive facets.

The following are some of the many important facets to describe, though of course not an exhaustive list:

- Data Model: **What/when/where** is it?
  - Spatio-temporal Frameworks (spatio-temporal schema & semantics)
  - Attribute Schema & Semantics
- Process: **How** was it made and thus how confident are we in it?
  - Quality (Accuracy & Uncertainty)
  - Provenance (lineage)
- Community: **Who** can/should use it? **Why** was it made?
  - Motivation
  - Access and licensing
  - Authority (Governance & Trustworthiness)

1.2 Background

The description of geographic data into distinct spatial, temporal, and thematic components (**where, when,** and **what**) pre-dates modern geographic information systems and goes back at least to Berry's geographic matrix [3]. This matrix has formed the basis for much of the conceptual modeling surrounding geographic data into logical systems for representing geographic units [17]. Conceptual modeling in GIScience has looked at many dimensions of geographic and spatial information, including the object/field distinction, spatial relations, temporal relations [27,33,45]. Representation of the semantics of attributes using object-oriented databases and formal semantics continues to be an active area of GIScience research [11,10,24]. However, by ignoring **how**, **who**, and **why** these models (explicitly or implicitly) take either an exclusive producer's view on what the data means, or attempt to describe a universal view; in either case without situatedness, or context. When context has been studied it has been operationalized in terms of weights on attributes for semantic similarity measurement — not in terms of process and community [36,39,23]. But we need this situatedness to allow us to differentiate between the perspectives that naturally arise with a community, for example between of the producer of the data and the eventual consumers, particularly the unexpected consumers [14].

Philosophical Foundations: Peirce's Firstness, Secondness, and Thirdness. The representation of the situatedness of information is a natural consequence of acknowledging that we do not all see things the same, or that meaning and utility can depend on the situation at hand. C.S. Peirce [32] first proposed such a model, broadly based on semiotics, to demonstrate how signs are created and interpreted in communication.

> Peirce's notion of sign was broad enough to include situations, contexts, propositions ... and their expression in any language, including English and logic. His notion of ground is crucial: it acknowledges that some agent's purpose, intention, or "conception" is essential for determining the scope of a situation or context.[44]

In Peirce's pragmatics, **firstness** refers to a concept that remains constant when viewed from different points of view; it simply "is," and requires no qualification. An example might be the fact that a city's population is 1.5 million people. Firstness could also include the thematic aspects of the data as articulated in the attribute fields of the data. Similarity of features based on geometry and much of the semantic similarity measurement work done in geosemantics falls under this category [38,39,23]. Databases and GISystems are well equipped for representing this kind of information.

Secondness. refers to concepts that require further description or explanation via first-order relations to other concepts, but without the need for further interpretation or qualification. For geographic data this means the relationships to scientific conceptual knowledge that informs the data, such as: the tasks and scientific processes that consume or produce the data or the semantic commitments of the domain knowledge. For example, a city is a kind of settlement, or a city is bigger than a town. Similarity based on secondness is the similarity of tasks and domain knowledge during acts of production and use of the data. Ontologies and workflows represent this kind of information well.

Thirdness. adds a qualifier: two things are brought into relation only within the context of a third (i.e., relations of relations). In the case of data, thirdness can, for example, represent the community of people who accept as true a certain set of attribute values and semantic commitments (statements of firstness and secondness, respectively). For example, a concept that only has relevance or acceptance among a specific group, such as *provisioning services*, which is a notion accepted by scientists studying ecosystem services but not widely accepted by other ecologists. Thirdness forms the basis of pragmatic reasoning, that data and relationships may not be true in all circumstances or to all participants, but may require interpretation in the light of experience or within a given situation. Thirdness measures of similarity are almost always overlooked but can prove valuable if one wants to find data that match community constraints. For example, data that fulfills a community's usage or personalizing data search based on matching user profiles [8,6].

Note that firstness, secondness, and thirdness are not necessarily fixed; at some time we may wish to assert a "fact," at other times we may challenge the same fact and wish to explore its foundation, or decide that it only applies within a specific context. Importantly, the nexus of interactions between community, scientific knowledge, and data can be examined from different perspectives [15,34]. Here we have focused on data as the immediate subject and looked at relations to that data. If we had taken the data producer as the subject, then firstness similarity would refer to qualities of the producer and the data could be modeled via thirdness relations that provide insight to the likely domain expertise of the producers.

2   Four Description Spaces: Data, Domain, Task, and Community

Here we propose a simple model for the pragmatic description of data that moves from *community* to *task* to *scientific domain knowledge* to *data description*, or visa versa. These four aspects of the pragmatics of geographic data provide a more complete context for understanding the meaning of data, or the fitness of data for various purposes, because they describe knowledge of community and knowledge of the underlying science along with the semantics and schemata of the data.

Using Peirce's categories as a guide we present a theory for comparing the similarity of geographic data based on firstness, secondness, and thirdness measures over these description spaces. We conceive of these spaces as having similarity metrics because that will allow us to define aspects of community, domain knowledge, task, and data as compact regions in the spaces. The similarity metric for the space can be defined in terms of categorical or set-theoretic similarity over a knowledge graph, such as a description logic ontology, or any of many other similarity strategies [13].

The important point is that each of these spaces has a number of facets that allow us to reason about the similarity of the instances in those spaces. The facets provide constraints by which we can match queries for data from a data consumer to the data objects that have been created by a producer in a potentially very different context. The facets can be as simple or as complex as needed, experience suggests that simpler is better because some descriptive information is better than none (because it is too demanding to supply). In our example below, simple ordinal statements implying a greater level of compliance to some agreed set of information goals might be a practical and useful approach for many tasks, although other approaches may be equally valid [7]. Table 1 lists two sample dimensions or facets for each of the four spaces we have described, which can be used to make compliance judgments for data to determine for example if it is fit-for-purpose. Each statement represents a progressively deeper commitment towards some ideal, and subsumes the previous commitments.

Figure 1 illustrates how two datasets can be represented across these spaces. The first dataset, represented by **o**, is historical monitoring data about water wells and aquifers

Table 1. This table shows eight ordinal facets that can be used to reason about compliance of data based on four pragmatic spaces

Community
Data Standing
0. No information
1. Intent behind the data is known (implies an understanding of the purpose beyond some threshold)
2. Data originates from a reputable source (implies community aspects are known beyond some threshold)
3. Peer review and repeated use has verified utility and quality of the data
4. Authoritative data source endorsed by community
Data Licensing and Openness
0. No information
1. Author publishes a link to the data
2. Data license and reuse terms are known and published with the data
3. Data is available via persistent URI
4. Data is registered with an open SDI or similar cataloging service

Task
Process / Workflow
0. No information
1. Some aspects of the task can be inferred from knowledge of the community (and/or the data)
2. A clear description of the task is provided as text
3. A formal description of the task is provided (such as via a task or application ontology)
4. A full, repeatable workflow and associated data are provided, that allow the task outcomes to be repeated
Intention
0. No information
1. Some aspects of the intent can be inferred from knowledge of the community (and/or the data)
2. Clear text statement of intent or scientific goals behind the task
3. Description of intention using a controlled vocabulary
4. Detailed description of meta-level science model

Domain Semantics
Formality of domain semantics
0. No information
1. Informal concept maps of domain are provided
2. Controlled vocabularies used to describe data
3. Lightweight (Web) Semantic schema and SPARQL end points provided
4. Uses appropriate domain ontologies to describe semantics
Completeness of domain semantics
0. No information
1. Upper-level domain ontology for broad concepts (such as SWEET) [35]
2. Anchored into top-level ontology (such as Dolce) [16]
3. Detailed domain ontology (such as GeoSciML) [40]

Data Syntax and Attributional Semantics
Data Schema
0. No information
1. Spatial data correctly geo-registers (we know the projection, coordinate system, etc.)
2. Attribute schema is published and correct (we can actually parse the data content!)
3. Data is published using relevant (open) standards
Metadata (beyond data schema)
0. No information
1. A minimal metadata standard is met
2. Full metadata is provided using relevant open standards.
3. Validated account of data collection and interpretation process is available (such as a geological field manual for a mapsheet)

made available as part of the National Groundwater database (NGWD) by Natural Resources Canada's Earth Sciences Sector Groundwater program.[2] In the community space this dataset is at the higher end of both dimensions as it is an authoritative source and it is registered and made available on an official website. In the task space it scores a 1 on the process/workflow dimension as some aspects of the data collection process can be inferred from the data. It scores a 2 on the intention dimension because there are clear descriptions of important uses of the data on the Environment Canada website. Along the domain semantics dimensions it scores highly, because the concepts are de-

---

[2] http://ngwd-bdnes.cits.nrcan.gc.ca/service/api_ngwds:def/en/
presentation.html

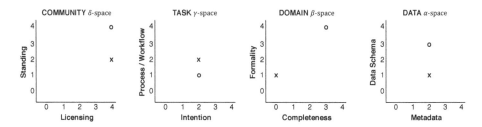

Fig. 1. Four domains of pragmatic and semantic description with example ordinal dimensions for reasoning about compliance. The **o** represents a dataset from NGWD and the **x** represents a dataset collected by the LTER network.

scribed using the GroundWater Markup Language (GWML) specification, an extension to GeoSciML [5]. Finally, it scores a 3 in the data schema dimension, because the data is easily parsed and linked to OGC open data standards, and it scores a 2 on the metadata dimension because the data attributes are fully described in the metadata.

The second example, represented by **x**, is a sample dataset of temperature and snow density data collected by a member of the Long Term Ecological Research (LTER) network [22]. The data originates from a reputable source (LTER), so scores 2 on data standing and is made freely available on the DataONE data network with DOI (`knb-lter-nwt.34.8`), so scores a 4 on the data licensing and openness [30]. It scores a 2 on both process/workflow and intention dimensions because the task and scientific goals are both clearly presented in the abstract associated with the dataset. It scores a 1 on the formality of domain semantics dimension because it is aligned with the LTER controlled vocabulary, but scores a 0 on completeness of domain semantics as that controlled vocabulary is described in SKOS, not a formal ontology. The geographic data schema correctly registers to WGS 84 coordinate system, so scores 1 on the data schema dimension. The attribute metadata dimension scores 2, because the metadata is described using the Ecological Metadata Language (EML) [12].

3   Generative Model for Geographic Data Creation

One goal of describing a model of geographic data semantics and pragmatics is to provide a mechanism to find data that are *fit-for-purpose*. The examples that follow assume this goal. From our point of view, we approach this goal by creating a representation of the communication act (both intentional and by implication) that is occurring between the producers and consumers of the data. This can be modeled using a graphical representation, as sets of relations. The generative process is illustrated in Figure 2 and demonstrates how a consumer and producer are indirectly linked to a data object via the description spaces of task and domain. This generative model is a sub-graph of the broader nexus of interactions that contribute to geographic (and other types of) understanding (see [14] for a more detailed discussion of this). Other models derived from this nexus are certainly possible, e.g., one might specify a direct edge between the community and domain space.

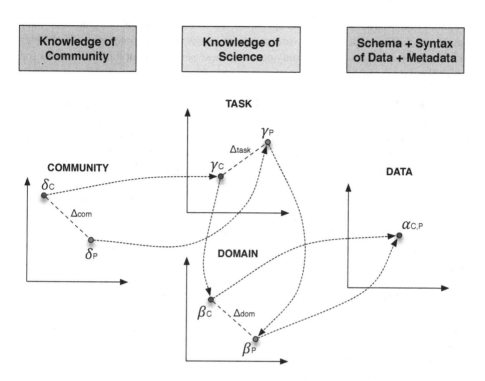

Fig. 2. Graphical model of data production and use via tasks and domain-specific semantic commitments. Consumer δ_C uses data object $\alpha_{C,P}$ with semantics β_C for task γ_C, although it was made by producer δ_P with semantic commitments γ_P for task β_P.

Data Model Space. Data in geographic information systems are often described in terms of geometry, other attributional characteristics, and occasionally temporal aspects. The data description space consists of dimensions that differentiate data along these respects. The GIS operations that transform data, e.g., projection and cartographic generalization, have the effect of moving a data instance from one point in this space to another. We can compare the similarity of two data objects based on their data description and this subsumes both traditional geometric matching, such as used in conflation algorithms, as well as similarity based on attribute value statistics. In our model all of these measures of similarity constitute *firstness* measures of similarity. They are based on characteristics of the geographic data artifacts themselves, divorced from interpretive modifiers. Within GIScience firstness measures of similarity have dominated the literature.

Domain Space. The domain-specific semantic commitments describe the semantics of the data in terms of a scientific domain. The interpretation of domain semantics can be restricted through the use of formal ontology, although the facets of this space do not necessarily need to be defined in this manner [20]. The work in geosemantics that looks at comparing the similarity of geographic concepts falls in this space and is a kind of *secondness* similarity. It remains an open question whether practical merging of domain ontologies and concept similarity measurement across multiple ontologies is solvable, thus we deem it important to not only consider these semantics but look at the tasks for which the data is intended [42].

Task Space. It may be well that the tasks that one wants to perform with the data is a better indicator of fitness-for-purpose than similarity measurement based on the data description. For example, if a user wishes to model wildfire, and knows that a specific vegetation coverage was created for exactly this purpose, it may well be useful to explore it further and, if necessary, adapt their own methodology or conceptual understanding to use it. It is also perhaps more likely that such a coverage will use data models and make ontological commitments that will be in keeping with those of the user: a vegetation coverage created to explore species diversity may not be so suitable. Note that this claim is not necessarily true for any specific example datasets. There will undoubtedly be counter-examples, but the principle applies in the sense of increased likelihood.

Community Space. The dimensions of the community space provide a means to describe the properties of both consumers and producers of data. Within this space we might recognize key themes and specializations that occur in the work of individuals and groups, constraints on information licensing and sharing, and governance issues about the authoritativeness of data. Based on usage, we may also be able to infer qualities such as trust and expertise [2].

3.1 What Variables do we Observe in the Graphical Model?

What we know about the pragmatics of geographic data will vary greatly from one data object to another. For example, it is possible that we might know nothing of the provenance of the data; we might only know the schema and attribute semantics of the data themselves. In other cases we might have information about the community, based on keeping track of use-cases, but no semantics or schema published. The model asserts that even when we do not have a full set of information available in relation to the data, the four description spaces can act as latent variables. For example, Figure 3a shows the case where we only have information about the data. Figures 3b–3d show cases where we know progressively more about the pragmatics of the data until we have a full picture with description of the producer in the community space, a description of the task in the task space, a description of the semantic commitments in the domain space, and a rich description of the data in the data model space. We explore later the question: To what extent can knowledge from one space provide insights into another?

3.2 Formalization

Formally, we define a *pragmatic data description* as a 4-tuple $< \delta, \gamma, \beta, \alpha >$, where δ is the community descriptor, γ is the task descriptor, β is the domain semantics descriptor, and α is the descriptor of data schema, spatio-temporal properties, and attribute semantics. The *context relation* $\boxdot$ is a binary relation between a symbolic data

(a) Case where we have a description of the data but know nothing of its provenance.

(b) Case where we have a description of the data and have a model of the producer in the community space.

(c) Case where we have a description of the data, have a model of the producer, and know the task for which the data is created.

(d) Ideal case where we have descriptions of the data, producer, task, and the semantic commitments from the scientific domain.

Fig. 3. Different degrees of observed pragmatics

object d (i.e., the actual digital encoding of the data) and its pragmatic data description: $\boxdot(d, < \delta, \gamma, \beta, \alpha >)$.

For simplicity's sake if we consider these spaces to be independent, then we can define fitness-for-purpose (ffp) as a compound distance measure across the firstness, secondness, and thirdness similarity spaces (Equation 1). In Section 4 we will discuss a probabilistic approach to measure the relatedness *between* the elements of the δ, γ, β, and α spaces.

$$ffp(i,j) = \Delta(\alpha_i, \alpha_j) + \Delta(\beta_i, \beta_j) + \Delta(\gamma_i, \gamma_j) + \Delta(\delta_i, \delta_j) \tag{1}$$

3.3 Consumer and Producer

This generative model starts at the producer and ends with data. We can use this model for the consumer as well if we swap the consumer in for the producer. We consider the problem of finding data that is fit-for-purpose as one of finding the ideal data object d^* given that we know the consumer's description within the community space, we know what task they want to perform with the data, and we understand the semantic commitments they have made. Thus, we want to find a data object d from the set of all data objects D that minimizes $ffp(d^*, d)$. (Practically speaking we want to find several

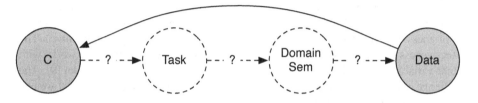

Fig. 4. The consumer wants fit-for-purpose data but the task and domain semantics are not-observable (latent variables in the generative model)

examples of data objects ordered by ffp, thus giving the user a few options to choose from.)

In many cases the task, the domain semantics, and even the data description will not be explicitly defined by the consumer and therefore these must be treated as latent variables in the model (see Figure 4). Realistically, we might be restricted to thirdness similarity measures in this case. For example, we might offer data from producers with a similar user profile.

4   Bayesian Interpretation for Learning and Prediction

The graphical model presented in the previous section points to mechanistic approaches to learn categories of geographic data by community, task, and domain semantics, in addition to traditional geosemantics. This is done by interpreting the graphical model described in the previous section as a Bayesian network, which provides significant statistical inferential power. A Bayesian network is a directed acyclic graph where nodes represent random variables and the edges represent their conditional dependencies. The directed edges between nodes are assigned probabilities and it satisfies the local Markov property that the variables are conditionally independent of other variables that are not parents in the graph [37]. The relationships between variables in a Bayesian network are often interpreted as causal relationships and can be used to model generative processes [31,4]. Thus, e.g., we can describe the probability that a producer δ_i will perform task γ_j. That task γ_j will entail domain semantics β_k and so on. Figure 5 shows an example of the Bayesian network that extends from a given producer and describes the probabilities of dependent tasks, domain semantics, and data descriptions.

Given a hypothesis space $\mathcal{H}$, we can use Bayes theorem to identify the most probable hypothesis, h, in that space to explain observed data, $\mathbf{d}$. $P(h)$ is the prior probability that a hypothesis is correct based on background knowledge. $P(\mathbf{d})$ is the prior probability that the data $\mathbf{d}$ is observed and $P(\mathbf{d}|h)$ is the probability that $\mathbf{d}$ is observed given that h is true. $P(h|\mathbf{d})$ is the posterior probability of h, i.e., what is the probability that the hypothesis holds given that $\mathbf{d}$ has been observed. We calculate this posterior probability by rewriting the denominator of Bayes' Rule (Equation 2).

$$P(h|\mathbf{d}) = \frac{P(\mathbf{d}|h)P(h)}{\sum_{h' \in \mathcal{H}} P(\mathbf{d}|h')P(h')} \qquad (2)$$

By parameterizing the dimensions in the community, scientific knowledge, and data description spaces and maximizing posterior probability given a set of training data, it opens the possibility of induction of new classification and prediction methods based on firstness, secondness, and thirdness categories (and compositions of all three). The effectiveness of Naïve Bayes classifiers and other hierarchical Bayesian networks with latent variables are well established and can be directly applied to this model [9]. The challenge moving ahead is articulating the dimensions of these spaces such that we can use these machine learning methods.

A probabilistic generative model gives us a way to describe the *potential* kinds of geographic data that a source can generate. Based on the probabilities in the graph we can

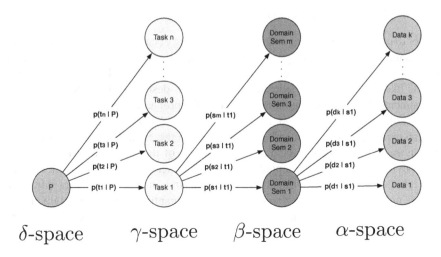

δ-space γ-space β-space α-space

Fig. 5. Data production graph as a Bayesian network

see the data that are likely to be generated by an individual source represented in community space. In addition to unsupervised learning of categories of tasks, communities, and semantic commitments, we can ask questions about latent variables in the model given some other knowledge that is available. For example, (a) probability of data given task, given domain semantic commitments and given producer; (b) probability of task given data; and (c) probability of community category given task.

Figure 6 provides a schematic of the kind of results we can anticipate given descriptions using facets such as those described in Table 1, which defines pragmatic description in terms of four two-dimensional spaces. This figure shows that a measure of 2 along both the Process/Workflow and Intention dimensions in task (β) space probabilistically implies certain values in other dimensions. Different descriptions of pragmatics will lead to different results. Importantly, this gives us the ability to experiment with different kinds of data description, changing facets and their dimensions and generative relationships to compute their utility via information gain measures [29].

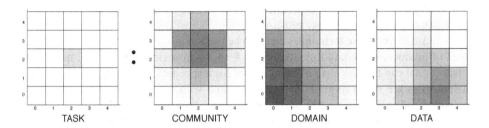

Fig. 6. Pixel-oriented visualization of probabilities of descriptions in the other three spaces, given a fixed point in Task space [25]. Darker colors represent higher probabilities.

4.1 Using Existing Descriptions

In order to perform this kind of unsupervised learning of categories of community, task, and domains; sufficient training data is required. But of course, much of the information we seek is not recorded directly in any current system. Large scale cyberinfrastructure projects like DataONE — a federated data network designed to enable discovery of environmental data — are beginning to address the problem of pragmatics [30]. Metadata describing purpose, method, authorship, rights holders, usage rights, and general abstracts written by the producers provide views to the pragmatics of the data, albeit often in unstructured natural language. A majority of the metadata that exist within DataONE have a geospatial component, but formal description of geosemantics and the geographic data model are virtually non-existent. In contrast, spatial data infrastructures are moving toward richer descriptions of geosemantics but broader pragmatics are largely lacking [24]. With some work, it is possible to assemble a rich enough description from which to begin.

As proof-of-concept of the network model, we downloaded 59,879 metadata descriptions from DataONE data objects that include geographic data. Although, the metadata do not describe pragmatics in the rich way we advocate earlier in this paper, we can demonstrate that by string matching terms that we associate with community members, methods and tasks, scientific domain knowledge and geographic representation we can find statistical pragmatic relationships. Figure 7 shows a small set of terms from this DataONE metadata mapped into a simple Bayesian network like the one shown in Figure 5. Since the DataONE metadata does not clearly differentiate between task and domain, we describe a simplified **science** description space, roughly covering both of the task and domain spaces as defined earlier in the paper. Once the network is built we use Markov chain Monte Carlo inference to find the likelihood of data given pragmatic evidence.

For example, we find that the probability of **fire**-related data given an **ecologist** producer is 8.0%, but when we add that the domain is **disturbance**, then the probability increases to 29.5%. Likewise, given a **climate scientist** producer there is only a 2.2% likelihood of precipitation data, but when the condition of **vegetation dynamics** is added then it rises to 58.5%. When scientific concepts are researched together, then it can imply high likelihood of data. For example, the probability of **tree** data given both **vegetation dynamics** and **disturbance** is quite high: 55.5%. The code and data for running these and similar experiments are available for download at https://wiki.auckland.ac.nz/x/mBKsAw.

To illustrate that these techniques can also be used to describe relationships between types of producers and data formats, Figure 8 shows how (in the DataONE network) a data format can be indicative of being useful for a specific community. For example, **hydrologists** are much more likely to do research with digital elevation model data (presumably due to their interest in catchment areas) than are **climatologists**. Whereas a NetCDF format strongly indicates relevance for a **climatologist**. Thus, a spatial data infrastructure that has user profiles of data consumers can provide a personalized data search service based on these results — e.g., suggesting DEMs if the system knows that the consumer is a hydrologist and so on.

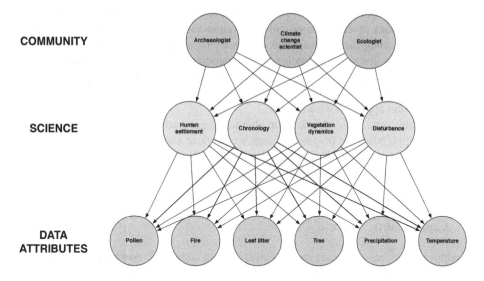

Fig. 7. Example of terms from DataONE metadata mapped to Bayesian network

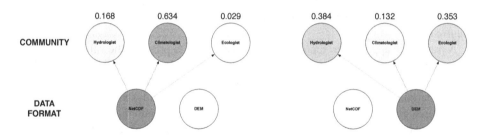

Fig. 8. Depending on the category of user different data formats will be more or less likely to be used in research

By finding the data with high relative likelihood, these probabilities can be used by data search applications to suggest potentially useful data for consumers who match community profiles, who are performing specific tasks, or working within specific scientific domains. Even with crude matching of terms to metadata text we begin to see value added in adopting this methodology. We anticipate being able to slowly build up richer descriptions of geospatial data, task and domain ontologies, and community space descriptions. Combined with Bayesian inference, we believe this holds great promise for new and better ways to find fit-for-purpose data.

4.2 Extending Toward Pragmatic Facets

Although the previous examples point to how we might use existing metadata to find potentially relevant and useful relationships between communities, tasks, scientific domains, and data schemata based on term matching; we contend that describing data using relatively simple descriptive dimensions, such as those listed in Table 1, that target the pragmatics of data will provide additional valuable information for data discovery. Values along these dimensions can easily be assigned by data providers, consumers, and also third-parties, such as data custodians of spatial data infrastructures. Four description spaces consisting of two dimensions each and five ordinal values per dimension (0..4) form a universe of 390,625 possible descriptions, a tractable number for the Bayesian approach we advocate.

Conceivably, one could also develop alternative generative models that combine the data description based on pragmatic facets we propose with other commonly used descriptions of schema and file format. By measuring the utility of these various models with data *in situ* we can begin to evaluate and refine our data description methods in a systematic way.

5   Conclusion

Modern approaches to science are providing us with additional, non-traditional ways to describe our data, such as the way they are used, and the community they originate from. Currently, we cannot use these descriptions because they don't fit in our conceptual data models. Yet for us, describing data well is still a very complex, perhaps untenable — and certainly impractical — proposition. To take full advantage of these new descriptions, we need to let go of the need to define data universally and objectively. This is not how we *use* data.

A pragmatic approach to representation can allow us to preserve the value of current facts and ontological commitments (Peirce's firstness and secondness), but add in the notion of context where it is needed to account for the fact that many things are true only in certain situations or to certain groups. This paper provides a workable and flexible pragmatic model to describe data, which can be reconfigured according to need. We have demonstrated how some of the (usually) opaque knowledge about community, task, and domain can be inferred from current meta-data text descriptions — thus bootstrapping the movement towards richer descriptions without placing additional burdens of description on the data producer. We have a pressing need to evaluate the utility and practicality of all such new descriptions, along with the old, so we can know with some confidence where to focus our efforts when it comes to providing data descriptions. Our next paper will provide a practical assessment of utility and practicality by measuring improvement in search results when pragmatic aspects are facilitated in the search process.

References

1. Adams, B., Gahegan, M.: Emerging data challenges for next-generation spatial data infrastructure. In: Winter, S., Rizos, C. (eds.) Research@Locate 2014, Canberra, Australia, April 7-9, pp. 118–129 (2014), http://ceur-ws.org
2. Artz, D., Gil, Y.: A survey of trust in computer science and the semantic web. Web Semantics: Science, Services and Agents on the World Wide Web 5(2), 58–71 (2007)
3. Berry, B.J.: Approaches to regional analysis: a synthesis. Annals of the Association of American Geographers 54(1), 2–11 (1964)

4. Bishop, C.M.: Pattern Recognition and Machine Learning. Springer, New York (2006)
5. Boisvert, E., Brodaric, B.: GroundWater Markup Language (GWML)-enabling groundwater data interoperability in spatial data infrastructures. Journal of Hydroinformatics 14(1), 93–107 (2012)
6. Carmel, D., Zwerdling, N., Guy, I., Ofek-Koifman, S., Har'el, N., Ronen, I., Uziel, E., Yogev, S., Chernov, S.: Personalized social search based on the user's social network. In: Proceedings of the 18th ACM Conference on Information and Knowledge Management, CIKM 2009, pp. 1227–1236. ACM, New York (2009)
7. Costello, M.J., Michener, W.K., Gahegan, M., Zhang, Z.Q., Bourne, P.E.: Biodiversity data should be published, cited, and peer reviewed. Trends in Ecology & Evolution 28(8), 454–461 (2013)
8. Crompvoets, J., Bregt, A., Rajabifard, A., Williamson, I.: Assessing the worldwide developments of national spatial data clearinghouses. International Journal of Geographical Information Science 18(7), 665–689 (2004)
9. Domingos, P., Pazzani, M.: On the optimality of the simple Bayesian classifier under zero-one loss. Machine Learning 29(2-3), 103–130 (1997)
10. Egenhofer, M.: Toward the semantic geospatial web. In: GIS 2002: Proceedings of the 10th ACM International Symposium on Advances in Geographic Information Systems, pp. 1–4. ACM, New York (2002)
11. Egenhofer, M.J., Frank, A.: Object-oriented modeling for GIS. Journal of the Urban and Regional Information Systems Association 4(2), 3–19 (1992)
12. Fegraus, E.H., Andelman, S., Jones, M.B., Schildhauer, M.: Maximizing the value of ecological data with structured metadata: an introduction to ecological metadata language (EML) and principles for metadata creation. Bulletin of the Ecological Society of America 86(3), 158–168 (2005)
13. Gahegan, M., Agrawal, R., Jaiswal, A., Luo, J., Soon, K.H.: A platform for visualizing and experimenting with measures of semantic similarity in ontologies and concept maps. Transactions in GIS 12(6), 713–732 (2008)
14. Gahegan, M., Luo, J., Weaver, S.D., Pike, W., Banchuen, T.: Connecting GEON: Making sense of the myriad resources, researchers and concepts that comprise a geoscience cyberinfrastructure. Computers & Geosciences 35(4), 836–854 (2009)
15. Gahegan, M., Pike, W.: A situated knowledge representation of geographical information. Transactions in GIS 10(5), 727–749 (2006)
16. Gangemi, A., Guarino, N., Masolo, C., Oltramari, A., Schneider, L.: Sweetening ontologies with DOLCE. In: Gómez-Pérez, A., Benjamins, V.R. (eds.) EKAW 2002. LNCS (LNAI), vol. 2473, pp. 166–181. Springer, Heidelberg (2002)
17. Goodchild, M.F.: Geographic data modeling. Computers and Geosciences 18(4), 401–408 (1992)
18. Goodchild, M.F.: Data models and data quality: problems and prospects. In: Environmental Modeling with GIS, pp. 94–104. Oxford University Press (1993)
19. Grira, J., Bédard, Y., Roche, S.: Spatial data uncertainty in the VGI world: Going from consumer to producer. Geomatica 64(1), 61–72 (2010)
20. Guarino, N.: Formal Ontology and Information Systems. In: Guarino, N. (ed.) International Conference on Formal Ontology in Information Systems (FOIS 1998), pp. 3–15. IOS Press, Trento (1998)
21. Heuvelink, G.B., Burrough, P.A., Stein, A.: Propagation of errors in spatial modelling with GIS. International Journal of Geographical Information System 3(4), 303–322 (1989)
22. Hobbie, J.E., Carpenter, S.R., Grimm, N.B., Gosz, J.R., Seastedt, T.R.: The US long term ecological research program. BioScience 53(1), 21–32 (2003)
23. Janowicz, K., Raubal, M., Kuhn, W.: The semantics of similarity in geographic information retrieval. Journal of Spatial Information Science (2), 29–57 (2011)

24. Janowicz, K., Schade, S., Bröring, A., Keßler, C., Maué, P., Stasch, C.: Semantic enablement for spatial data infrastructures. Transactions in GIS 14(2), 111–129 (2010)
25. Keim, D.A.: Designing pixel-oriented visualization techniques: Theory and applications. IEEE Transactions on Visualization and Computer Graphics 6(1), 59–78 (2000)
26. Kuhn, W.: Ontologies in support of activities in geographical space. International Journal of Geographical Information Science 15(7), 613–631 (2001)
27. Langran, G., Chrisman, N.R.: A framework for temporal geographic information. Cartographica: The International Journal for Geographic Information and Geovisualization 25(3), 1–14 (1988)
28. MacEachren, A.M., Robinson, A., Hopper, S., Gardner, S., Murray, R., Gahegan, M., Hetzler, E.: Visualizing geospatial information uncertainty: What we know and what we need to know. Cartography and Geographic Information Science 32(3), 139–160 (2005)
29. MacKay, D.J.C.: Information Theory, Inference, and Learning Algorithms, 7.2nd edn. Cambridge University Press, Cambridge (2003)
30. Michener, W., Vieglais, D., Vision, T.J., Kunze, J., Cruse, P., Janée, G.: Dataone: Data observation network for earth - preserving data and enabling innovation in the biological and environmental sciences. D-Lib Magazine 17(1/2) (2011)
31. Pearl, J.: Causality: Models, reasoning and inference. Cambridge University Press, Cambridge (2000)
32. Peirce, C.S.: The Collected Papers of Charles Sanders Peirce. Harvard University Press (1931)
33. Peuquet, D.J.: Representations of space and time. Guilford Press (2002)
34. Pike, W., Gahegan, M.: Beyond ontologies: Toward situated representations of scientific knowledge. International Journal of Human-Computer Studies 65(7), 674–688 (2007)
35. Raskin, R.G., Pan, M.J.: Knowledge representation in the semantic web for Earth and environmental terminology (SWEET). Computers & Geosciences 31(9), 1119–1125 (2005)
36. Raubal, M.: Formalizing conceptual spaces. In: Varzi, A.C., Vieu, L. (eds.) Formal Ontology in Information Systems, Proceedings of the Third International Conference (FOIS 2004), pp. 153–164. IOS Press (2004)
37. Russell, S., Norvig, P.: Artificial Intelligence: A Modern Approach, 3rd edn. Prentice Hall (2010)
38. Saalfeld, A.: Conflation automated map compilation. International Journal of Geographical Information System 2(3), 217–228 (1988)
39. Schwering, A.: Approaches to semantic similarity measurement for geo-spatial data: A survey. Transactions in GIS 12(1), 5–29 (2008)
40. Sen, M., Duffy, T.: GeoSciML: development of a generic geoscience markup language. Computers & Geosciences 31(9), 1095–1103 (2005)
41. Shi, W.: A generic statistical approach for modelling error of geometric features in GIS. International Journal of Geographical Information Science 12(2), 131–143 (1998)
42. Shvaiko, P., Euzenat, J.: Ontology matching: state of the art and future challenges. IEEE Transactions on Knowledge and Data Engineering 25(1), 158–176 (2013)
43. Sinton, D.: The inherent structure of information as a constraint to analysis: Mapped thematic data as a case study. Harvard Papers on Geographic Information Systems 7, 1–17 (1978)
44. Sowa, J.F.: Syntax, semantics, and pragmatics of contexts. In: Ellis, G., Rich, W., Levinson, R., Sowa, J.F. (eds.) ICCS 1995. LNCS, vol. 954, pp. 1–15. Springer, Heidelberg (1995)
45. Worboys, M.F., Duckham, M.: GIS: a computing perspective. CRC Press (2004)

Fields as a Generic Data Type
for Big Spatial Data

Gilberto Camara[1,2], Max J. Egenhofer[3], Karine Ferreira[1], Pedro Andrade[1],
Gilberto Queiroz[1], Alber Sanchez[2], Jim Jones[2], and Lubia Vinhas[1]

[1] Image Processing Division, National Institute for Space Research (INPE),
São José dos Campos, Brazil
[2] Institute for Geoinformatics (ifgi), University of Münster, Germany
[3] National Center for Geographic Information and Analysis and
School of Computing and Information Science, University of Maine, Orono, ME, USA

Abstract. This paper defines the Field data type for big spatial data. Most big spatial data sets provide information about properties of reality in continuous way, which leads to their representation as fields. We develop a generic data type for fields that can represent different types of spatiotemporal data, such as trajectories, time series, remote sensing and, climate data. To assess its power of generality, we show how to represent existing algebras for spatial data with the Fields data type. The paper also argues that array databases are the best support for processing big spatial data and shows how to use the Fields data type with array databases.

1 Introduction

One of the biggest changes in Geoinformatics in recent years arises from technologies that produce lots of data. Earth observation and navigation satellites, mobile devices, social networks, and smart sensors create large data sets with space and time references. Big spatial data enables real-time applications, such as tracking environmental changes, detecting health hazards, analyzing traffic, and managing emergencies. Big data sets allow researchers to ask new scientific questions, which is both an opportunity and a challenge [2]. However, there are currently no appropriate conceptual models for big spatial data. Lacking sound guidance, we risk building improvised and incompatible application, with much effort wasted.

A model for big spatial data should consider the nature of the data, which are records of measurements and events in space-time. Sensors measure properties of nature, such as temperature, soil moisture, and land surface reflectance, and human events, such as locations of people and cars. Since these sensors observe the world in real-time, we take big spatial data to be records of continuous phenomena.

The terms *fields* and *coverages* describe real-world phenomena that vary continuously in space and time [5,10,26]. Despite the abstract nature of the concept, most work on fields deals with concrete data structures (e.g., triangulations, cells, and contours). The OGC definition for coverages–"digital spatial information representing space-time varying phenomena" [25]–is similar to the definition of the Fields data type. Since OGC's coverages focus on describing operations on concrete spatial representations,

M. Duckham et al. (Eds.): GIScience 2014, LNCS 8728, pp. 159–172, 2014.

they add complexity and reduce generality [24,25]. Big spatial data, however, needs an inclusive model that starts with the measurements (i.e., the data collected) and builds on top of them a generic scheme for space-time analyses. The lack of such a high-level model is a serious impediment in analyses of large, complex, and diverse data sets. To avoid makeshift approaches, one needs a wide-ranging, yet simple model for fields at a higher abstraction level than implementation-specific solutions.

Early work on spatial data modeling viewed *fields* as four-dimensional functions $f(x, y, z, t)$ that describe positions in space-time [15,21]. This approach was later refined with *geo-atoms*, the minimal form common to all geographic information and a basis for modeling spatial phenomena [16]. A geo-atom combines a position in space-time and a property, expressed as a tuple $[\mathbf{x}, \mathbf{Z}, z(\mathbf{x})]$, where $\mathbf{x}$ is a position in spacetime, $\mathbf{Z}$ is a property, and $z(\mathbf{x})$ is the value of the property at that position. To represent fields, we take the idea of geo-atoms one step further and consider how one observes reality. Since one will never have complete information about external reality, one needs to make inferences about positions in space-time for which there are no observations [22]. Thus, field representations have to combine *observed* and *inferred* measures of a phenomenon. One needs to put together observations of properties of reality with a procedure that estimates values of these properties at non-observed positions [8].

This paper defines *fields* as *sets of geo-atoms* $\{[\mathbf{x}, \mathbf{Z}, z(\mathbf{x})]\}$ that are observations of a property Z in an space-time extent, and an *estimator function* that estimates values of this property in non-observed locations of this extent. A field has a *space-time* extent, a set of *positions* inside this extent, and a set of *values* observed or estimated for each position. We define a Field data type based on abstract specifications, following a line of research in Geoinformatics that considers formal definitions precede reliable system implementation [13,12,31].

Although the Field data type is not specific for dealing with big spatial data, it is particularly relevant for handling large data sets. Contemporary object-relational data models are built around layers, which slice the geographic reality in a particular area. The use of layers as a basis for spatial data organization comes from how data is organized in thematic and topographic maps. When applied to big spatial data, the organizing principle of geographic layers breaks down, however. Instead of a set of static spatial layers (each with its legend), big spatiotemporal data sets store information about changes in space and time. Conceiving such information as fields captures their inherent nature better than the traditional layer-oriented view.

After a brief discussion on generic programming and generic types in Section 2, we introduce the *Field data type* (Section 3). We show how to use the Field data type to represent time series, sensor networks, trajectories, collections of satellite images, and climate data, sharing common operations. Section 4 shows how to implement existing spatiotemporal algebras using the Field data type. Section 5 discusses the nature of big spatial data; we make a case for array databases as the best current support for handling these data sets. Section 6 shows how to use the Field data type in connection with array databases for processing large spatial data. The paper closes with a discussion of a road map for making the Field data type a tool for developing new types of GIS applications.

2 Generic Programming and Generic Types

The design of the Field data type is based on the ideas of generic programming. Generic programming uses *abstract data types*, which are formal tools that allow an objective evaluation of computer representations [3]. Abstract data type definitions have an externally viewable set of operations and a set of axioms applicable to them [17]. The operations are generic, so they work for different data structures and different implementations.

Generic programming is well-suited for building GIS [9]. Most spatial algorithms can be designed to be independent of spatial data structure, relying instead on basic properties that most of them provide. To find the mean value of an attribute in a spatial data set, it is irrelevant whether the data structure is a TIN, a grid, or a set of polygons. All one needs is to get from one data item to the next, and to compare two items. Even algorithms that depend on spatial properties can be expressed in an abstract form. One can define the local mean of a data set using an abstract definition of neighborhood, leaving the details to the implementation phase.

To define an abstract data type, we use the following notation. Type definitions and operations use a monospaced font. Type names are capitalized (e.g., Integer). Sets of instances of a type are included in curly braces, for instance, {Integer} is a set of variables of type Integer. We write an ordered pair of variables of types A and B as (TypeA, TypeB).

Generic types are indicated by T:GenericType where T is a placeholder for a concrete type. The notation I:Item defines a generic type of items, where the concrete type can be, for example, Integer or Real. Types that use other generic types are written as CompositeType [T:GenericType], so Stack[I:Item] defines a composite type Stack that handles instances of the generic type I:Item.

To associate concrete types to a generic type, we write T:GenericType ⊨ ConcreteTypeA, ConcreteTypeB. To point out that one can replace the generic type I:Item by concrete types Integer and Real, we write I:Item ⊨ Integer, Real.

Names of functions and operators begin with a lowercase letter. Examples are top, pop, and new. Function signatures point out their input types and the output type. The notation (TypeA x TypeB → TypeC) describes a function where TypeA and TypeB are the types of the input and TypeC is the type of the output. A factorial function has (Integer → Integer) as a signature. Functions can use generic types. A generic sum function has I:Item x I:Item → I:Item as a signature.

Consider a stack, a last-in, first-out data structure, whose specification is given in Fig. 1. It has three fundamental operations: push, pop, and top. The push operation adds an item to the top of the stack, pop removes the item from the top of the stack, while top returns the element at the top of the stack, without changing the stack. The Stack data type is defined independently of the data structures and algorithms that implement it. This specification provides support for implementing stacks of different concrete types (e.g., stacks of integers, stack of strings, or stacks of any other user-defined type including stacks of stacks).

```
Type Stack [I] uses I:Item
Functions
     new:        Stack
     push:       I x Stack → Stack[I]
     pop:        Stack[I] → Stack[I]
     isEmpty:    Stack[I] → Boolean
     top:        Stack[I] → I
Variables
     s: Stack
     i: Item
Axioms
     isEmpty (new ()) = true
     isEmpty (push(i, s)) = false
     top (new ()) = error
     top (push (i, s)) = i
     pop (push (i, s)) = s
```

Fig. 1. Abstract specification of the data type stack

3 Fields as Generic Types

What is in common between a time series of rainfall in Münster, the trajectory of a car in Highway 61, a satellite image of the Amazon, and a model of the Earth's climate? They share the same inherent structure. They all have a space-time extent, within which one measures values of a phenomenon, providing observations of reality. Within this extent, one can also compute the values of these phenomena at non-observed positions. We thus conceptualize these data sets as *fields*, made of *sets of geo-atoms* $\{[\mathbf{x}, \mathbf{Z}, z(\mathbf{x})]\}$ that are observations of a property Z in an space-time extent, and an *estimator function* that estimates values of this property in non-observed locations of this extent.

This definition of fields is a generalization of the traditional view of fields as functions that map elements of a bounded set of locations in space onto a value-set [14]. We extend this idea in two ways: (1) we consider different types of locations in space and time and (2) we consider that the elements of the value-set can also be positions in space-time. Thus, a field is a function whose domain is an extent of space-time, where one can measure the values of a property in all positions inside the extent.

The key step in this conceptualization is the generic definition of the concepts of *position* and *value*, shown in Fig. 2. In a time series of rainfall, positions are time instants, since space is fixed (the sensor's location), while values are the precipitation counts. In a remote sensing image, positions are samples in 2D space (the extent of the image), since time is fixed (the moment of image acquisition), while values are attributes, such as surface reflectance. Logistic and trajectory models record moving objects by taking positions as time instances, while their values are the objects' locations in space.

The generic type P:Position stands for positions in space-time. This type is mapped onto concrete types that express different time and space cases. Some non-exhaustive examples are Instant for time instants, 2DPoint and 3DPoint

```
P:Position ⊨  Instant, 2DPoint, 3DPoint,
               (2DPoint, Instant), (3DPoint, Instant)
V:Value    ⊨  Integer, Real, Boolean, String, P:Position
E:Extent   ⊨  [(3DCube, Interval)], [(3DPolygon, Interval)]
```

Fig. 2. Building blocks for the Fields data type

for purely spatial positions, and pairs (2DPoint, Instant) and (3DPoint, Instant) for space-time positions. The generic type V:Value stands for attribute values. Concrete types linked to V:Value include Integer, Real, String, Boolean and their combination. Values can also be associated to positions, as in the case of trajectories.

The formal description of a Field data type is shown in Fig. 3. Each field exists inside an extent of space-time, represented by the type E:Extent, whose instances are sets of 3D compact regions in space-time. Each field has an associated G:Estimator function that enables estimating values at positions inside its extent. This allows a field to infer measures at all positions inside the extent. The estimator function use the field's information and thus has a signature (F:Field x P:Position → V:Value).

The relationship between positions and extents is a key part of the model. All positions of a field are contained inside the extent. Thus, the possible concrete types for the generic type Position are those that can be topologically evaluated as being part of a space-time hypercube or a space-time polygon. The definition of an extent as a set of space-time hypercubes also avoids the problems with null values. Thus, there are no null values inside a field extent in this Field model.

The operations of the Field data type are:

New. Creates a new Field, given an extent and an estimator function.

Add. Adds one observation with a (position, value) pair to the Field.

Obs. Returns all observations associated to the Field.

Domain. Returns the full set of positions inside the Field's extent. The actual result of this operation depends on the Field's granularity, but the operation can be defined in a problem-independent way.

Extent. Returns the extent of the Field.

Value. Computes the value of a given position, using the estimator function. The estimator ensures that a field will represent a continuous property inside its extent.

Subfield. Returns a subset of the original Field according to an extent. This function is useful to retrieve part of a Field.

Filter. Returns a subset of the original Field that satisfies a restriction based on its values. Examples include functions such as "values greater than the average."

Map. Returns a new Field according to a function that maps values from the original Field to the field to be created. Examples of map include unary functions such as double and squareRoot. This function corresponds to a map in functional programming.

Combine. Creates a new Field combining two fields with the same extent, according to an operation to be applied for each element of the original Fields. Examples of combine include binary functions such as sum and difference.

```
Field [E, P, V, G] uses E:Extent, P:Position, V:Value, G:Estimator
Operations
new:          E x G → Field
add:          Field x (P, V) → Field
obs:          Field → {(P, V)}
domain:       Field → {P}
extent:       Field → E
value:        Field x P → V
subfield:     Field x E → Field
filter:       Field x (V → Bool) → Field
map:          Field x (V → V) → Field
combine:      Field x Field x (V x V → V) → Field
reduce:       Field x (V x V → V) → V
neigh:        Field x P x (P x P → Bool) → Field
Variables
f, f1, f2: Field
g: Estimator
p: Position
e: Extent
v: Value
Functions
uf: (V → V)          -- unary function on values
bf: (V x V → V)      -- binary function on values
ff: (V → Bool)       -- filter function on values
nf: (P x P → Bool) -- neighborhood function on positions
Axioms
-- basic fields axioms: a field is dense relative to its extent
∀ p ∈ extent(f)  ⟹  ∃ value(f, p) = g(f, p)
∀ p ∉ extent(f)  ⟹  value(f, p) = ∅
-- axioms on operation behavior
∀ f, domain(f) ⊆ extent (f)
subfield(f, e) ⊆ f  ⟺  e ⊆ extent (f)
filter(f, ff) ⊆ f
obs (new(e, g)) = ∅
obs (add (new(e, g)), (p, v))) =
   (p, v)  ⟺  p ⊂ e
subfield(f, extent(f)) = f
neigh (f, p, nf) ⊆ f, ∀ p ∈ extent (f)
value (map (f, uf), p) =
   uf (value (f, p)), ∀ p ∈ extent (f)
value (combine(f1, f2, bf), p) =
   bf (value (f1, p), value (f2, p))  ⟺
   p ∈ extent (f1) and p ∈ extent (f2)
reduce (f, bf) =
   bf (reduce (f1, bf), reduce(f2, bf))  ⟺
   f1 = subfield (f,e1) and f2 = subfield(f,e2) and
   e1 ∩ e2 = ∅ and e1 ∪ e2 = extent (f)
```

Fig. 3. Generic data type definition of Field

Reduce. Returns a value that is a combination of all the values of some positions the Field. Examples include statistical summary functions such as `maximum`, `minimum`, and `mean`.

Neigh. Returns the neighborhood of a position inside a Field. It uses a function that compares two positions and finds out whether they are neighbors. One example of the function is a proximity matrix where each position is associated to all its neighbors.

The Field definition is independent of granularity, which we take to be a problem-dependent issue. Each concrete field will have its spatial and temporal granularity that will determine how its operations are implemented. Temporal granularity will be represented by the concrete implementation of types `Interval` and `Instant`. The granularity of type Instant should be such that it is always possible to test whether an instant is inside an interval.

The Fields data type distinguishes between the extent and the domain of a field. The extent is the region of space-time where one is able to get a value for each position. The domain of a field is the set of positions it contains, whose granularity depends on how the field was constructed. For example, two fields may have the same extent and different domains. For the same extent, one field may have a set of scattered positions as its domain, while another may have its positions organized in a regular grid in space-time. One can perform operations between these fields without changing their granularities, since they adhere to the same operations.

4   Implementing Existing Algebras with the Fields Data Type

To show how to use the Fields data type, we consider how to express two existing algebras for spatial data using it: Tomlin's map algebra [30] and the STAlgebra [8]. Map Algebra is a set of procedures for handling continuous spatial distributions. It has been generalized to temporal and multidimensional settings [11,4,23]. Tomlin defines the following map algebra operations:

Local Functions: The value of a location in the output map is computed from the values of the same location in one or more input maps.

Focal Functions: The value of a location in the output map is computed from the values of the neighborhood of the same location in the input map.

Zonal Functions: The value of a location in the output map is computed from the values of a spatial neighborhood of the same location in an input map. This neighborhood is a restriction on a second input map.

Fig. 4 shows how to express Tomlin's map algebra functions with the Field data type.

To implement a generic map algebra, the *local unary* and *local binary* functions are mapped onto the `map` and `combine` operators, respectively. Local functions involving three or more maps can be broken down into unary and binary functions. A *focal function* uses the functions `neigh` and `reduce`. The `neigh` function returns a field with only those local values that are used by `reduce` to get a new value for the position in the output field. The same combination implements *zonal functions*. The difference is

```
Variables
f1, f2: Field                       -- input
f3: Field                           -- output
p, p1: Position
Functions
uf: (v:Value → v:Value)             -- unary function
bf: (v:Value x v:Value → v:Value)   -- binary function
nf: (p:Position x p:Position → Bool) -- neighborhood function
Operators
localUnary (f1, uf) = map (f1, uf)
localBinary (f1, f2, bf) = combine (f1, f2, bf)
focalFunction (f1, nf, bf) =
    ∀ p ∈ domain (f3)
      add(f3, (p, reduce (neigh (f1, p, nf), bf)))
zonalFunction (f1, f2, nf, bf) =
    ∀ p ∈ domain(f3)
      add(f3, (p, reduce (subfield (f1,
                          extent (neigh (f2, p, nf), bf)))))
```

Fig. 4. A generic map algebra

that the neighborhood function is defined on a second field. The extent of the neighborhood of the second field is used to extract a subfield of the first field. The function `reduce` then produces a unique value that is the new value of the position in the output field. The mapping is dimension-independent and can be used to implement not only Tomlin's 2D map algebra [30], but also a multidimensional map algebra [23] and a temporal map algebra [11].

A second example is STAlgebra [8], which takes observations as its basic building blocks. Based on Sinton's view of the inherent nature of geographical data [28], STAlgebra singles out different types for spatiotemporal data: `Coverage`, `CoverageSeries`, `TimeSeries`, and `Trajectories`. Operations on these types allow queries and inferences on space-time data. Instances of these types can be related to *events*. The mappings from the four spatiotemporal data types `TimeSeries`, `Trajectory`, `Coverage` and `CoverageSeries` onto the `Field` type are as follows:

Time Series. A *time series* represents the variation of a property over time in a fixed location. For example, a time series of rainfall has measured values of precipitation counts at some controlled times (e.g., hourly) at the sensors' locations. A `TimeSeries` type is mapped onto a `Field[E:Extent, Instant, V:Value, G:Estimator]` where positions are time instants.

Trajectory. A *trajectory* represents how locations or boundaries of an object evolve over time. For example, a trajectory of an animal, which has a fixed identification, is composed of measured spatial locations at controlled times (e.g., hourly). The `Trajectory` type of STAlgebra is mapped to a `Field[(3DPolygon, Interval), Instant, 2DPoint,`

G:Estimator] or to a Field[(3DPolygon,Interval), Instant, 3DPoint, G:Estimator], if the trajectory is taken in 2D or 3D space, respectively.

Coverage. A *coverage* represents the variation of a property within a spatial extent at a certain time. A remote sensing satellite image is an example of a coverage. It has a fixed time, the moment of the image acquisition, and measured values of surface reflectance at spatial locations. The Coverage type is mapped onto a Field[E:Extent, 2DPoint, V:Value, G:Estimator] whose positions are 2D spatial locations.

CoverageSeries. A *coverageseries* represents a time-ordered set of coverages that have the same boundary, as in the case of a sequence of remote sensing images over the same region. The CoverageSeries type has a fixed spatial extent and measured coverages at controlled times. It is mapped onto a Field[E:Extent, (2DPoint, Instant), V:Value, G:Estimator] whose positions have variable 2D spatial locations and times. The field's extent is composed of the coverage series' spatial extent and an interval that encloses all position instances.

5   Array Databases for Big Spatial Data

Big spatial data comes from many different sources and with different formats. Among those sources are Earth Observation satellites, GPS-enabled mobile devices and social media. For example, the LANDSAT data archive at the United States Geological Survey has more than 5 million images of data of the Earth's land surface, collected over 40 years, comprising about 1 PB of data. These data sets allow researchers to explore big data sets for innovative applications. One example is the world's first forest cover change map from 2000 to 2012 at a spatial resolution of 30 meters [18].

The challenge for handling big spatial data is to design a programming model that can be scaled up to petabyte data sets. Currently, most scientific data analysis methods for Earth observation data are file-based. Earth observation data providers offer data to their users as individual files. Scientific and application users download scenes one by one. For large-scale analyses, users need to obtain hundreds or even thousands of files. To analyze such large data sets, a program has to open each file, extract the relevant data and then move to the next file. The program can only begin its analysis when all the relevant data has been gathered in memory or in intermediate files. Data analysis on large datasets organized as individual files will run slower and slower as data volumes increase. This practice has put severe limits on the scientific uses of Earth Observation data.

To overcome these limitations, there is a need for a new type of information system that manages large Earth Observation data sets in an efficient way and allows remote access for data analysis and exploration. It should also allows existing spatial (image processing) and temporal (time series analysis) methods to be applied to large data sets, as well as enabling development and testing of new methods for space-time analyses of big data. After analyzing alternatives, such as MapReduce [7], we consider that array databases offer the best current solution for big spatial data handling. Array databases offer a model of programming that suits many of tasks for analysis of spatiotemporal data.

Array databases organize data as a collection of arrays, instead of tables used in object-relational DBMSs. Arrays are multidimensional and uniform, as each array cell holds the same user-defined number of attributes. Attributes can be of any primitive data type such as integers, floats, strings or date and time types. To achieve scalability, array databases strive for efficiency of data retrieval of individual cells. Examples of array databases include RasDaMan [1] and SciDB [29].

Array databases have no semantics, making no distinction between spatial and temporal indexes. Thus, to be used in spatial applications, one needs to extend them with types and operations that are specific for spatiotemporal data. That is where the Fields data type is particularly useful.

6   Fields Operations in Array Databases

This section shows how to map the fields data type onto the array database SciDB [29]. SciDB splits big arrays into chunks that are distributed among different servers; each server controls a local data storage. One of the instances in the cluster is the coordinator, responsible for mediating client communications and for orchestrating query executions. The other instances, called workers, participate in query processing. SciDB takes advantage of the underlying array data model to provide an efficient storage mechanism based on chunks and vertical partitions. Compared to object-relational databases, the SciDB solution provides significant performance gains. Benchmarks comparing object-relational databases and array databases for big scientific data have shown gains in performance of up to three orders of magnitude in favor of SciDB [6,27].

SciDB provides two query languages: an Array Query Language (AQL) that resembles SQL and an Array Functional Language (AFL) closely related to functional programming. There are two categories of functions:

Scalar Functions. Algebraic, comparison and temporal functions, that operate over scalar values.

Aggregates. Functions that operate on array level, like average, standard deviation, maximum and minimum values.

Natively, SciDB already supports some of the operations of the Fields data type. The operations of the Fields data type currently available in SciDB are described in Table 1. We tested these operations using arrays of different sizes, as discussed below.

Table 1. Fields model mapped onto SciDB

Field op	signature	SciDB op
map	`Field x (v:Value → v:Value)`	`apply`
subfield	`Field x e:Extent → Field`	`subarray`
filter	`Field x (v:Value → Bool)`	`filter`
reduce	`Field x (v:Value x v:Value → v:Value)`	`aggregate`

Our evaluation used a set of images from the MODIS sensor, which flies onboard NASA's Terra and Aqua remote sensing satellites. The MODIS instruments capture data in 36 spectral bands. Together the instruments image the entire Earth every 1 to 2 days. They are designed to provide measurements in large-scale global dynamics, including changes in the Earth's cloud cover, radiation budget, and processes occurring in the oceans, on land, and in the lower atmosphere [20].

We used the MODIS09 land product with three spectral bands (visible, near infrared, and quality). Each MODIS09 image is available for download at the NASA website as a tile covering 4,800 x 4,800 pixels in the Earth's surface at 250 meters x 250 meters ground resolution. We then combined more than ten years of data (544 time steps) of the 22 MODIS images that cover Brazil, giving a total of 11,968 images that were merged into an array of 2.75×10^{11} (275 billion) cells. Each cell contains three values, one for each band. This array was then loaded into SciDB for our experiment.

We first used the SciDB `subarray` function to select subsets of the large array for evaluation purposes. For each subarray, we used the SciDB `apply` function to calculate the enhanced vegetation index [19] associated to each cell and stored the results in a new subarray. Next, we used the `filter` operation to select from each resulting subarray those cells whose red value was greater than 100 and stored the results. Finally, we used the `aggregate` function to calculate the average of the one attribute of each subarray and store the results. Fig. 5 shows the test results as the average of 5 runs for the following number of cells: $46 * 1024^2, 46 * 2048^2, 46 * 3072^2, 46 * 4096^2, 46 * 5120^2, 46 * 6144^2, 46 * 7168^2, 46 * 8192^2, 46 * 9216^2, 46 * 10240^2, 46 * 11264^2, 46 * 12288^2, 46 * 13312^2, 46 * 14336^2$.

These results were obtained in a single Ubuntu server, having 1 Intel Xeon 2.00 GHz CPU, with 24 cores and 132 GB memory. The performance results are satisfactory,

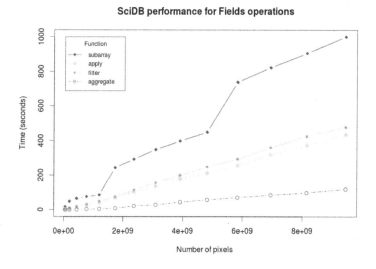

Fig. 5. Performance measures for Field operations in SciDB operations

since the processing time grew roughly linearly with array size. With a bigger server configuration, we can expect better results. These results have given us confidence that combining the Fields data type with array database is viable and likely to produce good results. As part of later work, we will implement the whole Fields data type in SciDB, making it a suitable environment for processing large spatial data.

Although array databases currently offer the most promising approach for handling big spatial data sets, they do not yet offer all of the support required by spatial applications. Most spatial applications need to combine field data sets with information about spatial objects, such as cities and farms. Also, array databases treat all dimensions equally. Therefore, developers of spatial applications need to provide additional support to use array databases effectively. This is a promising new research area that can lead to spatial information infrastructures that will make good use of large data sets.

7   Conclusions

This paper defined the Field abstract data type for representing continuous spatiotemporal data. The motivation was to provide a sound basis for applications that deal with big spatial data sets. These data sets can come for many different sources and have many purposes, yet they share common features: in all of them, one measures values at positions in space-time. The underlying conceptual view is that these data sets are measures of continuous phenomena, thus leading to fields. We showed that the Fields data type can represent data sets, such as maps, remote sensing images, trajectories of moving objects, and time series.

We also considered the problem of how to implement the Field data type operations in an environment suitable for handling large spatial data and argued that array databases are currently the best approach available. Some of the operations of the Field data type are already available in the open source array database SciDB, and our experiments showed that the performance of SciDB is encouraging. Given the results so far, we will implement the full set of the Field data type operations directly in SciDB to provide a full features of Field data type in array databases.

We anticipate that the combination of the Field data type and array databases can bring about a disruptive change in spatial information infrastructures. Consider the case of Earth Observation data. Currently, remote sensing data is retrieved from the data archives on a scene-by-scene basis and most applications use only one temporal instance per geographical reference. In an advanced infrastructure, researchers and institutions will break the image-as-a-snapshot paradigm, as entire collections of image data will be archived as single spatiotemporal arrays. Users will be able to develop algorithms that can span seamless partitions in space, time, and spectral dimensions, and arbitrary combinations of those. These algorithms will provide new insights into changes in the landscape.

We believe that the combination of simple, yet powerful data types with new technologies for spatial data management will bring about large changes in the use of spatial information, especially for data that promotes the public good. Data management of large data sets will be done in petascale centers. Users will have the means to perform analysis and queries on these data sets. Petascale centers that promote open data policies and open data analysis will get large benefits from increased awareness of the value of spatial information for society.

Acknowledgments. Gilberto Camara thanks for the support of the Brazilian research agencies FAPESP (grant 2008/58112-0) and CNPq (grant 04752/2010-0). This work was written while Gilberto Camara was the holder of the Brazil Chair at the University of Münster, supported by the Brazilian agency CAPES (grant 23038.007569/ 2012-16). Gilberto also received substantial financial and logistical support provided by the Institute of Geoinformatics at the University of Münster, Germany. Max Egenhofer's work was partially supported by NSF Grant IIS-1016740.

References

1. Baumann, P., Dehmel, A., Furtado, P., Ritsch, R., Widmann, N.: Spatio-temporal retrieval with RasDaMan. In: Proceedings of the 25th International Conference on Very Large Data Bases, VLDB 1999, pp. 746–749 (1999)
2. Campbell, P.: Editorial on special issue on big data: Community cleverness required. Nature 455(7209), 1 (2008)
3. Cardelli, L., Wegner, P.: On understanding type, data abstraction, and polymorphism. ACM Computing Surveys 17(4), 471–552 (1985)
4. Cordeiro, J., Camara, G., Freitas, U., Almeida, F.: Yet another map algebra. Geoinformatica 13(2), 183–202 (2009)
5. Couclelis, H.: People manipulate objects (but cultivate fields): Beyond the raster-vector debate in GIS. In: Frank, A.U., Formentini, U., Campari, I. (eds.) GIS 1992. LNCS, vol. 639, pp. 65–77. Springer, Heidelberg (1992)
6. Cudre-Mauroux, P., Kimura, H., Lim, K.T., Rogers, J., Madden, S., Stonebraker, M., Zdonik, S., Brown, P.: SS-DB: A standard science DBMS benchmark. In: XLDB 2010 - Extremely Large Databases Conference (2012)
7. Dean, J., Ghemawat, S.: MapReduce: Simplified data processing on large clusters. Communications ACM 51(1), 107–113 (2008)
8. Ferreira, K., Camara, G., Monteiro, A.: An algebra for spatiotemporal data: From observations to events. Transactions in GIS 18(2), 253–269 (2014)
9. Frank, A.: One step up the abstraction ladder: Combining algebras - from functional pieces to a whole. In: Freksa, C., Mark, D.M. (eds.) COSIT 1999. LNCS, vol. 1661, pp. 95–108. Springer, Heidelberg (1999)
10. Frank, A.: Tiers of ontology and consistency constraints in geographic information systems. International Journal of Geographical Information Science 15(7), 667–678 (2001)
11. Frank, A.: Map algebra extended with functors for temporal data. In: Akoka, J., et al. (eds.) ER Workshops 2005. LNCS, vol. 3770, pp. 194–207. Springer, Heidelberg (2005)
12. Frank, A.: GIS theory - the fundamental principles in GIScience: A mathematical approach. In: Harvey, F.J. (ed.) Are there Fundamental Principles in Geographic Information Science?, pp. 12–41 (2012)
13. Frank, A., Kuhn, W.: Specifying Open GIS with functional languages. In: Egenhofer, M., Herring, J.R. (eds.) SSD 1995. LNCS, vol. 951, pp. 184–195. Springer, Heidelberg (1995)
14. Galton, A.: Fields and objects in space, time and space-time. Spatial Cognition and Computation 4 (2004)
15. Goodchild, M.: Geographical data modeling. Computers and Geosciences 18(4), 401–408 (1992)

16. Goodchild, M., Yuan, M., Cova, T.: Towards a general theory of geographic representation in GIS. International Journal of Geographical Information Science 21(3), 239–260 (2007)
17. Guttag, J., Horowitz, E., Musser, D.: Abstract data types and software validation. Communications of the ACM 21(12), 1048–1064 (1978)
18. Hansen, M., Potapov, P., Moore, R., Hancher, M., Turubanova, S., Tyukavina, A., Thau, D., Stehman, S., Goetz, S., Loveland, T., Kommareddy, A., Egorov, A., Chini, L., Justice, C., Townshend, J.: High-resolution global maps of 21st-century forest cover change. Science 342(6160), 850–853 (2013)
19. Jiang, Z., Huete, A., Didan, K., Miura, T.: Development of a two-band enhanced vegetation index without a blue band. Remote Sensing of Environment 112(10), 3833–3845 (2008)
20. Justice, C., Townshend, J., Vermote, E., Masuoka, E., Wolfe, R., Saleous, N., Roy, D., Morisette, J.: An overview of MODIS land data processing and product status. Remote Sensing of Environment 83(1), 3–15 (2002)
21. Kemp, K.: Fields as a framework for integrating GIS and environmental process models. part one: Representing spatial continuity. Transactions in GIS 1(3), 219–234 (1997)
22. Kuhn, W.: Geospatial semantics: Why, of what, and how? Journal of Data Semantics 3, 1–24 (2005)
23. Mennis, J.: Multidimensional map algebra: Design and implementation of a spatiotemporal GIS processing language. Transactions in GIS 14(1), 1–21 (2010)
24. OGC: The OpenGIS abstract specification - Topic 6: Schema for coverage geometry and functions (Tech. Rep. OGC 07-011). Tech. rep., Open Geospatial Consortium, Inc. (2007)
25. OGC: OGC web coverage service (WCS) interface standard - Core (OGC 09-110r3). Tech. rep., Open Geospatial Consortium, Inc. (2010)
26. Peuquet, D.: Representations of geographic space: Toward a conceptual synthesis. Annals of the Association of American Geographers 78(3), 375–394 (1988)
27. Planthaber, G., Stonebraker, M., Frew, J.: EarthDB: scalable analysis of MODIS data using SciDB. In: Proceedings of the 1st ACM SIGSPATIAL International Workshop on Analytics for Big Geospatial Data, pp. 11–19. ACM (2012)
28. Sinton, D.: The inherent structure of information as a constraint to analysis: Mapped thematic data as a case study. In: Dutton, G. (ed.) Harvard Papers on Geographic Information Systems, vol. 7, pp. 1–17. Addison-Wesley, Reading (1978)
29. Stonebraker, M., Brown, P., Zhang, D., Becla, J.: SciDB: A database management system for applications with complex analytics. Computing in Science & Engineering 15(3), 54–62 (2013)
30. Tomlin, C.: Geographic Information Systems and Cartographic Modeling. Prentice-Hall, Englewood Cliffs (1990)
31. Winter, S., Nittel, S.: Formal information modelling for standardisation in the spatial domain. International Journal of Geographical Information Science 17(8), 721–741 (2003)

Linked Data - A Paradigm Shift for Geographic Information Science

Werner Kuhn[1,2], Tomi Kauppinen[3,4], and Krzysztof Janowicz[2]

[1] Center for Spatial Studies, University of California, Santa Barbara CA, USA
[2] Department of Geography, University of California, Santa Barbara CA, USA
[3] Cognitive Systems Group, Universität Bremen, Germany
[4] Department of Media Technology, Aalto University School of Science, Finland

Abstract. The Linked Data paradigm has made significant inroads into research and practice around spatial information and it is time to reflect on what this means for GIScience. Technically, Linked Data is just data in the simplest possible data model (that of triples), allowing for linking records or data sets anywhere across the web using controlled semantics. Conceptually, Linked Data offers radically new ways of thinking about, structuring, publishing, discovering, accessing, and integrating data. It is of particular novelty and value to the producers and users of geographic data, as these are commonly thought to require more complex data models. The paper explains the main innovations brought about by Linked Data and demonstrates them with examples. It concludes that many longstanding problems in GIScience have become approachable in novel ways, while new and more specific research challenges emerge.

1 Introduction

Linked Data is an example of a technological innovation that transforms the way we think about information and its role in society, in particular geographic information. However, discussions on it tend to focus on technical aspects, such as how to convert existing data sets or how to deal with semantics in shared vocabularies and ontologies. This paper explains why the difference between traditional data holdings and Linked Data repositories is more than one of formats and is largest for sophisticated types of information, in particular information with spatial and temporal components.

With the adoption of Linked Data, the familiar complexities of conceptual database schemata for spatial data can safely remain internal to organizations, from where they have been too hard to share anyway. Their externally relevant contents get streamlined into the open and more manageable form of vocabulary definitions. Users of Linked Data do not need to be aware of complex schema information to use data adequately, but "only" of the semantics of types and predicates (such as `isLocatedIn`) occurring in the data. While many questions remain to be answered about how to produce and maintain vocabulary specifications [1], the elaborate layering of syntactic, schematic, and semantic interoperability issues [2] has simplified to a single common syntax (RDF), the irrelevance of traditional schema information outside a database, and a focus on specifying and sharing vocabularies.

M. Duckham et al. (Eds.): GIScience 2014, LNCS 8728, pp. 173–186, 2014.

This simplification is more dramatic for spatially and temporally referenced data (with their complexities in the form of geometries and scale hierarchies) than it is for, say, financial data. The resulting paradigm shift, from distributed complex databases accessed through web services that expose schemata to knowledge represented as graphs, whose links can be given well-defined meaning, radically changes some of the long-standing problems of GIScience and GIS practice. This paper attempts to raise the level of discussion on Linked Data from the "how" to the "why" by describing the changes in perspective on some deeper issues of GIScience. It also summarises new problems and research questions arising from the paradigm shift.

The discussion should be seen against the broader background of commonly identified limitations of existing data models for spatial and spatio-temporal data, including that

- access to data is software-dependent
- metadata and schemata are separate entities
- data models mix concerns of semantics and data management
- the semantics of terms remains implicit or hard to share and reason with
- data are seen as provider-independent truths, though they often contradict
- there are too few simultaneously accessible viewpoints or versions of data
- global, unique identifiers are hard to obtain and not encouraged
- valuable data sets remain isolated and hard to integrate
- the emphasis on consistency and quality restricts data availability
- data about a particular topic, place, or period are hard to find
- incentives for producing metadata and enabling reuse are lacking.

Linked Data is not a panacea removing these obstacles to producing, accessing, and using spatial data. But it is more than a set of new technologies, as it substantially changes how we deal with these problems. For example, geographic information has long been recognized as a powerful "glue" to integrate information across domains, but putting this vision in place was often too hard. With Linked Data, the gluing function of spatial and temporal referencing has finally become reality [3], as the Linked Data Cloud[1] allows for linking any data to geographically referenced data, such as linked geodata[2].

After introducing the basics of Linked Data in the next section, sections three to nine discuss the impact of Linked Data on how we understand spatial and spatio-temporal information and the issues surrounding it: provenance, consistency, metadata, semantics, maintenance, data publishing, and data integration. We conclude with some new or revised research challenges and a summary of the points made.

2   Linked Data in a Nutshell

Linked Data is the name for a collection of design principles and technologies centered around a novel paradigm to publish, retrieve, reuse, and

---

[1] http://datahub.io/group/lodcloud
[2] http://linkedgeodata.org/

integrate data on the Web. In contrast to the Document Web, the Web of Linked Data aims at establishing named and directed links between typed data. For example, a normal Web page about Portsmouth (such as http://en.wikipedia.org/wiki/Portsmouth) may link to another page about Hampshire (such as http://en.wikipedia.org/wiki/Hampshire). For a machine, the intended meaning of such links is difficult to interpret and the Web pages can only be consumed as integral units of text or other media. On the Linked Data Web, by contrast, the link between Portsmouth and Hampshire would be directed and labeled, for example, forming the statement that Portsmouth is located in Hampshire. Additionally, the two places would be typed, e.g., as city and county, jointly leading to the statement that the city of Portsmouth is located in the county of Hampshire. Finally, the predicate isLocatedIn could be defined as a transitive relation in an ontology. Thus, in conjunction with a statement that Hampshire county is located in the UK, one could automatically derive the new statement that Portsmouth is located in the UK.

Given that three elements constitute each piece of information in Linked Data, one refers to such statements as *triples*, consisting of a subject (Portsmouth), a predicate (isLocatedIn), and an object (Hampshire). This syntax, which happens to be the simplest form in which statements can be made in natural language, has thus been carried over to the world of data. The data model for triples is the so-called Resource Description Framework (RDF).

Tim Berners-Lee established four principles of Linked Data [4]:

- Uniform Resource Identifiers (URIs) should be used to denote things.
- HTTP URIs should be used so that these things can be referred to and dereferenced (looked up) by human users and software agents.
- W3C standards such as RDF or OWL should be used to provide information about the things when their URIs are dereferenced.
- Data about anything should link out to other data, using their URIs to create a densely interconnected graph of knowledge (the so-called Linked Data Cloud).

As these principles are expressed in the technical jargon of the Web, it helps to relate them to spatial data and entities in geographic space. According to the principles, an entity in the physical world, such as the historic ship HMS Victory, should be identified by a globally unique URI. As the HMS Victory is not an information resource, one cannot directly retrieve information about it using a browser or Linked Data tools. However, when visiting the ship's URI, the responding Web server can redirect (HTTP code 303) the visitor to an information resource, such as an RDF document containing statements about HMS Victory or an HTML page that renders RDF in a human readable way[3]. (As a URI for HMS Victory, we use http://dbpedia.org/page/HMS_Victory, which can be abbreviated to dbpedia:HMS_Victory, thanks to the predefined name space dbpedia[4]).

Linked Data can be queried using SPARQL[5], a query language for RDF. GeoSPARQL adds the possibility to query over topological relations and thus enriches

---

[3] See http://live.dbpedia.org/page/HMS_Victory

[4] http://dbpedia.org/sparql?nsdecl

[5] DBPedia has an open SPARQL endpoint at http://live.dbpedia.org/sparql

SPARQL by quantitative reasoning. So far, GeoSPARQL support is limited but some reference implementations, such as Parliament, have been proposed and implemented as free and open source [5]. The following SPARQL query example retrieves all predicates and objects of statements that have dbpedia:HMS_Victory as a subject.

```
SELECT * WHERE {
  dbpedia:HMS_Victory ?predicate ?object
}
```

The queried statements might contain information about the ship, e,g., when it was laid down or the battles it participated in. While the first case can be represented by a single date, e.g., using XSD date type, the linked historic battles could themselves be represented by URIs[6], linking to actors involved in the battles, such as Vice-Admiral Horatio Nelson. A triple may state that the HMS Victory is located at Portsmouth, UK, which in turn leads to more resources about that city, its population, and so forth, contributing to the more and more densely interconnected Linked Data Cloud.

Linked Data is usually stored in so-called triple stores and accessed via so-called SPARQL endpoints. The ontologies that allow human users and machines to understand which concepts and predicates can be queried, and how they are formally defined, are described using languages such as the Web Ontology Language (OWL).

Listing 1.1 shows five RDF triples in Turtle syntax[7]. Some of them are assertions, e.g., that Horatio Nelson died in the Battle of Trafalgar, while others are taxonomic, e,g., that naval battles are special battles. One can derive new statements from those triples; for instance, one can automatically infer that Nelson died in a battle.

```
ex:HoratioNelson ex:diedIn ex:BattleOfTrafalgar .
ex:BattleOfTrafalgar ex:during ex:NapoleonicWars ;
                     rdf:type ex:NavalBattle .
ex:NavalBattle rdfs:subClassOf ex:Battle .
ex:diedIn rdfs:subPropertyOf ex:participatedIn .
```

Listing 1.1. Example of RDF statements

The discussion in this paper focuses on data exposed as Linked Data, regardless of a possible co-existence of the same information in other data models (e.g., in GIS, databases, or file systems) and regardless of whether the data are considered open or not. Linked Data is in fact most typically treated as (but not limited to) a secondary, openly available, exposure of contents that are also held in other formats, typically enabled through government subsidies or open source communities.

---

[6] Though DBPedia represents the battles only as literal values, not enabling queries to follow links.
[7] http://en.wikipedia.org/wiki/Turtle_(syntax)

3   Data Are Statements

Linked Data express claims [3], i.e., statements made by somebody somewhere at some time. Meta-information about this context is obtainable through the owner of the URI at which the statements reside and the place and time of their publication. This idea holds the key for solving provenance, currency, and versioning problems. Provenance or lineage (the origin and chronology of data) remains a thorny issue in theory and practice around geographic data and data in general. Linked Data deals well with provenance, but remains somewhat weak in dealing with the temporal evolution of knowledge.

For an example of how to obtain provenance information, consider the triple stating that Portsmouth is located in Hampshire. The author of such a statement is typically made explicit only at the level of whole data sets. The statement about Portsmouth could be (but actually is not) part of Ordnance Survey's publicly available linked geodata[8]. According to http://data.ordnancesurvey.co.uk/datasets/os-linked-data, Ordnance Survey did make the taxonomic statement, however, that the City of Portsmouth is a Borough, on October 25th 2010, and confirmed it on May 10th, 2013.

The nature of triples as statements, as obvious as it is from their subject-predicate-object syntax, tends to be mixed up with the conventional view of data as facts, even in the specialized literature discussing how to reason with Linked Data. Triples are still too often seen as single, objective, eternal truths about some contents, irrespective of authorship or date or other contextual and quality aspects. The implicit assumption is that they will conveniently be overwritten or forgotten if better data become available.

When triples are instead seen as expressions of beliefs held by individuals or organizations at some point in time, the Linked Data paradigm easily admits reviewing, commenting, revising, and extending information. If Ordnance Survey, for example, makes statements about the geography of the UK, these come with a significant level of authority and trustworthiness. But they still express a view of the world held by a (professional, authoritative) organization at some point in time, subject to disagreement, revision, and improvement.

Trust and reputation are naturally attached to the authors of data, propagated to their statements, and calibrated through links and their weights in search engines. The Linked Data paradigm is ideal to express and reason with such information. There are many solutions for handling and increasing trust: 1) professional and volunteered curation of data, 2) inconsistency checks via automated reasoning, 3) applications and services using the data. These processes can also take the form of syntactic checks. Similarly, reputation should increase based on communication, i.e., sharing evidence about the level of trust: for instance communicating that the data in some source is curated.

Thanks to the Linked Data paradigm, digital maps and other geospatial models can now be seen as sets of statements made by authors with some reputation at some well-defined points in time [6]. Ideally, these statements will never be removed, because it is valuable (and sometimes legally required or otherwise essential) to know about previous world views or states of affairs. New insights may be gained, the world or the ways to describe it may change, and statements can simply be added to the Linked Data Cloud.

---

[8] http://data.ordnancesurvey.co.uk/

From a technical point of view, with RDF 1.1, the possibility to group statements in data sets attributed to a well-defined (by URI) provenance has now become a standard solution (using multiple graphs)[9].

4   Statements Can Contradict

Another long standing concern in GIScience is that of the consistency of its models. Following Allemang and Hendler's AAA slogan that "Anyone can say Anything about Any topic" [7], consistency is not a key concern for Web-scale systems. Given the variety of sources, perspectives, granularities, and data creation and curation procedures, the collection of statements in the Web can, do, and will contain contradictions. Early on, the Semantic Web community decided to handle this challenge not by stricter models and model checking but by adopting the Open World Assumption (where the absence of a statement does not imply its falsehood). It focuses on inferential semantics, deferring consistency checking to the level of ontology engineering.

To give a concrete example for the consequences of this decision, consider the death of a prominent person. This information may be updated at Wikipedia (and thus DBpedia live[10]) within minutes but may take longer to be integrated into other data repositories. News stations may decide to wait for independent confirmation of the event. Thus, the global knowledge graph will contain information that the given person is alive and dead at the same time. A Web-scale system such as the global graph of Linked Data still has to be able to function despite such apparent inconsistencies.

On the terminological level, the Open World Assumption ensures that a lack of knowledge or the temporal unavailability of a certain data repository does not imply that a statement is false. In contrast, under the Closed World Assumption statements that are not known to be true, are assumed to be false. In the example, the absence of a date of death does not imply that the person is still alive, just that it is unknown whether the person is alive or dead (until a statement clarifies this either way).

In addition to such temporal aspects, space and place also play a key role in interpreting statements. For instance, news agencies from two different countries may have diverging opinions (on a cause of death or anything else). These regional (as well as temporal) differences can go as far as authorities disagreeing at the terminological level. The case of "freedom fighters" versus "terrorists" used to label the same people over time is an infamous example. Two data providers at the same time, or the same provider at two different times, may *classify* a particular individual in different ways. This leads to an extended version of the AAA slogan to an AAAAA view of "Anyone can say Anything about Any topic at Any time and Anywhere" [8]. The lesson from such examples is that spatial and temporal indexing of statements is important for reasoning, even where the topics reported do not seem to be spatial or temporal.

Additionally, and in clear contrast to previous knowledge engineering paradigms, the Linked Data community has taken the stance that there is no need for linking to abstract top-level or domain-level ontologies (while retaining the benefits of designing

---

[9] https://dvcs.w3.org/hg/rdf/raw-file/default/rdf-dataset/index.html
[10] http://live.dbpedia.org

ontologies along well-defined upper level distinctions). It has invested instead into research on ontology alignment, data-driven knowledge patterns, and query federation. Consequently, what appears to be the same statement in two triple stores can have vastly different interpretations. Consider, for instance, a triple stating that Horatio Nelson died on the deck of the HMS Victory, and a second triple that the HMS Victory is located at Portsmouth, UK. Depending on the choice of ontology axioms used to interpret these statements, including decisions on how to assess the temporal validity of statements, one can infer that Nelson's place of death is in the UK or not. Whether this is an unintended logical consequence is left to decide at the terminological level, i.e., by considering and adapting the ontologies used. This attitude fosters reusability and integration. Its downside is that the (mis)use of so-called co-reference resolution (e.g., via `owl:sameAs`) may hamper conflation; whether two resources actually refer to the same entity is often a very complex decision to make; see Haalpin et al. [9] for a detailed discussion.

5   Metadata Are Data

GIScience and GIS practice distinguish data (such as on geographic features and their attributes) from metadata (such as on the creation dates and names of creators of the data). The two types of data tend to be captured and kept in separate locations, models, formats, granularities, and business models. Linked Data blurs this largely artificial distinction, as each statement can be semantically typed and annotated. It is still possible to create provenance information about collections of statements (RDF documents), or SPARQL endpoints serving certain data. This can, for example, be done using the Prov ontology [10]. Provenance statements, however, are statements of the same nature and form as all other statements in RDF (which itself evolved from a metadata format). It is possible to follow links from data to metadata and back. It is also possible to define what is considered metadata in some application scenario and what is considered data. Finally, on the terminological level, ontologies developed in OWL or the RDF Schema language (RDFS) can also be queried to retrieve data about individuals as well as about concepts (which would be considered metadata) and one can pose queries that filter the results by criteria about concepts and data.

Metadata about statements (spatial or not) are themselves often spatial and temporal: when did the information become known? when and where did an event occur? what was its duration? what else happened during that time or at that place? These spatial and temporal aspects of metadata make spatio-temporal computing attractive for information in general, beyond properly geographic data. They can be better exploited now that the data about them is part of content data. Through the tight integration of data and metadata in a single and simple format (RDF), we are now in fact applying GIScience methods to information infrastructures in general [8].

6   Semantics Is in Predicates, Not Schemata

One of the biggest innovations of Linked Data is the way semantics gets handled. Traditional geographic data modeling wisdom (to be found in any GIScience textbook or course) has it that a conceptual database schema, together with a data dictionary, is the best way to capture what is meant by the data. Ontologies may be used to expose semantics and allow for machine reasoning about it, but they are typically not seen as having reached the expressive power or practicality of database schemata and dictionaries. A database schema, in this sense, describes the structure of data, while an ontology provides a specification of the intended meaning of terms. Thus, the triple structure is a schema, while ontologies (expressed in OWL or other languages) specify the types and predicates used in triples.

One problem with the database schema approach to semantics is that data often leave their native environments and get repackaged in forms which may or may not capture the intended semantics adequately. Supplying them with some schema information (say, in XML-based form) is normally not enough. Database schemata, even in their layered form standardized by ANSI and ISO (conceptual - logical - physical), fail to separate the concerns of data organization from those of semantics. They do a great job on the former, but a notoriously poor one on the latter task, especially when data leave their native environments. Semantics has very little to do with how data are structured, and much more with how terms are being used in the data (and their structuring).

Linked Data provides this missing semantic link for spatial (as well as any other types of) data by connecting statements to definitions of the predicates used in the triples. Since their conceptual schema is the simplest possible one, that of a triple with a subject, predicate and object, nothing else needs to be stated about structure or schemata and one can concentrate all semantic efforts on capturing the intended meaning of the terms used in these three elements. Semantics that was traditionally captured in schema form (say, about cardinalities of relations) can be restated in shared vocabularies. In this form, it will be explicit and accessible to inspection and revision anywhere within and outside the organization producing or holding the data. As syntactic interoperability is largely handled by common standards such as RDF, we can entirely focus on addressing semantic interoperability.

7   Maintenance without Deletion

GIScience still relies massively on static world views—or at least on an implicit assumption that changes can be recorded through sequences of snap shots. For instance, changes related to weather (say, temperature or amount of rain) or other environmental conditions (sea level or river width) are monitored and recorded as time series of attribute values. The underlying assumption is that the essence of world knowledge is static: values of temperature change but the temperature concept itself remains unchanged and only current values of such concepts need to be stored. Many counter examples show the need for modeling changes in more sophisticated ways:

– natural and administrative regions change (they split, merge or otherwise change their form) due to human decisions or natural processes;
– connections between things change (e.g., a person moves from one affiliation to another)
– concepts change over time or acquire multiple senses (like the famous example by Frege about the concepts of *evening star* and *morning star*)

There are many such changes in the world [11] and in our way of talking about it. When formalized, they can be used for inferences. An example of this is to study [12] what topological inferences can be made when a region is cut into pieces. Linked Data allows for storing and sharing both the data used as input and the inference results. This way, inference results become links in RDF and provide a way to traverse longer paths, for instance from contemporary place names to historic ones [13].

Ideally, with Linked Data as statements, monotonicity is regained. A true statement never needs to be retracted, but the time span in which it is considered true needs to be made explicit. For place-based information, this means to understand and state the beginnings and ends of the validity of place names. The time span for the resource representing the place can be defined, for instance, as a the time when the borders of an administrative place remained unchanged [13]. This way borders can be linked to the right place, and spatial relations—like overlap with historic or contemporary regions—can be traced over time.

New challenges emerge for this approach, such as how to deal with imprecision and uncertainty—for instance how to infer when a certain statement is valid if the beginning/end of its validity is not known exactly. A Linked Data solution for this problem is to define fuzzy temporal intervals. Instances of this concept can then be annotated with the fuzzy begin, begin, end, and fuzzy end predicates of the interval. This allows for computing with the validity of statements even if the validity itself is imprecise. Maintenance also calls for documented provenance: about who has created a statement, when it was made, or when it became obsolete [14].

8   Data Publishing and Sharing by URI

The principles of Linked Data are built around URIs. Things should be named with URIs, preferably with HTTP URIs. Accessing data by URI allows for individual statement retrieval, in contrast to the need for always downloading complete, often large datasets. SPARQL can be used to query just that part of the data that one needs for a given task, down to a single piece of information (such as what county contains Portsmouth).

It is notable in this context that SPARQL endpoints are themselves identifiable via URIs. This allows for automatic querying of data provided by endpoints, and gathering of large-scale documentation of available data of different types. However, it calls for research on representation mechanisms for spatial accuracy, resolution, and other data quality aspects. Linked data thus provides a transparent way for building future Spatial Data Infrastructures (SDI) [15].

The availability of data about all parts of a statement (i.e., subject, predicate and object) differs considerably from traditional SDI, where predicates are not described in a machine-understandable fashion. For instance, the following excerpt describes the predicate DEFOR_2008[11] used to describe "new deforestation in 2008" in the Linked Brazilian Amazon Rainforest Data [16]:

---

[11] Full example is accessible as Linked Data at http://spatial.linkedscience.org/context/amazon/DEFOR_2008

```
@prefix amazon:<http://spatial.linkedscience.org/context/amazon/> .

amazon:DEFOR_2008
    rdfs:label "Percentage of new deforestation in 2008";
    amazon:aggregation amazon:Pixel ;
    amazon:columnnumber amazon:c10 ;
    amazon:source amazon:INPE ;
    amazon:timeperiod amazon:year2008 ;
    amazon:unit amazon:percent ;
    amazon:variabletype amazon:LandUse .
```

Deferencing by URI provides a way to check what statements are currently served by that URI. By accessing the URI of a 25kmx25km grid cell about the Brazilian Amazon Rainforest one gets the following kinds of statements[12] —i.e., aggregated information about that cell.

```
amazon:AMZ_LINKED_25K_1000
    rdfs:label "Cell 1000";
    amazon:DEFOR_2002 "0.039"^^xsd:double ;
    amazon:DEFOR_2003 "0.0030"^^xsd:double ;
    amazon:DEFOR_2004 "0.031"^^xsd:double ;
    amazon:DEFOR_2005 "0.042"^^xsd:double ;
    amazon:DEFOR_2006 "0.0050"^^xsd:double ;
    amazon:DEFOR_2007 "0.012"^^xsd:double ;
    amazon:DEFOR_2008 "0.0040"^^xsd:double .
```

Another example of Linked Data publishing is spatial@linkedscience [17] which contains Linked Data about papers published in the GIScience, COSIT, ACM GIS (SIGSPATIAL), and AGILE conference series. Each paper, author, and affiliation is assigned a URI. Accessing the data is again done via URI[13] to retrieve an RDF version of the data.

Communities ranging from libraries to environmental scientists have been seeking a way to identify the information resources they are dealing with. The solutions—such as Digital Object Identifiers (DOIs) for identifying outlets and papers—call for establishing registries to maintain identifiers and their mutual mappings. Linked Data facilitates such registries and the trust in them by tracing the hubs of data, similarly to search engines using HTML pages to find trusted hubs of information.

9   Data Integration by Linking

Linking enriches not only source data (that links to destination data), but also the destination data. For instance, referencing digital cultural heritage data to places creates rich

---

[12] Full example is available at http://spatial.linkedscience.org/context/amazon/AMZ_LINKED_25K_1000

[13] For instance
http://spatial.linkedscience.org/context/acmgis/paper/doi10.1145/1653771.1653787

descriptions of the places themselves. This allows for studying, for example, connections between places and the culture of a region.

URIs enable integration of different kinds of data not only online, but also locally. Sensitive data can make use of openly available Linked Data by sharing its URIs. This supports providing of the context (e.g., via spatial or temporal references) for the sensitive data in question, while the sensitive data itself can remain private.

Federated queries allow for accessing data from different SPARQL endpoints, i.e., to combine results from multiple sources. For instance the following statement documents that a grid cell is partially overlapping a municipality.

```
amazon:AMZ_LINKED_25K_1000
  tisc:partiallyOverlapping
    amazon:BRAZIL_MUNICIPALITY_1508407 .
```

By further requesting statements about the municipality[14] `amazon:BRAZIL_MUNICIPALITY_1508407` one can retrieve not only its name (Xinguara), but also a link to `dbpedia:Para_State`[15] representing the State in which Xinguara is located. This way one can navigate from one resource to another, independently of which Linked Data repository each happens to be stored at.

Given the challenges of sharing and agreeing about meanings of predicates, the procedure of linking is not straightforward. Automatic linking can easily create inadequate links, but manual linking is often too time consuming [18]. A key research task is to support identity resolution, i.e., when two things denoted by two URIs are the same and when they are not. Linking also tends to have context-dependent outcomes. For example, information retrieval by a tourist can accept more loosely defined links (say, on partonomical relations) between places than retrieval for administrative tasks of authorities.

Specifying and publishing link types (i.e., predicates) encourages others to reuse them. For instance, the Citation Typing Ontology[16] [19] lists over 80 different types of citations (such as *cites as evidence*, *conforms* or *critiques*). If the GIScience community considers typing of citations between its publications, it will support deeper understanding of the impact of its work.

Furthermore, the sharing and reuse of spatial and temporal relations[17] as Linked Data by the community would support the large scale reuse of GIScience methods and applications. If two data sets use the same URIs for predicates and concepts, then queries and reasoning procedures tested with one data set will also work for the other.

---

[14] See `http://spatial.linkedscience.org/context/amazon/ BRAZIL_MUNICIPALITY_1508407`

[15] `http://dbpedia.org/resource/Para_State`

[16] `http://purl.org/spar/cito`

[17] Such as predicates defined by the Open Time and Space Core Vocabulary, see `http://observedchange.com/tisc/ns/`

10   New Challenges

With the adoption of Linked Data as a paradigm, new challenges emerge for research and practice. We briefly list some of them here, suggesting research directions for Geographic and other Information Sciences.

1. How to deal with *raster data*: if one separates pure rasters from their interpretations into object concepts and treats both as Linked Data, what problems remain to be solved [20]?
2. How to deal with *time* in its many forms of relevance to linked data [21,22]?
3. How to exploit (recently standardized) RDF notions like multiple graphs for spatial data sets?
4. How to scope statement *validity* temporally [23] and spatially?
5. How to talk about statements themselves in a logically clean form, providing meta information (for a new and promising proposal, see [24]).
6. How to use such meta-statements in *trust and reputation* models [25]?
7. How to determine which *ontologies* are needed for geodata, how to reuse them, how to align them with other ontologies, and how to ensure community buy-in[18]?
8. How to deal with *real-time* or near-real-time streams of data? [26]
9. How to extend Application Programming Interfaces (APIs) to serve Linked Data, in addition to their typical CSV, JSON, and XML outputs [3]?
10. How do the large volumes of simply structured Linked Data affect *efficiency* in accessing and analyzing geographic data, in comparison with database systems and web services [27]?
11. How to better handle co-reference resolution to enable geo-data *conflation*?
12. Where are the hard limits, if any, of the triple data model for spatial data and which data should not be triplified (e.g., Well-Known-Text)?

The characteristics of the Linked Data paradigm, as described in this paper, provide a strong basis for addressing these and other challenges. In particular, the recognition that data are statements made by somebody somewhere at some time has already proven[19] to be one of the most powerful ideas when it comes to dealing with geographic information and when using spatial and temporal references as glues for information in general [8].

11   Conclusions

We discussed the paradigm shift afforded by Linked Data and Semantic Web technologies, highlighting its impacts on key questions of GIScience. Many of these impacts have to do with the changing role of database schemata, conventionally thought to be essential for modeling geographic data. This role needs to be revisited in the light of the triple model and the outsourcing of semantics into explicitly specified and shared vocabularies. Another set of impacts has to do with space and time as efficient integrators of data. The Linked Data approach makes this capacity explicit by enabling a global identification and publication of spatial and temporal references.

---

[18] These questions are being addressed by the so-called GeoVoCamps, see
    `http://vocamp.org/wiki/Main_Page`
[19] in projects such as `http://lodum.de/life/`

We argued that creating data in the form of Linked Data statements produces several benefits: statements can contradict and their validity can be time stamped, provenance information can be combined with the data itself, and semantics can be defined explicitly. With Linked Data, these benefits are built into the general Semantic Web infrastructure, creating a large-scale distributed (and spatially enabled) information infrastructure. We illustrated the Linked Data approach via examples ranging from mundane geographic facts through historical battles and environmental observations to bibliographic data. We ended with a list of a dozen research questions around novel challenges posed by Linked Data in GIScience.

The paper has focused on Linked Data, rather than Open or Linked Open Data. Yet, beyond the technical aspects discussed here, Linked Data has become an important vehicle for transparency in society. The paradigm of Open Government [28], popularized through national efforts in Brasil, the UK, the US, and other countries has rapidly spread and is reaching municipal levels in some countries [29]. With a wide range of available tools and a growing choice of vocabularies to convert data to Linked Data, anybody who wants (or is mandated) to open up geodata can and should now do so. Technical hurdles will no longer serve as an excuse to keep geodata hidden where there are no real reasons to do so. As for Open Data and Linked Data in general, the GIScience community has a great opportunity to help exploit location as an integrator across platforms, domains, and disciplines.

References

1. Gangemi, A.: Ontology Design Patterns for Semantic Web content. In: Gil, Y., Motta, E., Benjamins, V.R., Musen, M.A. (eds.) ISWC 2005. LNCS, vol. 3729, pp. 262–276. Springer, Heidelberg (2005)
2. Bishr, Y.: Overcoming the Semantic and Other Barriers to GIS Interoperability. International Journal of Geographic Information Science 12(4), 299–314 (1998)
3. Hart, G., Dolbear, C.: Linked Data: A Geographic Perspective. Taylor & Francis (2013)
4. Berners-Lee, T.: Linked Data – Design Issues (2009),
   http://www.w3.org/DesignIssues/LinkedData.html
   (accessed February 3, 2014)
5. Battle, R., Kolas, D.: Enabling the geospatial semantic web with parliament and geosparql. Semantic Web 3(4), 355–370 (2012)
6. Scheider, S., Jones, J., Sanchez, A., Keß ler, C.: Encoding and querying historic map content. In: AGILE (in press, 2014)
7. Allemang, D., Hendler, J.: Semantic Web for the Working Ontologist: Effective Modeling in RDFS and OWL. Morgan Kaufmann Publishers Inc., San Francisco (2008)
8. Janowicz, K.: The role of space and time for knowledge organization on the semantic web. Semantic Web 1(1), 25–32 (2010)
9. Halpin, H., Hayes, P.J., McCusker, J.P., McGuinness, D.L., Thompson, H.S.: When owl:sameas isn't the same: An analysis of identity in linked data. In: Patel-Schneider, P.F., Pan, Y., Hitzler, P., Mika, P., Zhang, L., Pan, J.Z., Horrocks, I., Glimm, B. (eds.) ISWC 2010, Part I. LNCS, vol. 6496, pp. 305–320. Springer, Heidelberg (2010)
10. Lebo, T., Sahoo, S., McGuinness, D., Belhajjame, K., Cheney, J., Corsar, D., Garijo, D., Soiland-Reyes, S., Zednik, S., Zhao, J.: Prov-o: The prov ontology. W3C Recommendation (April 30, 2013)

11. Smith, B., Brogaard, B.: Sixteen days. The Journal of Medicine and Philosophy 28, 45–78 (2003)
12. Egenhofer, M., Wilmsen, D.: Changes in topological relations when splitting and merging regions. In: 12th International Symposium on Spatial Data Handling, pp. 339–352. Springer, Vienna (2006)
13. Kauppinen, T., Hyvönen, E.: Modeling and reasoning about changes in ontology time series. In: Kishore, R., Ramesh, R., Sharman, R. (eds.) Ontologies: A Handbook of Principles, Concepts and Applications in Information Systems. Integrated Series in Information Systems, pp. 319–338. Springer, New York (2007)
14. Zhao, J., Klyne, G., Shotton, D.: Provenance and linked data in biological data webs. In: The 17th International World Wide Web Conference (LDOW 2008) (2008)
15. Diaz, L., Remke, A., Kauppinen, T., Degbelo, A., Foerster, T., Stasch, C., Rieke, M., Schaeffer, B., Baranski, B., Broering, A., Wytzisk, A.: Future SDI—Impulses from Geoinformatics Research and IT Trends. International Journal of Spatial Data Infrastructures Research 7, 378–410 (2012)
16. Kauppinen, T., de Espindola, G.M., Jones, J., Sánchez, A., Gräler, B., Bartoschek, T.: Linked Brazilian Amazon Rainforest Data. Semantic Web Journal 5(2) (2014)
17. Keßler, C., Janowicz, K., Kauppinen, T.: spatial@linkedscience – Exploring the Research Field of GIScience with Linked Data. In: Xiao, N., Kwan, M.-P., Goodchild, M.F., Shekhar, S. (eds.) GIScience 2012. LNCS, vol. 7478, pp. 102–115. Springer, Heidelberg (2012)
18. Goodwin, J., Dolbear, C., Hart, G.: Geographical Linked Data: The Administrative Geography of Great Britain on the Semantic Web. Transactions in GIS 12, 19–30 (2008)
19. Shotton, D.: CiTO, the Citation Typing Ontology. Journal of Biomedical Semantics 1(suppl. 1), S6 (2010)
20. Scharrenbach, T., Bischof, S., Fleischli, S., Weibel, R.: Linked raster data. In: Xiao, N., Kwan, M.P., Goodchild, M.F., Shekhar, S. (eds.) Geographic Information Science. LNCS, vol. 7478, Springer, Heidelberg (2012)
21. Kauppinen, T., Mantegari, G., Paakkarinen, P., Kuittinen, H., Hyvönen, E., Bandini, S.: Determining relevance of imprecise temporal intervals for cultural heritage information retrieval. Int. J. Hum.-Comput. Stud. 68(9), 549–560 (2010)
22. Trame, J., Keßler, C., Kuhn, W.: Linked data and time–modeling researcher life lines by events. In: Tenbrink, T., Stell, J., Galton, A., Wood, Z. (eds.) COSIT 2013. LNCS, vol. 8116, pp. 205–223. Springer, Heidelberg (2013)
23. Gutierrez, C., Hurtado, C.A., Vaisman, A.A.: Temporal RDF. In: Gómez-Pérez, A., Euzenat, J. (eds.) ESWC 2005. LNCS, vol. 3532, pp. 93–107. Springer, Heidelberg (2005)
24. Nguyen, V., Bodenreider, O., Sheth, A.: Don't like RDF reification? creating meta triples describing triples using singleton property
25. Bishr, M., Kuhn, W.: Trust and Reputation Models for Quality Assessment of Human Sensor Observations. In: Tenbrink, T., Stell, J., Galton, A., Wood, Z. (eds.) COSIT 2013. LNCS, vol. 8116, pp. 53–73. Springer, Heidelberg (2013)
26. Calbimonte, J.P., Jeung, H., Corcho, Ó., Aberer, K.: Enabling query technologies for the semantic sensor web. International Journal on Semantic Web and Information Systems 8(1), 43–63 (2012)
27. Jones, J., Kuhn, W., Keßler, C., Scheider, S.: Making the web of data available via web feature services. In: Proceedings of the 17th AGILE Conference on Geographic Information Science, Castellón, Spain (forthcoming, June 2014)
28. Lathrop, D., Ruma, L.: Open government: Collaboration, transparency, and participation in practice. O'Reilly Media, Inc. (2010)
29. Consoli, S., Aldo Gangemi, A.G.N., Silvio Peroni, V.P., Recupero, D.R., Spampinato, D.: Geolinked Open Data for the Municipality of Catania. In: Proceedings of the 4th International Conference on Web Intelligence, Mining and Semantics, Thessaloniki, Greece (forthcoming, June 2014)

An Ontology Design Pattern for Surface Water Features

Gaurav Sinha[1], David Mark[2], Dave Kolas[3], Dalia Varanka[4], Boleslo E. Romero[5],
Chen-Chieh Feng[6], E. Lynn Usery[4], Joshua Liebermann[7], and Alexandre Sorokine[8]

[1] Department of Geography, Ohio University, Athens, OH, USA
[2] Department of Geography, University at Buffalo, Buffalo, NY, USA
[3] Raytheon BBN Technologies, Columbia, MD, USA
[4] U.S. Geological Survey, Rolla, MO, USA
[5] Department of Geography, University of California, Santa Barbara, CA, USA
[6] Department of Geography, National University of Singapore, Singapore
[7] Tumbling Walls LLC, 85 High Street, Newton, MA 02464
[8] Oak Ridge National Laboratory, Oak Ridge, TN, USA

Abstract. Surface water is a primary concept of human experience but concepts are captured in cultures and languages in many different ways. Still, many commonalities exist due to the physical basis of many of the properties and categories. An abstract ontology of surface water features based only on those physical properties of landscape features has the best potential for serving as a foundational domain ontology for other more context-dependent ontologies. The *Surface Water* ontology design pattern was developed both for domain knowledge distillation and to serve as a conceptual building-block for more complex or specialized surface water ontologies. A fundamental distinction is made in this ontology between landscape features that act as containers (e.g., stream channels, basins) and the bodies of water (e.g., rivers, lakes) that occupy those containers. Concave (container) landforms semantics are specified in a *Dry* module and the semantics of contained bodies of water in a *Wet* module. The pattern is implemented in OWL, but Description Logic axioms and a detailed explanation is provided in this paper. The OWL ontology will be an important contribution to Semantic Web vocabulary for annotating surface water feature datasets. Also provided is a discussion of why there is a need to complement the pattern with other ontologies, especially the previously developed *Surface Network* pattern. Finally, the practical value of the pattern in semantic querying of surface water datasets is illustrated through an annotated geospatial dataset and sample queries using the classes of the *Surface Water* pattern.

1 Introduction and Motivation

Surface water refers to water that exists on a surface at a greater mass than just detectable moisture on the earth's surface. It is a critical natural resource for life on earth, and a primary category of environmental reality within the realm of human experience. Yet, given the immensely rich and varied contexts of our experiences, it is not surprising that features associated with surface water (and the landscape, in general), are perceived and lexicalized quite differently, depending on which surface water

M. Duckham et al. (Eds.): GIScience 2014, LNCS 8728, pp. 187–203, 2014.
© Springer International Publishing Switzerland 2014

characteristics are recognized and emphasized by a culture or language. Similar, though less extreme, differences are also found among scientists and other professions. The innovative multidisciplinary field called *Ethnophysiography* has emerged recently for exploring these nuances arising from the intersection of language, culture, and cognition as they affect the interpretation of the landscape [21-22].

The variability implies that there exist several ways for classifying and relating features, and that there may be loss of information due to semantic interoperability between different conceptual systems. Researchers have explored topographic gazetteers and spatial data standards from countries across the world, such as GNIS[1], SDTS[2], and Geonet Names Server[3] (all from USA), INSPIRE[4] (Europe), TTDMS[5] (Taiwan), and topographic map standards from the Russian Federation (as discussed in [8]), to show the varied way topographic concepts are understood and formalized [5, 8, 20, 24]. Even formally developed systems such as WordNet[6] and EnvO[7], or SWEET[8] are inconsistent with each other—and with common sense conceptualizations of geospatial phenomena. Part of the problem is that different aspects of the landscape may be preferentially paid attention to in different cultures, languages and professions [31-34]. Hence, categories for landscape, including surface water, may be difficult to generalize, and terms may not have one-to-one correspondence with terms in other languages [21-22, 34].

Such dissimilarities should not, however, distract from the fact that people do communicate successfully across cultural and linguistic barriers and standards and ontologies can also be rendered interoperable (albeit with some information loss). For example, comparison of Russian, Taiwanese (Mandarin language) and US (English) geospatial standards revealed several terrain and hydrographic qualities, relations and categories that are shared and can be used for concept matching between the national geospatial standards [8]. As another example, the ambitious European INSPIRE spatial data infrastructure initiative can cater to the variation across European countries by capturing localized, country-specific geographic semantics in separate microtheories, and then allow inference of a subset of shared conceptualizations to enable semantic interoperability at the global European level [5]. Yet, such interoperability driven studies only hint at what may be some of the underlying principles of different conceptualization systems. For a comprehensive understanding, substantial research on geographic cognition [26], nature of geographic categories [31], and naïve geography [6] will be needed to discover general principles. It is safe to assume that such theories are unlikely to emerge anytime soon, since many languages, cultures and contexts would have to be investigated to identify truly stable categories and properties.

---

[1] Geographic Names Information System (GNIS): http://geonames.usgs.gov

[2] Spatial Data Transfer Standard (SDTS): http://mcmcweb.er.usgs.gov/sdts

[3] GEOnet Names Server (GNS): http://earth-info.nga.mil/gns/html

[4] INSPIRE Directive: http://inspire.jrc.ec.europa.eu

[5] Taiwan Topographic Map Data Standard (TTMDS): http://fas.harvard.edu/chgis/work/coding/feat_types_tw.htm

[6] Wordnet: A Lexical Database for English: http://wordnet.princeton.edu

[7] Environmental Ontology: http://environmentontology.org

[8] SWEET ontologies: http://sweet.jpl.nasa.gov

In the shorter term, scientific ontologies and well-known theoretical frameworks, including Gibson's theory of environmental affordances [10], Horton's primary theory [16], Lakoff and Johnson's ideas of experiential realism and embodied cognition [18-19], and Hayes' naïve physics manifesto [14, 15] offer reasonable justification for making some basic assumptions about how people experience the world. The consensus from these theoretical frameworks seems to be that there exist some physical precepts of the landscape that all human beings (with similar physical sensory capabilities) are able to perceive and experience, irrespective of their background and context. Identifying the minimal components of that *commonly experienced* landscape would allow the design of foundational landscape ontologies [29]. Such ontologies will serve best if conceptually grounded in basic, easily generalizable experiences of physical reality. This paper contributes to this research agenda by presenting a new *Surface Water* ontology that captures the essential semantics of discrete surface water features and their physical connection to the earth's surface. The semantics were primarily derived from considerations of physically observable properties and features in the landscape.

The success of any foundational domain ontology rests on it being relatively abstract and sparse in terms of how many categories and properties it specifies to avoid over-commitments and be useful across domains. It should also be easily extensible across geographic scales, and provide clear criteria for adding more specialized domain concepts. One popular and effective semantic engineering solution is to create small ontology design patterns to specify conceptualizations pertaining to a particular slice of a domain [9]. The *Surface Water* ontology presented here is such a small, easily comprehensible, and generalized ontology design pattern focused on some simple concepts from the domain of surface water. It is intended to serve both as a conceptual building-block for guiding the design of more complex surface water ontologies, and also as a self-contained knowledge representation unit capturing essential surface water semantics reusable in any domain with equal validity.

The fundamental design principle used for this pattern is to explicitly separate landscape features that act as containers for water to flow or collect, from the flowing and standing bodies of water that occupy those containers. The categories modeled by this pattern are abstract enough to function as "meta" categories closely corresponding to (but not equivalent to) basic categories encountered in hydrology, a field-observation driven geoscience domain that already offers a stable system of surface water feature types, and also to categories often encountered in natural languages. The pattern's categories reflect distinctions driven by observable physical properties (e.g., shape, size, depth, flow of liquids), and thus they are compatible with Horton's primary theory [16]. That leads to the surmise that such distinctions should also be inducing recognition of similar categories, albeit with additional properties and at different conceptual granularities, in most cultures and natural languages.

The *Surface Water* pattern captures semantics that arise from the object view of the surface water domain, but cannot capture non-channelized flows directly on the surface during floods and runoff. Hence, this paper also includes a brief discussion of another pattern (developed earlier by the authors), called the *Surface Network* pattern which is based on a well-known theory for conceptualizing any surface as an abstract network of simple shape elements [4, 25]. The *Surface Network* pattern can help incorporate surface flow semantics *anywhere* on the surface, not just in channels or basins. For complete representation of surface water semantics, several other intuitive object, network and field based ontologies will also be needed.

The rest of this paper is organized as follows. Section 2 presents the methodology and conceptual motivation for designing the pattern. The conceptual foundations and all the axioms of the OWL ontology are presented in Section 3. The practical utility of the pattern for semantic querying and annotation is discussed in Section 4. Section 5 briefly discusses how the *Surface Network* pattern applies to the domain of surface water, and Section 6 wraps up the paper with some general conclusions.

2   Pattern Design

2.1   Methodology

The *Surface Water* ontology pattern is supposed to function as a core surface water domain ontology (complemented by the *Surface Network* pattern), that is sufficiently abstract to be applied to more specific geospatial ontology applications. As mentioned above, ontology design patterns are small ontologies capturing essential, reusable qualities of a theme, and acting as building blocks in more complex ontologies. They reduce duplicated work and the core elements of the pattern facilitate data integration since they are designed to remain consistent when reused within different applications [9]. A key requirement for pattern design is that both domain and ontology engineering knowledge experts need to understand each other's perspectives. An increasing number of patterns are being designed at Geo-Vocabulary Camps (GeoVoCamps), which are a bottom-up, participatory approach to pattern design, achieved through 2-3 day working sessions of domain experts and ontology engineers to discuss and implement patterns for the geospatial domain. Philosophically motivated debates, and extensive discussions about the practical scope of the pattern and which domain entities and properties should be selected, characterize the GeoVoCamp workshops. Semantic engineering principles and implementation method determine the final form of the pattern which is generally available online and sometimes also documented as research publications [3, 17, 30].

The *Surface Water* pattern was developed at GeoVoCampDC2013[9] by the authors of this paper, most of whom also worked together to develop the *Surface Network* pattern at an earlier GeoVoCampSOCoP2012[10]. Both workshops were organized by members of Spatial Ontology Community of Practice (SOCoP).[11] There is no single authoritative resource that can be cited for the *Surface Water* pattern. As is the case

---

[9] GeoVocampDC2013: http://vocamp.org/wiki/GeoVoCampDC2013
[10] GeoVocampSOCoP2012: http://vocamp.org/wiki/GeoVoCampSOCoP2012
[11] Spatial Ontology Community of Practice (SOCoP): http://socop.org

for most patterns, the insights came from the collective research and practice of the authors. A wide variety of resources on surface water concepts was also known to them and considered more than sufficient as background knowledge for making decisions about both surface water domain and pattern design issues. The following sources of knowledge informed the *Surface Water* pattern design: natural language texts, multilingual dictionaries, encyclopedias, geospatial data standards (GNIS, SDTS), geoscientific reference texts, lexical databases (NGA GNS, WordNet), geoscience ontologies (SWEET, EnvO) and prior geographic and formal ontology research on scientific, legal, and folk concepts of surface water or closely related concepts [1-2, 5, 7-8, 12-13, 20-22, 24, 28-34].

2.2   Conceptual Background

A pattern needs to be generic enough to find recurring use in diverse contexts [9]. A well-established method for designing and motivating patterns is identification of a set of competency questions that refine the general use case and illustrate the types of semantic queries that can be addressed by implementing the pattern in more domain-specific contexts [11]. Some typical questions that best illustrate the generality and scope of the *Surface Water* pattern in a wide variety of contexts are listed below.

Q1. "Find all standing water bodies that are completely located in region X."
Q2. "Find all direct tributaries flowing into river X."
Q3. "Find all types of streams that originate from and also terminate in a basin."
Q4. "Find all valleys draining into a lake X."
Q5. "Find all streams which drain into lakes that do not fill their basins."

These queries can be relevant in a wide range of domains such as topographic mapping and querying, hydrological analysis, digital terrain analysis, pollution transport modeling, navigation, habitat analysis, natural resource conservation, disaster planning etc.. For example, a water body is abstract enough to resolve to different entity types (lakes, ponds, reservoirs) in different contexts, including different geographic scales (Q1). Tributaries of a river could be queried for determining navigation, or tracing pollution pathways, or to assess stream volumes (Q2). Streams, rivers, creeks, runs and many other flowing water features can be all treated as specializations of a single abstract type of channelized flow, and yet be distinguished from each other when needed in different contexts (Q3). Similarly, hydrographic, terrain, and other databases can be integrated and queried collectively by creating ontologies that explicitly capture the physical relationship between the land surface, (concave) land forms, and surface water (Q3-5). These types of competency questions helped identify the essential classes and properties of the *Surface Water* pattern.

3   Formalization of the Surface Water Pattern

3.1   General Principles

The following conceptualization underlies the *Surface Water* pattern: *There are locations or regions on the surface of the earth that host concave landforms, many of which interconnect, and act as containers for water to collect and/or flow through in dominant amounts under the influence of gravity.* The pattern, therefore, distinguishes between the terrain feature that acts as a container and the body of water, both considered to be overlapping in the same space. This is the most fundamental idea recognized in this pattern. The pattern is, thus, divided into two conceptual parts or modules: *Dry* and *Wet* to capture the two types of semantics separately, and to allow focused specializations in the future. The *Dry* module captures the semantics of concave landscape features (channel, depression and interface), which can exist regardless of the presence of surface water, but do act as containers for sustained water flow and storage. The *Wet* module is dependent on and reuses the *Dry* module features to capture the semantics of hydro features (stream segment, water body, and fluence) that occupy the features whose semantics are defined in the *Dry* module. Note that there are several types of snow and ice formations that may not be properly addressed by this pattern, which has been designed for the typical cases of liquid surface water features and contained in channels and depressions. The classes and properties within the *Dry* and *Wet* modules are formally encoded using Web Ontology Language (OWL), and available online.[12, 13] All semantics of the modules and how they interrelate is also captured schematically in Fig. 1.

Some general issues related to pattern design and how they are discussed in this paper need to be clarified as well. First and foremost, Description Logic (DL) notation is used for presenting axioms in this paper since it is much more compact than OWL. Names of properties are simplified to not begin with "has", but they should be easy to track because class names begin with capital letters, and property names begin with small letters. Global domain and range declarations over properties are not used because that is known to reduce interoperability—all domain-range declarations for properties are defined only in the context of specific classes. All classes of the pattern are declared to be pairwise disjoint because they do not cover overlapping categories. Disjointness declaration is the recommended practice in OWL for improving inference about domain concepts. The DL axioms for disjointness are not presented below for lack of enough space. Property axioms are also not included for space constraints, but their intended purpose should be evident from Fig. 1, and the discussion of the axioms below.

---

[12] *Dry* module URI:
http://purl.org/geovocamp/ontology/SurfaceWater_Dry
[13] *Wet* module URI:
http://purl.org/geovocamp/ontology/SurfaceWater_Wet

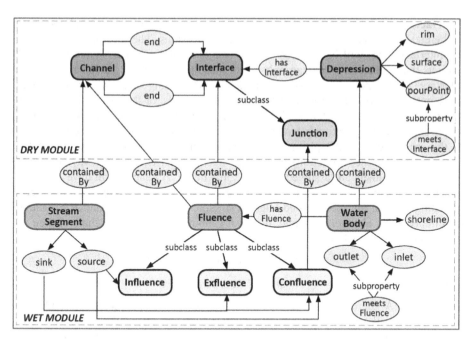

Fig. 1. Surface *Water* pattern's *Dry* and *Wet* module classes (brown/blue) and properties (grey)

3.2   Dry Module Semantics

The *Dry* module has two primary classes, *Channel* and *Depression*, while a third supporting class *Interface* formalizes concepts of spatial connections between surface water features in the terrain. These classes represent three-dimensional terrain features, where water occurs dominantly and their purpose is to be a foundation for specifying the semantics of classes of the *Wet* module.

Channel. The *Channel* class captures the semantics of a linear conduit with two ends, which is located on or is a natural part of the earth's surface, and as a consequence of its shape, it acts as a container where water can collect and flow in dominant amounts between the two ends of the conduit. Specifying that a channel has exactly two ends, each of which is formally represented through the *interface* class is considered sufficient to support flow semantics (specified in the *Wet* module). Axiom A1 encodes this logic. There are some properties (*lowerEnd, upperEnd, bed,* and *bank*) that are only included in the pattern to support future specializations of the *Channel* class, but not used in this pattern to maintain its generality. For example, the distinction between upper and lower ends of a channel is not made since it would preclude channels where flows reverse temporarily. Similarly, only well-defined channels may be deemed to have a bed and bank, but not many minor channels transporting thin rivulets of water.

$$Channel \sqsubseteq (\leq 2end.Interface \sqcap \geq 2end.Interface) \tag{A1}$$

Interface. An *interface* is a conceptual abstraction to represent "transition" locations at the end of channels or on the boundary of depressions. The most common use of the interface class will be to represent the two ends of a channel. While axiom A1 restricts that a channel's end can only be an interface, axiom A2 further restricts the interface to be the end of only a channel, and nothing else. An interface that represents the physical merger or bifurcation of channels, or the merger of a channel and depression, is a *Junction* (A3). When an interface represents the end of just a single channel, or a junction between a channel and a depression, it can be thought of as the cross sectional 2D planar area through which flow would enter or exit the channel. An interface (junction) involving three channel ends will always be a volume, since channels are assumed to have some depth. A channel can have only two interfaces, at each of its two ends, but a depression can have any number of interfaces, including none at all, as specified later (A7). Note that axioms A1 and A3 together also imply that a channel is atomic, in that it cannot contain another channel (i.e., no sub-channels are possible).

$$Interface \sqsubseteq \forall endOf.Channel \qquad (A2)$$

$$Junction \sqsubseteq Interface \qquad (A3)$$

Depression. A depression is a concavity in the earth's surface that is surrounded by higher ground all around and which can contain water by virtue of its shape and material surface. Depressions can be as large as ocean basins or as small as holes found in a channel bed. A depression is defined as being spatially enclosed by or having an upper bound marked by its rim (A4). The rim is the highest elevation line (a contour) that encloses the depression. Functionally, the rim denotes the level below which water can stay contained in the depression without overflowing. Formal specification of this definition of rim would require specification of multiple mathematical and spatial concepts, and is beyond the scope of this simple pattern. A depression also has the property of having exactly one pour point (A5), which marks the lowest location from where water would exit naturally if the depression was maximally filled, up to the level of its rim. The pour point is at the same elevation as and touches the rim (not formalized in OWL). A depression must also have a surface so that it can support a water body (A6). This means that this pattern allows only those individual depressions, whose surfaces allow containment of water bodies to be members of this class. Finally, the *meetsInterface* property is used to specify that a depression is connected to other channels, the outside region, or in rare cases, to another depression, through only *interfaces* (A7). As illustrated in Fig. 1, the *pourPoint* property is also declared as a subproperty of *meetsInterface*, since it must connect the depression to an interface which will then connect to other features or the region outside of the depression. The pour point is always on the rim, but interfaces with some channels which bring inflow may be on the rim or below it inside the depression (not formalized in OWL).

$$Depression \sqsubseteq (\leq 1\,rim \sqcap \geq 1\,rim) \qquad (A4)$$

$$Depression \sqsubseteq (\leq 1\,pourPoint \sqcap \geq 1\,pourPoint) \qquad (A5)$$

$$Depression \sqsubseteq (\leq 1\,surface \sqcap \geq 1\,surface) \qquad (A6)$$

$$Depression \sqsubseteq \forall meetsInterface.Interface \qquad (A7)$$

3.3   Wet Module Semantics

The *Wet* module reuses the classes of the *Dry* module to specify the semantics of surface water flow and collection in channels and depressions, respectively. There are three classes in the *Wet* module: *fluence, stream segment,* and *water body.* Instances of the latter are always contained by the channel and depression, respectively, while fluences may be contained in interfaces, junctions or channels.

Fluence. A fluence is the transitional water entering or leaving a stream segment or a water body. For stream segments within channels, the fluence can be either within a channel interface if flow starts or ends at the channel end, or outside the interface and within the channel, if flow starts or ends not at the end, but within a channel somewhere. For a water body, the fluence can either within be the interface to a channel or inside the depression containing the water body. The fluence class is further specialized through three (pairwise disjoint) subclasses: *influence, exfluence* and *confluence* (A8-A10) to capture all the ways flow can start or end for a stream segment or enter or exit from a water body. The influence is the source, the exfluence, the sink, and the confluence can be both the source and sink of (different) stream segments or a stream segment and a water body. The confluence is a type of fluence signifying water merging or transitioning from one stream segment into another stream segment or water body, or from a water body to a stream segment. It is always contained within the junction of channels, since stream segments must exit a channel to meet other stream segments (A11). If the stream flows through the entire channel, then the influence and exfluence are contained in the interfaces at the end of the channel, otherwise they are contained within the channel somewhere (A12-A13). Note that axioms A8-A13 do not specifically preclude the *fluence* classes from being re-used to cover non-channelized flow semantics. However, this pattern was designed to focus only on channelized flows, and broader than intended interpretation of the semantics is not recommended.

$$Influence \sqsubseteq Fluence \tag{A8}$$

$$Exfluence \sqsubseteq Fluence \tag{A9}$$

$$Confluence \sqsubseteq Fluence \tag{A10}$$

$$Confluence \sqsubseteq (\leq 1\,containedBy.Junction \sqcap \geq 1\,containedBy.Junction) \tag{A11}$$

$$Influence \sqsubseteq (\leq 1\,containedBy.Interface \sqcap \geq 1\,containedBy.Interface) \sqcup \\ (\leq 1\,containedBy.Channel \sqcap \geq 1\,containedBy.Channel) \tag{A12}$$

$$Exfluence \sqsubseteq (\leq 1\,containedBy.\,Interface \sqcap \geq 1\,containedBy.Interface) \sqcup \\ (\leq 1\,containedBy.Channel \sqcap \geq 1\,containedByChannel) \tag{A13}$$

Stream Segment. A stream segment is contained (and flows) within the channel (A14). Every stream segment has two flow related properties: a *source* and a *sink* (A15-A16), which mark the inflow and outflow ends of a stream segment. Flow as a process is too complex to be explicitly formalized in OWL. Instead, it is implied indirectly as directed from the source to the sink. Stream segments can only meet at a confluence. Note that multiple sources and sinks of merging or diverging streams are all 'resolved' to the *same* physical confluence within the junction that contains it. The use of source and sink properties prevents the unrealistic situation of multiple coincident fluence and interface entities, when stream segments meet. If a stream segment's source receives water from another stream segment or water body, a source will be a confluence, or if the source is neither a stream segment or water body it will be an influence, but never an exfluence (A17). Similarly, a stream segment's sink can never be an influence, and only either a confluence (if the stream segment loses water to another stream segment or water body) or exfluence, otherwise. The axioms do not preclude the possibility of more than one stream segment contained in a channel, if a stream segment does not flow end to end, but it should be a rare possibility, if at all.

$$StreamSegment \sqsubseteq (\leq 1\, containedBy.Channel \; \sqcap \geq 1\, containedBy.Channel) \qquad (A14)$$

$$StreamSegment \sqsubseteq (\leq 1\, source \; \sqcap \geq 1\, source) \qquad (A15)$$

$$StreamSegment \sqsubseteq (\leq 1\, sink \; \sqcap \geq 1\, sink) \qquad (A16)$$

$$StreamSegment \sqsubseteq (\leq 1\, source.(Influence \sqcup Confluence) \; \sqcap \\ \geq 1\, source.(Influence \sqcup Confluence)) \qquad (A17)$$

$$StreamSegment \sqsubseteq (\leq 1\, sink.(Exfluence \sqcup Confluence) \; \sqcap \\ \geq 1\, sink.(Exfluence \sqcup Confluence)) \qquad (A18)$$

Water Body. This class represents a standing collection of water contained within a depression (A19). The water is contained due to the (impermeable) depression surface and between the depression's surface anywhere up to the rim of the depression. Every water body, therefore, has a shoreline (A20). The shoreline achieves the highest level of the rim only when the depression is full (e.g. a full lake basin), otherwise the shoreline is at a lower level (typical in arid areas). This relationship between the shoreline and rim is not specified explicitly in OWL because it would need more complex axioms and incorporation of too many extra mathematical and spatial properties and entities. A water body meets stream segments that flow into or out of the water body. The *meetsFluence* property specifies that a water body connects to other bodies of water only through fluences (A21). If the water body fills a depression completely, it has an *outlet*, otherwise not (A22). The outlet can be either a confluence or exfluence (A23). It will be considered a confluence if the water body flows out to form a stream segment contained in a channel. Otherwise, if the water body loses water through the outlet in such a way that no sustained stream segment and channel are found connecting to the interface representing the pour point, then the outlet is considered to be an exfluence. A water body may have any number of *inlets*, including none at all, depending on how many stream segments or other discrete sources (e.g., underground springs) introduce net inflow into the water body. If an *inlet* exists, then it is a confluence if the inflow is from a stream segment in a channel, or an influence if its flow is not confined in a channel (A24). As shown in Fig. 1, *outlet* and *inlet* are also subproperties of *meetsFluence* because the inlet and outlet of a water body will always be represented by a fluence.

$$WaterBody \sqsubseteq (\leq\!1\, containedBy.Depression \sqcap \geq\!1\, containedBy.Depression) \quad (A19)$$

$$WaterBody \sqsubseteq (\leq\!1\, shoreline \sqcap \geq\!1\, shoreline) \quad (A20)$$

$$WaterBody \sqsubseteq \forall meetsFluence.Fluence \quad (A21)$$

$$WaterBody \sqsubseteq \leq\!1\, outlet \quad (A22)$$

$$WaterBody \sqsubseteq \forall outlet.(Exfluence \sqcup Confluence) \quad (A23)$$

$$WaterBody \sqsubseteq \forall inlet.(Influence \sqcup Confluence) \quad (A24)$$

3.4 Discussion

The *Surface Water* pattern was designed as a minimalist ontology modeling, and should not be expected to address all possible surface water types and cases. The designed classes cover a limited set of categories that are likely to be widely, if not universally, shared by most people. These categories are only supposed to serve as basic building blocks, similar to foundation ontology categories, for more specialized and context dependent categories. For example, semantics of braided streams may require a more complex channel type to be introduced. Wetlands (e.g., marsh, swamp, fen) forming over permeable lands and/or not contained in depressions may be modeled as another unrelated pattern, or as an extension to this pattern (e.g., as a special type of water body that supports substantial vegetation and/or is characterized by certain soil types). This pattern is also not designed to support semantics of processes that lead to changes in the physical properties of surface water features due to action of water contained inside. However, the pattern's classes should be able to describe the instances of surface water features at different times or stages of evolution, which will allow sharing information about the spatiotemporal behavior (albeit only in terms of snapshot states) of specific features and/or geographic areas where the features are located.

4 Applications of the Surface Water Ontology Pattern

4.1 Aligning Geo-Databases and Annotating Mapped Features

The *Surface Water* pattern has theoretical value as an encapsulation of fundamental domain categories and properties. The pattern is also an ontology for the Semantic Web, and a practical guide for making hydro-GIS datasets interoperable at a generalized level. Surface water features get represented in different ways in geospatial databases depending on intended use. A primary use of this pattern would be to make databases interoperable by interpreting the features as instances of the basic categories of the *Surface Water* pattern. As an example, Fig. 2 shows mapped representations of real world surface water features for a small study area. The left diagram in Fig.2 maps instances of Dry module classes, and the right diagram maps instances of the Wet module classes for the same study area. In this example, all depressions contain water bodies, and all junctions contain confluences. However, not all channels contain stream segments, and not all interfaces contain fluences. All fluences are confluences, except one which is an influence that can be inferred by visual comparison of the left and right diagrams to be located not within the interface at the channel's end, but within the channel itself. This implies that the stream segment does not fully traverse the channel end to end, but instead has a source starting somewhere downstream from the upper (flow) end of the channel. This example, thus, clearly underscores the importance of distinguishing between spatially overlapping instances of different surface water feature categories.

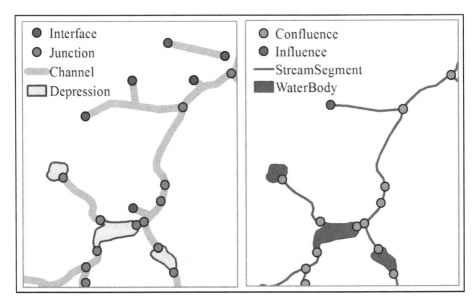

Fig. 2. An example of how surface water features can be described and mapped as instances of classes defined in the *Dry* (left) and *Wet* (right) modules of the *Surface Water* pattern

4.2   Querying Geo-Databases

A quick test of practicability is to check if the pattern is useful in answering the competency questions listed in Section 2. The queries listed below confirm and show how the pattern can be used to construct domain queries using terms (italicized below) corresponding to the pattern's categories and their properties. The returned features can be references to instances of surface water features shown above in Fig. 2.

Q1. "Find all *water bodies* which are *containedBy depressions* whose *rims* are completely contained within the spatial extent of region X."

Q2. "Find all *stream segments* whose *sinks* are *confluences* with any *stream segment* with name X."

Q3. "Find all stream segments whose *sources* and *sinks* are *confluences containedBy junctions* that are *interfaces* with a *depression*."

Q4. "Find all *channels* with *junctions* which *contains confluences* with the *water body* named X."

Q5. "Find all *stream segments* having *confluences* with *water bodies* that do not have an *outlet*."

As can be seen, semantics related to *both* bodies of water and their containing landforms are needed to correctly frame the queries. To make these sorts of queries possible, especially on the Semantic Web, we are also developing software to convert hydro-GIS datasets into RDF triples (standard graph data model for the Semantic Web), which can then be queried using the GeoSPARQL[14] semantic query language. Fortunately, the GeoSPARQL ontology also offers built-in support for geospatial data querying, as will be needed for most surface water datasets. There are, obviously, many other questions of interest for the domain that are not answerable with this simple pattern, and many complex geospatial queries that cannot be handled by GeoSPARQL. Also, data on "dry" stream channels and depressions are not as common as datasets that map only streams and water bodies. Establishing topological connections to find junctions and fluences, and distinguishing the types of fluences based on flow direction is only possible with advanced geospatial data models.

5   Integration with Other Ontologies

Several sources informed the design of the *Surface Water* pattern. The formalization of voids in hydrogeology [12] was identified as a mid-level foundational ontology before subsequently designing the *Surface Water* pattern, especially because it formalizes the container schema, and is itself aligned with the DOLCE foundational ontology [23]. However, it presented some immediate problems due to formalization in Common Logic and a focus on hydrogeology. These are not insurmountable issues, but still prevented a quick alignment process. Additionally, the immediate goal of this work is to integrate the *Surface Water* and *Surface Network* patterns since the latter addresses surface flow semantics not encoded in the former. The *Surface Network* pattern is unlikely to integrate well with the ontology of hydrogeological voids [12], which is another reason for not yet investing in alignment with the ontology of voids. In the following paragraphs the discussion is focused on the *Surface Network* pattern's utility as an abstract and reusable surface water domain ontology.

A surface network mathematically describes the global spatial shape of a (twice differentiable, smooth) surface in terms of a topological network of critical points (peaks, saddle points, and pits) and lines (ridge, course, slope, and contour lines), which together provide a generalized representation of the global surface shape [4, 25, 27]. The theory also includes two types of areal districts, hills and dales, which are bound by course lines and ridge lines, respectively, and exhaustively partition the surface into morphological parts, independent of each other. Three other areal feature types can be recognized, although only implicitly referred to in the literature: territory (area of overlap between exactly one hill and one dale), hilltop (enclosing area around

---

[14] GeoSPARQL: http://www.opengeospatial.org/standards/geosparql

a peak down to the contour of the highest saddle point connected to the peak via a ridge line), and basin (enclosing area around a pit up to the contour of the lowest saddle point connected to the pit via a course line). Surface networks offer a discrete, feature based abstraction of surfaces, which themselves lack searchable objects. This inspired the design of the *Surface Network* and the *Geospatial Surface Network* patterns, the former for topological abstractions of the surface, and the latter extending the former with support for metric geospatial surfaces. The two patterns are available as OWL ontologies[15, 16] and detailed in another manuscript [30].

These patterns are relevant to surface water semantics and are mentioned here because they formalize surface shape semantics. Intuitively, the earth's surface already provides the potential for surface water collection and flow. Water flows along channels under the influence of gravity, collects and flows along lines of steepest descent, often forming well-developed stream channels, and at other times just downhill anywhere on the surface, as during a rainfall event or a flood. Areas that drain together to a pit or basin form drainage basins (i.e., watersheds), and are bound and demarcated by drainage divides (i.e., ridges). Water naturally flows toward the lowest points available in a drainage basin, where it starts to collect and fill basins, which are the lowest areas around the pit within the drainage basin.

Keeping the above statements in mind, it emerges that if the earth's surface is abstracted as a surface network, its shape elements will easily capture the above semantics, albeit at an abstract level. Some shape elements correspond closely to categories of the *Dry* module of the *Surface Water* pattern, and some others capture additional surface semantics. For example, a surface network basin would be equivalent to a depression, the pit will model the lowest point in a depression, the pale will correspond exactly to the pour point of a depression, and the contour passing through the pale would be the rim of the depression. Surface networks also add concepts of drainage basins (dales) and drainage divides (ridge lines) missing in the *Surface Water* pattern, and course lines which abstract locations where water flows consistently, and therefore are conceptually quite similar to channels. The *Geospatial Surface Network* can be used to conceptualize hydrological stream networks, and non-channelized overland/sheet flows of water—semantics which are missing from the *Surface Water* pattern.

The reason for designing a *Surface Water* pattern, separate from the *Surface Network* pattern is that the mathematically abstract surface network elements do not correspond perfectly with the features of the real world. Because course lines must extend strictly between saddle points to pits, only a subset of flow channels can be made to correspond to course lines. Also, course lines technically never meet, so channel junctions cannot be modeled without making some theoretical adjustments to surface network theory. Furthermore, because pits and basins are often generalized in dry land focused digital elevation models, pits, depressions and the pales and channels connected to them practically never get recognized as part of surface networks. Still,

---

there are some benefits to using the *Surface Network* pattern as outlined above, and their possible integration into a larger pattern. The authors are currently planning to address the integration and re-evaluation of both patterns in a future GeoVoCamp workshop.

6   Conclusions

Ontology patterns allow us to specify our concepts in knowledge modules and force us to extract the most essential concepts of a domain. The *Surface Water* pattern was created at a GeoVoCamp, by domain experts and knowledge engineers. The pattern follows general design principles and includes only highly generalized categories based on physically observable characteristics. Since the pattern is simple and formalized in OWL, it can be used on the Semantic Web to share surface water feature datasets. The pattern should be easily reusable in different domains to annotate and implement semantic querying of surface water feature datasets. It can also be used to implement GIS data models that would be compatible with the naïve geography approach to GIS design and querying. As discussed above, the pattern is useful for intuitive querying and mapping of terrain and hydrographic GIS datasets.

The *Surface Water* pattern is abstract to be generalizable. On the other hand, that also means that it must be combined and or extended with other topography related ontologies, and also aligned with other foundational ontologies, to help realize its true potential. In that respect, the separation of the *Dry* and *Wet* modules is well suited conceptually to link and develop specialized ontologies for terrain and surface water in tandem. The *Dry* module should be specialized to add surface water semantics pertaining to morphology (e.g., size, shape, topology) and terrain composition (e.g., based on types of rocks, soils, vegetation), while the *Wet* module should be specialized to capture categories and properties related specifically to hydrologic characteristics (e.g. flow volume, flow frequency, source, and quality). Finally, it should be noted that although this pattern is designed and discussed for surface water semantics, it should be usable for modeling basic semantics of flow possibility of other liquids (e.g., polluted plumes, lava flows) on terrain or other physical surfaces, including that of other planets.

Acknowledgments. This manuscript has been prepared in part by Oak Ridge National Laboratory, P.O. Box 2008, Oak Ridge, Tennessee 37831-6285, managed by UT-Battelle, LLC for the U.S. Department of Energy under contract no. DEAC05-00OR22725. Accordingly, the publisher, by accepting the article for publication acknowledges that the United States Government retains a non-exclusive, paid-up, irrevocable, world-wide license to publish or reproduce the published form of this manuscript, or allow others to do so, for U.S. Government purposes. Any use of trade, product, or firm names is for descriptive purposes only and does not imply endorsement by the U.S. Government. The authors are grateful to Gary-Berg Cross and Spatial Ontology Community of Practice (SOCoP) for organizing GeoVoCampDC2013, and the USGS and three anonymous reviewers for providing comments.

References

1. Bennett, B.: Application of Supervaluation Semantics to Vaguely Defined Spatial Concepts. In: Montello, D.R. (ed.) COSIT 2001. LNCS, vol. 2205, pp. 108–123. Springer, Heidelberg (2001)
2. Bromhead, H.: Ethnogeographical Categories in English and Pitjantjatjara/Yankunytjatjara. Language Sciences 33, 58–75 (2011)
3. Carral, D., Scheider, S., Janowicz, K., Vardeman, C., Krisnadhi, A.A., Hitzler, P.: An Ontology Design Pattern for Cartographic Map Scaling. In: Cimiano, P., Corcho, O., Presutti, V., Hollink, L., Rudolph, S. (eds.) ESWC 2013. LNCS, vol. 7882, pp. 76–93. Springer, Heidelberg (2013)
4. Cayley, A.: On Contour Lines and Slope Lines. London, Edinburgh, and Dublin Philosophical Magazine and Journal of Science 18, 264–268 (1859)
5. Duce, S., Janowicz, K.: Microtheories for Spatial Data Infrastructures – Accounting for Diversity of Local Conceptualizations at a Global Level. In: Fabrikant, S.I., Reichenbacher, T., van Kreveld, M., Schlieder, C. (eds.) GIScience 2010. LNCS, vol. 6292, pp. 27–41. Springer, Heidelberg (2010)
6. Egenhofer, M., Mark, D.M.: Naïve Geography. In: Kuhn, W., Frank, A.U. (eds.) COSIT 1995. LNCS, vol. 988, pp. 1–15. Springer, Heidelberg (1995)
7. Feng, C.-C., Bittner, T., Flewelling, D.M.: Modeling Surface Hydrology Concepts with Endurance and Perdurance. In: Egenhofer, M., Freksa, C., Miller, H.J. (eds.) GIScience 2004. LNCS, vol. 3234, pp. 67–80. Springer, Heidelberg (2004)
8. Feng, C.-C., Sorokine, A.: Comparing English, Mandarin, and Russian Hydrographic and Terrain Categories. International Journal of Geographical Information Science (2013) (in press), doi:http://dx.doi.org/10.1080/13658816.2013.831420
9. Gangemi, A., Presutti, V.: Towards a Pattern Science for the Semantic Web. Semantic Web 1(1-2), 61–68 (2010)
10. Gibson, J.J.: The Ecological Approach to Visual Perception, Houghton Mifflin, USA (1979)
11. Gruninger, M., Fox, M.S.: The Role of Competency Questions in Enterprise Engineering. In: Proceedings of the IFIP WG5.7, Workshop on Benchmarking – Theory and Practice, pp. 212–221 (1994)
12. Hahmann, T., Brodaric, B.: The Void in Hydro Ontology. Frontiers in Artificial Intelligence and Applications: Formal Ontology in Information Systems 239, 45–58 (2012)
13. Hart, G., Dolbear, C., Goodwin, J., Feliciter, L.: Using Structured English to Represent a Topographic Hydrology Ontology. In: OWL Experiences and Directions Workshop (2007)
14. Hayes, P.: The Naïve Physics Manifesto. In: Michie, D. (ed.) Expert Systems in the Micro-Electronic Age, pp. 242–270. Edinburgh University Press (1979)
15. Hayes, P.: Naive Physics I: Ontology of Liquids. In: Hobbs, J., Moore, R. (eds.) Formal Theories of the Commonsense World, pp. 71–108. Ablex, Norwood (1985)
16. Horton, R.: Tradition and Modernity Revisited. In: Hollis, M., Lukes, S. (eds.) Rationality and Relativism, pp. 201–260. Basil Blackwell, Oxford (1982)
17. Hu, Y., Janowicz, K., Carral, D., Scheider, S., Kuhn, W., Berg-Cross, G., Hitzler, P., Dean, M., Kolas, D.: A Geo-Ontology Design Pattern for Semantic Trajectories. In: Tenbrink, T., Stell, J., Galton, A., Wood, Z. (eds.) COSIT 2013. LNCS, vol. 8116, pp. 438–456. Springer, Heidelberg (2013)
18. Lakoff, G.: Women, Fire, and Dangerous Things: What Categories Reveal about the Mind. University of Chicago Press, Chicago (1987)

19. Lakoff, G., Johnson, M.: Philosophy in the Flesh. University of Chicago Press, Chicago (1986)
20. Mark, D.M., Smith, B.: A Science of Topography: From Qualitative Ontology to Digital Representations. In: Bishop, M.P., Shroder, J.F. (eds.) Geographic Information Science and Mountain Geomorphology, pp. 75–100. Springer-Praxis, Chichester (2004)
21. Mark, D.M., Turk, A.G., Stea, D.: Progress on Yindjibarndi Ethnophysiography. In: Winter, S., Duckham, M., Kulik, L., Kuipers, B. (eds.) COSIT 2007. LNCS, vol. 4736, pp. 1–19. Springer, Heidelberg (2007)
22. Mark, D.M., Turk, A.G.: Landscape Categories in Yindjibarndi: Ontology, Environment, and Language. In: Kuhn, W., Worboys, M.F., Timpf, S. (eds.) COSIT 2003. LNCS, vol. 2825, pp. 28–45. Springer, Heidelberg (2003)
23. Masolo, C., Borgo, S., Gangemi, A., Guarino, N., Oltramari, A.: WonderWeb Deliverable D18 Ontology Library (final) (2003), http://wonderweb.semanticweb.org/deliverables/documents/D18.pdf (last accessed May 15, 2014)
24. Masser, I.: Building European Spatial Data Infrastructures, 2nd edn. ESRI Press, Redlands (2010)
25. Maxwell, J.C.: On Hills and Dales. The London, Edinburgh and Dublin Philosophical Magazine and Journal of Science 40, 421–427 (1870)
26. Montello, D., Freundschuh, S.: Cognition of Geographic Information. In: McMaster, R., Usery, L.E. (eds.) A Research Agenda for Geographic Information Science, pp. 61–91. CRC Press (2005)
27. Rana, S.S. (ed.): Topological Data Structures: An Introduction to Geographical Information Science. John Wiley & Sons, Ltd., Chichester (2004)
28. Santos, P., Bennett, B., Sakellariou, G.: Supervaluation Semantics for an Inland Water Feature Ontology. In: Proceedings of International Joint Conference on Artificial Intelligence, IJCAI 2005, pp. 564–569 (2005)
29. Sinha, G., Mark, D.M.: Toward a Foundational Ontology of the Landscape. In: Fabrikant, S.I., Reichenbacher, T., Kreveld, M., Schlieder, C. (eds.) GIScience 2010, Extended Abstracts, Zürich, Switzerland, September 14-17 (2010)
30. Sinha, G., Kolas, D., Mark, D.M., Romero, B.E., Usery, L.E., Berg-Cross, G., Padmanabhan, A.: Surface Network Ontology Design Patterns for Linked Topographic Data. Manuscript submitted to Semantic Web for review
31. Smith, B., Mark, D.M.: Geographic Categories: An Ontological Investigation. International Journal of Geographical Information Science 15 (7), 591–612 (2001)
32. Taylor, M., Stokes, R.: Up the creek. What is Wrong with the Definition of a River in New South Wales? Environment and Planning Law Journal 22(3), 193–211 (2005)
33. Taylor, M.P., Stokes, R.: When is a River not a River? Consideration of the Legal Definition of a River for Geomorphologists Practising in New South Wales, Australia. Australian Geographer 36(2), 183–200 (2005)
34. Wellen, C.C., Sieber, R.E.: Toward an Inclusive Semantic Interoperability: The Case of Cree Hydrographic Features. International Journal of Geographical Information Science 27(1), 168–191 (2013)

An Indoor Navigation Ontology for Production Assets in a Production Environment

Johannes Scholz[1] and Stefan Schabus[2]

[1] Research Studios Austria, Studio iSPACE, Salzburg Austria
[2] Carinthia University of Applied Sciences, School of Geoinformation, Villach, Austria

Abstract. This article highlights an indoor navigation ontology for an indoor production environment. The ontology focuses on the movement of production assets in an indoor environment, to support autonomous navigation in the indoor space. Due to the fact that production environments have a different layout than ordinary indoor spaces, like buildings for office or residential use, an ontology focusing on indoor navigation looks different than ontologies in recent publications. Hence, rooms, corridors and doors to separate rooms and corridors are hardly present in an indoor production environment. Furthermore, indoor spaces for production purposes are likely to change in terms of physical layout and in terms of equipment location. The indoor navigation ontology highlighted in this paper utilizes an affordance based approach, which can be exploited for navigation purposes. A brief explanation of the routing methodology based on affordances is given in this paper, to justify the need for an indoor navigation ontology.

1 Introduction

Spatial information systems concentrate on the outdoor space, while humans and things reside indoors and outdoors. Publications show, that an average person spends approximately 90% of their time inside buildings [1]. Compared with the developments for outdoor space, indoor space applications are quite behind and recently got into focus of research and development activities. Worboys [2] highlights the ubiquitous availability of satellite technology (GPS) and aerial photography as utilities used for data collection and positioning in an outdoor space. Due to the emergence and mass market availability of location-based service applications, there is a growing demand for such applications in an indoor environment. Location-based applications in an indoor environment are intended to support people in indoor decision processes – e.g. orientation, navigation and guidance.

The context of a production environment is a special indoor space, as the indoor space is laid out in order to support the production processes best. Hence, a production indoor layout looks different than a piece of architecture constructed for office or residential use. Due to the fact that the purpose of the production indoor space is solely devoted to support efficient production processes there are few fine grained architectural entities that are distinguishable – like rooms. Hence, theory has to cope with non-standard indoor entities that are subject of this paper. Additionally, the positions

M. Duckham et al. (Eds.): GIScience 2014, LNCS 8728, pp. 204–220, 2014.
© Springer International Publishing Switzerland 2014

of equipment can be reordered which alters the layout of the indoor space. This holds especially true for the use case semiconductor industry, which forms the application context of this paper. Due to the fact that any semiconductor production is done in a cleanroom environment, there are several constraints in terms of movements. Not every production asset is allowed to go anywhere in the production line due to cleanroom restrictions, and/or certain production processes which have to be separated due to contamination risks.

In order to support production processes accordingly, there is a need to locate two distinct object classes in the indoor environment: production assets that will undergo several production steps, and production equipment that processes the assets accordingly. In a flexible production environment, like the semiconductor industry, equipment and their positions might change. Either the tool itself is replaced by a new one or the location of a piece of equipment is altered. Additionally, the "production line" is not fulfilling a conveyor belt metaphor with a fixed processing chain. The semiconductor production line is a highly flexible and complex system, due to the following reasons:

- Overall processing time (from raw wafer to electronic chip) of a single production artifact can last from several days to a couple of weeks depending on the product.
- Several hundred production steps necessary until the production is finished.
- High number of different products that require different production steps.
- Each production step can be carried out on several tools which are sometimes geographically dispersed over several production halls – also with varying processing time and quality depending on the equipment used.
- High number of production assets – in different degrees of completion – present in the indoor production line.

The overarching goal is to support the transport processes of production assets in an indoor production environment. With such an approach the current production processes can be supported and an optimized physical layout of the indoor space could be computed by conducting specific simulation runs. In this paper we focus on the navigation and autonomous movement of production assets that shall be supported by means of Geographic Information Science and Technology. Autonomous in this context refers to the ability that each production asset knows explicitly where to go next after a completed processing step. Additionally, the indoor informatics system should be resilient in terms of changes to equipment and indoor spaces. The initial goal is to understand and model the movement of production assets in an indoor production environment. In order to model the movement of production assets an ontology is created that describes indoor space, indoor movements and navigation tasks. Both – indoor space and indoor movements – are necessary in order to fully understand the movement processes possible in the indoor production environment. The ontology is based on the work of Yang and Worboys [3] and Worboys [2].

In this paper we focus on the modeling of movements of production assets in an indoor production environment in order to support autonomous navigation in the indoor space. The environment "production line", which differs from ordinary indoor spaces by the unstable behavior of the indoor entities, requires the movement ontology to look different than in current literature. In order to support autonomous routing in an indoor production environment we utilize the concept of affordances.

The remainder of the paper is structured as follows. In section two the relevant literature is presented, followed by a description of the indoor production environment. This is followed by a section elaborating on the movement behavior of production assets in an indoor production environment, which depends on the description of the indoor production space. Consecutively, we present the indoor movement ontology and extend it towards affordance based routing in an indoor environment in the subsequent section. In the last section we summarize the paper, discuss the results and future work.

2   Relevant Work

This section covers the relevant literature for the paper. First we the highlight relevant work covering indoor geography and switch to indoor geography and production line processes with spatio-temporal data mining in an indoor environment. Additionally, this section covers some literature on affordance-based ontologies for navigation purposes.

A significant number of research activities were carried out over the last decades in the context of modeling outdoor space, providing a rich set of methods high level of structuring and applications. However, indoor geography related research has attained increasing attention during the last years due to the fact that an average person spends about 90% inside a building [1, 4]. Early research works on indoor wayfinding include Raubal and Worboys [5] and Raubal [6]. The work in [6] uses an airport as example of an indoor environment and presents an agent-based indoor wayfinding simulation.

In order to model indoor spaces there exist several approaches that use topology, where the indoor space is "reduced" to a graph [5, 7, 8, 9]. Jensen et al. [10] employ a graph based model to track entities in an indoor environment by placing sensors in the indoor space. To model the 3D geometry of buildings Building Information Systems are used, which do not support navigation and routing in general [11]. Worboys [2] mentions hybrid models that include geometrical and topological features, which are well studied in literature [12, 13, 14]. Other approaches provide different levels of granularity of the indoor space. Hence, the user can rely on more details for important points on a journey which requires route generation and visualization in one application [15, 16, 17].

Production line processes represent a challenging research and application field for indoor geography. Due to the fact that any optimization of production processes is depending on allocation and sequencing of production processes. Such optimization can increase the efficiency of production processes and therefore provide an interesting option for cost savings based on an increase of performance and productivity [18, 19, 20]. An increase of productivity can also be realized by analyzing spatio-temporal data, which are generated by storing historical information on production processes. Data mining methods are appropriate to analyze spatio-temporal data accordingly [21]. In order to create maps to visually analyze such data, geovisual analytics can be employed [22]. The main advantage is that a person has the ability to recognize visual patterns [23].

In order to model indoor movement of production assets we use ontologies to formally describe the behavior. Ontologies try to determine the "various types and categories of objects and relations in all realms of being" [24]. A domain ontology describes what is in the specific domain in a general way, resulting in a formal description of the content and the behavior of a part of the physical world [6]. Davis [25] lists the elements of a domain ontology: entities, relations and the rules applied. The theory of affordances is used to model routing and navigation of production assets, as they should be able to move in an autonomous manner, requiring the detection of the best possible path with respect to given constraints. The term "affordances" is coined by Gibson [26, 27]. Affordances and ontologies have been subject to research in outdoor and indoor environments [28, 29, 30]. While Anagnostopoulos et al. [31] and Tsetsos et al. [32] develop an indoor space ontology focusing on navigation, Yang and Worboys [3] develop an ontology for indoor-outdoor space. They separate different "microworlds" by distinguishing between the upper level ontology, domain ontology and a task ontology. The navigation ontology developed in this paper inherits elements describing the indoor space in order partially integrate indoor space entities in the navigation ontology. Hence, the approach in this paper includes a task and domain ontology – indoor space – with respect to Yang and Worboys [3]. Hence, the work here can be related and integrated in the upper as well as the indoor space and task ontology published in [3].

3   Indoor Production Environment

This section describes the indoor production environment under review. As previously mentioned, the objective of this paper is the modeling of production assets in a semiconductor fabrication. Such an indoor environment has several peculiarities that distinguish it from other production environments and ordinary indoor spaces. This section is based on the work of Geng [33], Osswald et al. [34] and personal experience.

Any semiconductor fabrication has to be operated in a clean room environment that ensures a low proportion of contaminating particles – both in size and quantity. Due to the fact that clean room space is expensive to construct and maintain, clean rooms are designed to be as compact as possible for the chosen equipment to be placed inside. Hence, the space dedicated to movement (people and production assets) and storage of production assets is limited. In addition, different quality classes of clean rooms exist, that are distinguishable by air quality (particles per m^3 air). Generally, the changeover between different clean room quality classes – often adjacent – is not easily possible. While it is allowed to switch to a clean room of lower quality at any time through doors, the switch to a clean room of higher quality is only possible through special airlock. This is especially true for the process of entering a clean room environment, which is only possible via specific airlocks. Hence, any humans – i.e. operators – can only leave and enter a production line using the airlocks. Similar, production assets can only enter the clean room at a specific airlock designed for production assets and are thoroughly cleaned thereafter, in order to prevent any contamination in the main production line.

The movement of operators and production assets is additionally restricted to other quality issues. Specific production asset types are prone to contamination due to chemical processes which are a result of certain production processes. Hence, selected production assets are not allowed to enter or leave a certain area of the production line to prevent them from contamination. As the production is located on different floors there are several possibilities to switch floors. Some staircases can be used by operators carrying production assets, while others can only be used by operators. In general production assets change floors by using elevators.

The indoor space under review is highly unstable, due to constant change of market demand and, thus altered production necessities. Hence, equipment has to be relocated, removed or new equipment is brought into the production facility. These processes can result in an altered layout of the indoor space, as corridors might change according to the space needed for certain equipment. This has consequences for the navigation of production assets as the "best" paths connecting two devices are altered. Generally, the layout of the production hall differs from classical production environments and ordinary indoor environments. Office or residential buildings' indoor space can be divided into rooms and corridors that are connected by doors. In a semiconductor environment, rooms are hardly present due to the fact that the indoor space is organized in distinguishable corridors with considerable length (see Fig. 1).

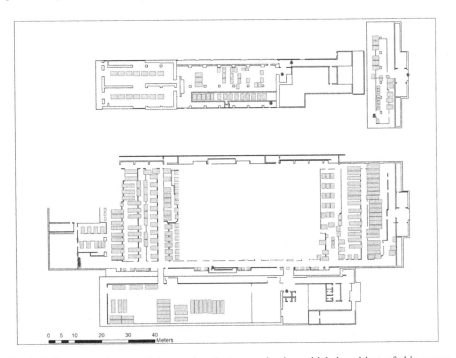

Fig. 1. Indoor space layout of the semiconductor production which is subject of this paper. Yellow rectangles represent devices in the clean room, and red dots represent transfer nodes. The white spaces are intentionally to disguise the complete production layout.

The production of microchips is a complex process chain that involves several hundred different production steps not aligned on a conveyor belt. Hence, there movement processes have a multifaceted structure due to a multitude of different microchip types having different production process chains. Additionally, each production step can possibly be done on several tools which increases the flexibility in terms of production, and increases the complexity of the movement behavior. In addition, the equipment suitable for a certain production step may be geographically dispersed. Nevertheless, each microchip type has a specific production plan that defines the process chain. Hence, each production asset in the clean room has a certain grade of completion and the next production step can easily be determined.

The indoor production line under review consists of one production hall of an Austrian semiconductor manufacturer. The layout of the indoor space is depicted in Fig. 1, showing the equipment positions as yellow and blue rectangles. In order to track production assets accordingly, an indoor tracking system called LotTrack is employed that relies on RFID and ultrasound technology. A detailed description of the system, the rationale behind the utilized technology and the application itself is found in [35].

4   Movement of Production Assets

In order to model the movement of production assets in an indoor environment, we start with a monitoring of the current in-situ "behavior" of production assets. The evaluation of trajectories collected gives insight in the behavior and helps shaping the navigation ontology accordingly. Thus, the following section elaborates on the movement behavior of production assets in the indoor environment. It is intended to show that we can model the movement of the agents using a graph, consisting of edges and nodes respectively.

The hypothesis regarding the movement is that production assets are moving along the corridors, most probably along the centerline of a corridor. Hence, the positions of production assets are compared with a graph consisting of corridor center lines and connection lines to equipment only in areas that are traversable by humans and production assets (see Fig. 2). To evaluate the spatial nearness between gathered asset positions and the graph a 1m buffer around the graph is created. In total a number of 41097 position recordings are tested (see Fig. 3) with respect to the buffer zone. In total 97.3% of the positions are inside the network buffer of 1m.

Problematic in this respect is the position of the antennas used to gather the production assets' position. The positioning antennas are placed on the ceiling with special rails and the positioning algorithm of LotTrack snaps positions to the nearest antenna rail. Hence, any tracked positions are generally shifted.

The evaluation of tracked positions of production assets as well as the layout of the indoor space – i.e. corridors – gives evidence that movements can be modeled utilizing a graph [7, 8, 9]. The graph used to model the movement of assets comprises of nodes and edges, which are described in detail in the navigation ontology in section 5.

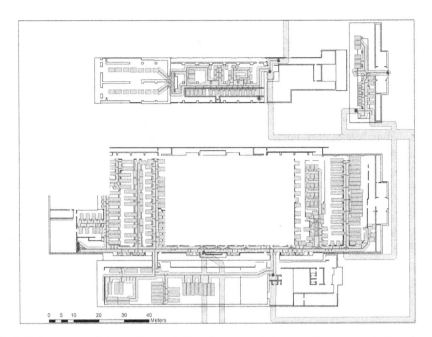

Fig. 2. Indoor space of the production hall under review. The green line represents the network that is traversable by humans and production assets, whereas the blue areas mark a 1m buffer around the network. Blue areas without green network lines are intentionally created, and represent the "virtual" connection of transfer nodes. The white spaces are intentionally to disguise the complete production layout.

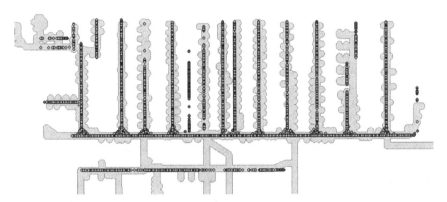

Fig. 3. Tracked production asset positions (approx. 41000) in relation to the 1m buffer around the network (marked in purple). The network positions are marked in green if they are inside the buffer and red if outside.

5   Affordance-Based Indoor Navigation Ontology for Production Environments

Creating the navigation ontology for production assets is closely related to the work of Yang and Worboys [3]. The navigation ontology developed in this paper inherits also elements describing the indoor space in order to have an integration of the navigation ontology and indoor space entities. The ontology developed here is based on affordance theory [26, 27] which can be used to establish connections between indoor and outdoor space. In addition, we employ the theory proposed by Jonietz and Timpf [36] of an affordance-based simulation framework for spatial suitability for navigation purposes.

5.1   Indoor Navigation Ontology

The indoor navigation ontology for production assets is presented in the following section. The ontology is depicted in Fig. 4 providing an overview of the model itself. The definitions of the concepts are given in this section.

Production Unit: A production unit represents the whole equipment of a production line. For example a Facility or a Device that are used during the various production steps. The subclasses are *ProductionUnit_Facility* and *ProductionUnit_Device*.

— *ProductionUnit_Device:* A device is the production unit used for the processing of goods. The device has a fixed position in the production line.
— *ProductionUnit_Facility:* The facility supports transport processes in the production line. The goods can be placed on shelves or tables if they are waiting to be processed or transported. The subclasses of a facility are *ProductionUnit_Facility_Moveable* and *ProductionUnit_Facility_Fixed*.
  • *ProductionUnit_Facility_Moveable:* A moveable facility is used to support a high stock of goods in the production line. They are e.g. bottleneck shelves used to store an extra amount of production assets. Such objects are removed if the stock in the production line is decreasing.
  • *ProductionUnit_Facility_Fixed:* Fixed facilities represent tables, shelves and other not moveable equipment in the production line.

Barrier: A barrier is limiting the transportation or movement behavior in the production line. The subclasses are *Barrier_Fixed* and *Barrier_Moveable*.

— *Barrier_Fixed:* A fixed barrier is limiting the movement behavior and cannot be changed very easily. Subclasses are *Barrier_Fixed_Wall*, *Barrier_Fixed_ProductionDevice* and *Barrier_Fixed_AirQuality,*
  • *Barrier_Fixed_Wall:* A wall is a fixed barrier. It is limiting the transport behavior within a production line.
  • *Barrier_Fixed_ProductionDevice:* The device in a production unit is linked with several infrastructure items such as electricity and gas lines and is regarded as a fixed or not easily changeable barrier.
  • *Barrier_Fixed_AirQuality:* For several production goods the air quality in a clean room is of importance and is also a barrier for the transport and movement behavior.

— Barrier_Moveable: Moveable barriers represent mainly barriers that can change over time very easily. The subclasses are *Barrier_Moveable_ProductionFacility* and *Barrier_Moveable_Contamination*.
  • *Barrier_Moveable_Contamination:* A contamination is a barrier over time. Hence, a certain production good is not allowed to enter a specific area of the production line.
  • *Barrier_Moveable_ProductionFacility:* Any production facility can impede movement as it is limiting the space for transportation. E.g. The position of shelves may easily be changed if they are not necessary anymore.

AccessNode: An AccessNode is linking outdoor and indoor space or vice versa. The subclasses are *AccessNode_Outdoor2Indoor*, *AccessNode_Indoor2Indoor* and *AccessNode_Indoor2IndoorTransfer*.

— *AccessNode_Outdoor2Indoor:* The connection from outdoor geography into the indoor environment. Therefore, the subclasses *Entrance*, *Exit* and *EntranceExit* are necessary.
  • *AccessNode_Outdoor2Indoor_Exit:* The exit is representing the way from an indoor geography back to the outdoor geography. This is necessary as there exist designated doors for leaving a production line (especially true for a production environment with clean rooms)
  • *AccessNode_Outdoor2Indoor_EntranceExit:* The EntranceExit represents both the way from outdoor geography to indoor geography and backwards.
  • *AccessNode_Outdoor2Indoor_Entrance:* The entrance enables the interaction and movement from outdoor into the indoor space.
— *AccessNode_Indoor2IndoorTransfer:* The transfer indoor is representing the connection in the same indoor space, thus connecting e.g. different floors.
  • *AccessNode_Indoor2IndoorTransfer_Elevator:* The transfer of production assets with an elevator in order to change the floor level.
    ○ *AccessNode_Indoor2IndoorTransfer_Elevator_TimeDependend:* The time dependence of an elevator is used in order to integrate the average waiting time until an elevator is available, due to the fact that elevators are mostly not available instantaneously.
  • *AccessNode_Indoor2IndoorTransfer_Stair:* A stair enables the transfer between different floors in an indoor space.
    ○ *AccessNode_Indoor2IndoorTransfer_Stair_NonRestricted:* Traversing a stair is allowed for all production asset types.
    ○ *AccessNode_Indoor2IndoorTransfer_Stair_Restricted:* The traversal of a stair is not allowed for certain production asset types.
— *AccessNode_Indoor2Indoor:* This class represents the transfer between different indoor spaces – e.g. different production halls.
  • *AccessNode_Indoor2Indoor_QualityCheckpoint:* A quality check such as an e.g. air quality check with an airlock.
  • *AccessNode_Indoor2Indoor_SecurityCheckpoint:* The entrance to certain areas can be restricted.

Corridor: A corridor is describing and including the ways where an operator – i.e. human being – can walk and transport the production goods in the production line. The subclasses are *Corridor_Node*, *Corridor_Passage* and *Corridor_Entrance*.

— *Corridor_Node:* Corridor nodes include the starting point, end point or interaction point of a navigation process.
  • *Corridor_Node_ProductionFacility:* A start point, end point or interaction point can be a production facility. For example a good has to be brought to a shelf because something has to be controlled.
  • *Corridor_Node_ProductionDevice:* A production device is mainly a start or end point for the transportation or navigation as the production goods are processed here.
— *Corridor_Passage:* The passage itself is representing the way between two consecutive navigation tasks.
  • *Corridor_Passage_Edge:* An edge is used between the different nodes and is combined to a passage along the corridor.
— *Corridor_Entrance:* Corridors need entrance points to the network for navigation and transportation in the production line.
  • *Corridor_Entrance_AccessNode:* The access node is one opportunity where operators or production assets are accessing the transportation network.
  • *Corridor_Entrance_Node:* Entrance nodes can also be production devices or facilities.

Navigation_Event: Any navigation task is described through the classes *Navigation_End*, *Navigation_Start* and *Navigation_Turn*.

— *Navigation_End:* This class represents the destination of a transportation or navigation task.
  • *Navigation_End_AccessNode:* An access node is the destination node of the navigation process if e.g. a production asset leaves the production line.
  • *Navigation_End_ProductionUnit:* The transportation between devices or facilities implies that a production facility or device is the end of the navigation task.
— *Navigation_Start:* The navigation start is representing the start of a navigation task, which can either be an *AccessNode* or a *ProductionUnit*.
  • *Navigation_Start_AccessNode:* An access node is the start of the navigation if a production asset is entering the production line.
  • *Navigation_Start_ProductionUnit:* The production unit is a starting point for the navigation.
— *Navigation_Turn:* During the navigation a production asset can perform several actions. These actions are the subclasses *Navigation_Turn_Right*, *Navigation_Turn_Left*, *Navigation_Turn_Backward* and *Navigation_Turn_Forward*.
  • *Navigation_Turn_Right:* The production asset turns right.
  • *Navigation_Turn_Left:* Represents a turn to the left.
  • *Navigation_Turn_Backward:* This event is a turn backward or represents backwards moving.
  • *Navigation_Turn_Forward:* This is a move forward.

Navigation_Agent: The agent that is navigating through the indoor space.

— *Production_Asset:* This class represents the navigation agent, and encompasses various types of production assets with different properties that have an influence on the suitability of a certain route and the choice of a certain route.

Navigation_Structure: This class contains generic entities that are necessary for route calculation proposes. A sequence of instances of the subclasses *Navigation_Node* and *Navigation_Edge* on which an agent moves defines a *Navigation_Path*. The objects of the class *Navigation_Structure* are help to specify the indoor space entities in terms of representation in a graph with nodes and edges.

5.2   Affordance-Based Routing

The navigation of production assets is based on affordances offered by the objects in indoor space with an approach similar to [36]. Affordances, initially coined Gibson [26, 27], describes a concept where an object offers its meaning. Gibson [27] further specifies the concept, that an affordance is not only defined by attributes of an object, but also by the abilities and properties of the interacting object [36]. In this context this approach is applied to the relations of machines and production assets with respect to their properties respectively.

For the case of production assets, several types of assets with specific properties exist that have to be respected when navigating. In addition, in order to define a navigation task the determination of a destination point – i.e. equipment offering a certain production process – and the selection of an appropriate path has to be carried out. This section gives only a rough overview of the algorithm in order to give an impression on the usage of the indoor navigation ontology.

In order to facilitate autonomous navigation of production assets in a semiconductor production environment each instance of the class *ProductionAsset* has certain characteristics:

— Product type: The product type reveals information on possible means of transport (e.g. thin wafer shall be carefully handled [i.e. only elevator, no stairs], 300mm wafers can withstand a low quality clean room due to a specialized plastic enclosure, with 300mm wafers it is not possible to open doors due to the weight of wafers including the plastic enclosure). In addition, the product type reveals information on barriers (quality, contamination) applicable that impede movement.
— List of production processes: This holds information on the sequence of production processes that have to be carried out. Due to the fact that certain processes can be done on several machines, with different processing results in terms of quality, each production asset has to select the piece of equipment that fulfills the requirements "best".

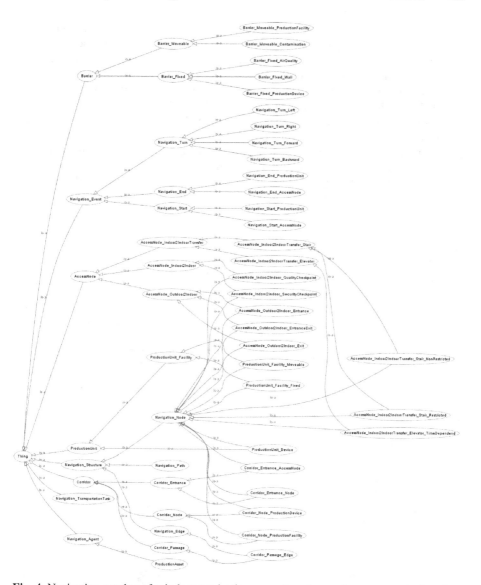

Fig. 4. Navigation ontology for indoor production space focusing on the movement of production assets

To support navigation processes in an indoor production space we apply the framework laid out in Fig. 5, which shares similarities with the approach of Jonietz and Timpf [36]. The methodology comprises of the collection of actions of a single production asset – e.g. move to the next production step "cleaning" starting from equipment "etcher_12". In order to determine the sub-actions contained in an action, the framework starts to analyze the destination production step of the action and moves towards the start point until the starting point is reached. For the action 'move to the next production step "cleaning" starting from equipment "etcher_12"' the approach starts to find indoor entities offering the production step "cleaning". If there is one piece of equipment affording the process of "cleaning" the algorithm analyzes the properties of the cleaning equipment, the start equipment "etcher_12" and the production asset. This results in differences in terms of indoor location – e.g. equipment located on different floors – and/or additional properties that have to be respected – e.g. thin wafers, where no stairs are allowed. Based on the differences and properties of indoor space entities and production assets the sub-actions are determined, starting from the destination equipment towards the start node. Based on the sub-actions found, the algorithm determines the nodes offering the required movement processes. E.g. a sub-action 'change from floor 1 to floor 2 with an elevator' searches for a node offering a connecting floor 1 and 2 by an elevator. This process finally results in a set of candidate nodes that are the basis for the navigation of the production assets.

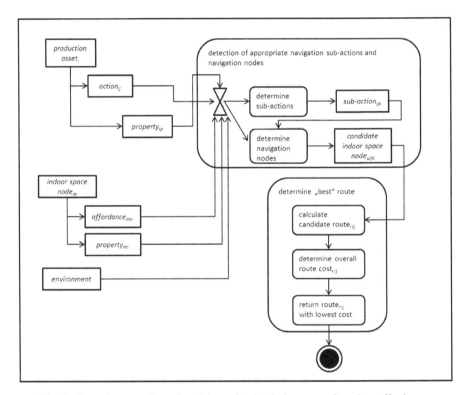

Fig. 5. General approach employed in navigating indoor space based on affordances

Based on the set of candidate nodes a routing algorithm calculates the "best" route which will be traversed by the production asset. First, candidate routes from start node to target node are determined and evaluated regarding overall route cost. Costs in this respect could be time, overall path length, or any other metric applied. Finally, the route with the lowest cost is returned.

Fig. 6 shows an application prototype for affordance based routing in the indoor production environment. There a production asset starts at an entrance node – labeled with 1 – and has 5 actions to perform, i.e. navigate to five devices in a certain order, where equipment 6 is located on a different floor. In addition, the production asset requires to be moved with care, thus the transition between the floors must be done with an elevator.

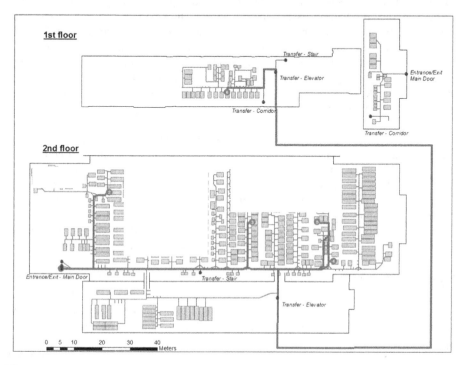

Fig. 6. Prototype application for affordance based routing in an indoor production environment. The red lines represent the traversable graph, and the green lines the route for the production asset. Five actions starting from the main entrance exist that have to be carried out, which are labeled with numbers in ascending order (start node is labeled with 1, final end node is labeled with 6). Of interest is the mandatory transfer from floor 2 to floor 1 by elevator. The white spaces are intentionally to disguise the complete production layout.

6   Conclusion and Discussion

The article elaborates on an ontology for indoor navigation in a production environment – semiconductor manufacturing. The agents moving in the indoor space are production assets that undergo several production processes, which are not aligned sequentially on a conveyor belt. Hence, any production assets should autonomously navigate from one production step to the next with respect to properties of the production asset and the indoor environment. The ontology describing indoor navigation processes is affordance based and includes a description of the indoor space. Based on the results an affordance based routing methodology is outlined and applied in a pro-totypical application.

The indoor ontology of a production indoor space looks different than current approaches [3] because the indoor space of production environments has different entities than ordinary indoor spaces. Ordinary indoor spaces comprise of rooms, corridors, doors, etc. while the production environment in semiconductor operates in a cleanroom and consists of mainly corridors without e.g. doors or distinct rooms. Due to the fact that production assets should be able to navigate between production equipment, machinery present in the indoor space, barriers (fixed and temporary) impeding movement, and any transfer between different floors are part of the ontology. In addition, the traversable space is modeled as graph that connects elements present in the indoor space. For navigation purposes an affordance based approach is proposed, that identifies required actions and detects nodes that afford the requirements, i.e. transfer from floor 1 to floor 2.

Future research directions include connections between indoor and outdoor space – already mentioned in [3]. In addition, the navigation and movement patterns in an indoor production environment are subject to further research that can be used to evaluate the navigation ontology. To do so we intend to use the concept of Self-organizing Maps [38, 39] and spatio-temporal data mining methods for trajectory pattern mining. Furthermore, we plan to use SOM and analysis of the geographic and attribute space applying the TRI-space approach [37]. In order to focus on the affordance-based routing approach presented in this paper a study highlighting general results of affordance-based routing in comparison to contemporary routing methods.

Acknowledgments. The work has been performed in the project EPPL, co-funded by grants from Austria, Germany, The Netherlands, Italy, France, Portugal- ENIAC member States and the ENIAC Joint Undertaking. Furthermore, the authors are kindly supported by Karl-Heinrich Anders and Gernot Paulus from Carinthia University of Applied Sciences.

References

1. Jenkins, P., Phillips, T., Mulberg, E., Hui, S.: Activity patterns of Californians: Use of and proximity to indoor pollutant sources. Atmospheric Environment — Part A General Topics 26A(12), 2141–2148 (1992)
2. Worboys, M.: Modeling indoor space. In: Proceedings of the 3rd ACM SIGSPATIAL International Workshop on Indoor Spatial Awareness, pp. 1–6. ACM (2011)
3. Yang, L., Worboys, M.: A navigation ontology for outdoor-indoor space (work-in-progress). In: Proceedings of the 3rd ACM SIGSPATIAL International Workshop on Indoor Spatial Awareness, pp. 31–34. ACM (2011)
4. Klepeis, N., Nelson, W., Ott, W., Robinson, J., Tsang, A., Switzer, P., Behar, J., Hern, S., Engelmann, W.: The National Human Activity Pattern Survey (NHAPS): A resource for assessing exposure to environmental pollutants. Journal of Exposure Analysis and Environmental Epidemiology 11(3), 231–252 (2001)
5. Raubal, M., Worboys, M.: A formal model of the process of wayfinding in built environments. In: Freksa, C., Mark, D.M. (eds.) COSIT 1999. LNCS, vol. 1661, pp. 381–399. Springer, Heidelberg (1999)

6. Raubal, M.: Ontology and epistemology for agent-based wayfinding simulation. International Journal of Geographical Information Science 15(7), 653–665 (2001)
7. Goetz, M.: Using Crowdsourced Indoor Geodata for the Creation of a Three-Dimensional Indoor Routing Web Application. Future Internet 4, 575–591 (2012)
8. Goetz, M., Zipf, A.: Formal definition of a user-adaptive and length-optimal routing graph for complex indoor environments. Geo-Spatial Information Science 14(2), 119–128 (2011)
9. Meijers, M., Zlatanova, S., Preifer, N.: 3D geoinformation indoors: structuring for evaluation. In: Proceedings of the Next Generation 3D City Models, Bonn, Germany, pp. 11–16 (2005)
10. Jensen, C.S., Lu, H., Yang, B.: Graph model based indoor tracking. In: Tenth International Conference on Mobile Data Management: Systems, Services and Middleware, pp. 122–131. IEEE (2009)
11. Howell, I., Batcheler, B.: Building Information Modeling Two Years Later – Huge Potential, Some Success and Several Limitations. The Laiserin Letter 22 (2005), http://www.laiserin.com/features/bim/newforma_bim.pdf (last accessed: December 7, 2013
12. Stoffel, E.P., Schoder, K., Ohlbach, H.J.: Applying hierarchical graphs to pedestrian indoor navigation. In: Proceedings of the 16th ACM SIG Spatial International Conference on Advances in Geographic Information Systems (2008)
13. Lorenz, B., Ohlbach, H.J., Stoffel, E.-P.: A hybrid spatial model for representing indoor environments. In: Carswell, J.D., Tezuka, T. (eds.) W2GIS 2006. LNCS, vol. 4295, pp. 102–112. Springer, Heidelberg (2006)
14. Becker, T., Nagel, C., Kolbe, T.: A multilayered space-event model for navigation in indoor spaces. In: 3D Geo-Information Sciences, pp. 61–77. Springer, Berlin (2009)
15. Hagedorn, B., Trapp, M., Glander, T., Döllner, J.: Towards an indoor level-of-detail model for route visualization. In: Tenth International Conference on Mobile Data Management: Systems, Services and Middleware, pp. 692–697 (2009)
16. Stoffel, E.P., Lorenz, B., Ohlbach, H.J.: Towards a semantic spatial model for pedestrian indoor navigation. In: Hainaut, J.-L., et al. (eds.) ER Workshops 2007. LNCS, vol. 4802, pp. 328–337. Springer, Heidelberg (2007)
17. Richter, K.F., Winter, S., Rüetschi, U.J.: Constructing hierarchical representations of indoor spaces. In: Tenth International Conference on Mobile Data Management: Systems, Services and Middleware, pp. 686–691. IEEE (2009)
18. Niebel, B.W., Freivalds, A.: Methods, standards, and work design. McGraw-Hill (2003)
19. Nyström, R.H., Harjunkoski, I., Kroll, A.: Production optimization for continuously operated processes with optimal operation and scheduling of multiple units. Computers & Chemical Engineering 30(3), 392–406 (2006)
20. Scholl, A., Becker, C.: State-of-the-art exact and heuristic solution procedures for simple assembly line balancing. European Journal of Operational Research 168(3), 666–693 (2006)
21. Bogorny, V., Palma, A.T., Engel, P., Alvares, L.O.: Weka-gdpm: Integrating classical data mining toolkit to geographic information systems. In: SBBD Workshop on Data Mining Algorithms and Aplications (WAAMD 2006), Florianopolis, Brasil, pp. 16–20 (2006)
22. Andrienko, G., Andrienko, N., Jankowski, P., Keim, D., Kraak, M.J., MacEachren, A., Wrobel, S.: Geovisual analytics for spatial decision support: Setting the research agenda. International International Journal of Geographic Information Science 21(8), 839–857 (2007)

23. Compieta, P., Marion, D.S., Bertolotto, M., Ferrucci, F., Kechadi, T.: Exploratory spatio-temporal data mining and visualization. Journal of Visual Languages and Computing 18, 255–279 (2007)

24. Smith, B.: Objects and their environments: from Aristotle to ecological ontology. In: Frank, A., Raper, J., Cheylan, J.P. (eds.) Life and Motion of Socio-economic Units, Taylor & Francis, London, pp. 79–97. Taylor & Francis, Abington (2001)

25. Davis, E.: Representations of Commonsense Knowledge. Representation and Reasoning. Morgan Kaufmann Publishers (1990)

26. Gibson, J.J.: The theory of affordances. In: Shaw, R., Bransford, J. (eds.) Perceiving, Acting, and Knowing, pp. 67–82. Lawrence Erlbaum (1977)

27. Gibson, J.J.: The Ecological Approach to Visual Perception. Houghton Mifflin Company (1979)

28. Raubal, M., Moratz, R.: A functional model for affordance-based agents. In: Rome, E., Hertzberg, J., Dorffner, G. (eds.) Towards Affordance-Based Robot Control. LNCS (LNAI), vol. 4760, pp. 91–105. Springer, Heidelberg (2008)

29. Turner, A., Penn, A.: Encoding natural movement as an agent-based system: an investigation into human pedestrian behaviour in the built environment. Environment and Planning B: Planning and Design 29, 473–490 (2002)

30. Kapadia, M., Singh, S., Hewlett, B., Faloutsos, P.: Egocentric Affordance Fields in Pedestrian Steering. In: Proceedings of the 2009 Symposium on Interactive 3D Graphics and Games (2009)

31. Anagnostopoulos, C., Tsetsos, V., Kikiras, P., Hadjiefthymiades, S.: OntoNav: A semantic indoor navigation system. In: 1st Workshop on Semantics in Mobile Environments (SME 2005), Cyprus (2005)

32. Tsetsos, V., Anagnostopoulos, C., Kikiras, P., Hadjiefthymiades, S.: Semantically enriched navigation for indoor environments. International Journal of Web and Grid Services 2(4), 453–478 (2006)

33. Geng, H. (ed.): Semiconductor manufacturing handbook. McGraw-Hill (2005)

34. Osswald, S., Weiss, A., Tscheligi, M.: Designing wearable devices for the factory: Rapid contextual experience prototyping. In: International Conference on Collaboration Technologies and Systems (CTS), pp. 517–521. IEEE (2013)

35. Thiesse, F., Fleisch, E., Dierkes, M.: LotTrack: RFID-based process control in the semiconductor industry. IEEE Pervasive Computing 5(1), 47–53 (2006)

36. Jonietz, D., Timpf, S.: An Affordance-Based Simulation Framework for Assessing Spatial Suitability. In: Tenbrink, T., Stell, J., Galton, A., Wood, Z. (eds.) COSIT 2013. LNCS, vol. 8116, pp. 169–184. Springer, Heidelberg (2013)

37. Skupin, A.: Tri-space: Conceptualization, transformation, visualization. In: Proceedings of Sixth International Conference on Geographic Information Science, Zurich, pp. 14–17 (2010)

38. Skupin, A., Esperbé, A.: An alternative map of the united states based on an n-dimensional model of geographic space. Journal of Visual Languages & Computing 22(4), 290–304 (2011)

39. Kohonen, T.: The self-organizing map. Neurocomputing 21(1), 1–6 (1998)

Wayfinding Decision Situations: A Conceptual Model and Evaluation

Ioannis Giannopoulos[1], Peter Kiefer[1], Martin Raubal[1],
Kai-Florian Richter[2], and Tyler Thrash[3]

[1] Institute of Cartography and Geoinformation, ETH Zürich, Zürich, Switzerland
[2] Department of Geography, University of Zürich, Zürich, Switzerland
[3] Department of Humanities, Social and Political Science, Chair of Cognitive Science,
ETH Zürich, Zürich, Switzerland

Abstract. Humans engage in wayfinding many times a day. We try to find our way in urban environments when walking towards our work places or when visiting a city as tourists. In order to reach the targeted destination, we have to make a series of wayfinding decisions of varying complexity. Previous research has focused on classifying the complexity of these wayfinding decisions, primarily looking at the complexity of the decision point itself (e.g., the number of possible routes or branches). In this paper, we proceed one step further by incorporating the user, instructions, and environmental factors into a model that assesses the complexity of a wayfinding decision. We constructed and evaluated three models using data collected from an outdoor wayfinding study. Our results suggest that additional factors approximate the complexity of a wayfinding decision better than the simple model using only the number of branches as a criterion.

1 Introduction

Successful wayfinding (i.e., our ability to find a distal destination from some origin; [23]) depends on several factors, including the complexity of the environment in which wayfinding occurs. The layout of an environment influences the ease with which a corresponding mental representation is formed [5,32]. In addition, familiarity with and structure of the environment help determine which strategies are used to find the way [7,14].

During wayfinding, the layout of the path network (e.g., the street network of an outdoor environment) is of particular importance. In these networks, path segments (the streets) meet at intersections where wayfinding decisions must be made. Accordingly, these intersections and their configuration are a main contributor to route complexity. In the dynamic context of wayfinding, they are often referred to as decision points (e.g., [16]).

One simple measure for establishing a decision point's complexity is the InterConnection Density (ICD; [24]). The ICD of a network is the average number of path segments meeting at an intersection. In other words, in O'Neill's terms, the complexity of a decision point is determined by the number of options to continue one's way.

However, this measure ignores certain dynamics of wayfinding [16]. For example, continuing straight at an intersection is arguably easier than turning left or right. These dynamics are reflected in Mark's measure of route complexity [21]. In this measure,

M. Duckham et al. (Eds.): GIScience 2014, LNCS 8728, pp. 221–234, 2014.

slot values are attributed to wayfinding situations, depending on the complexity of an intersection (e.g., whether the intersection is a T-intersection or the convergence of six different streets) and the corresponding, possible actions (e.g., continuing straight or turning left). Higher slot values denote higher complexity.

Ambiguity in the decision situation also needs to be considered. For example, executing the instruction "turn left" becomes more complex when there are several options to turn left compared to when there is only one path segment heading in that direction [11]. Landmarks may help to reduce ambiguity (and thus complexity). References to salient geographic objects (e.g., "turn left at the post office") anchor actions in space [4]. They signal crucial actions to perform and support identifying the right spot at which to perform them [20].

During route following (i.e., instructed wayfinding) the interplay between instructions and environment also become important. Good instructions may ease wayfinding considerably even in highly complex environments; bad instructions on the other hand may make wayfinding nearly impossible even in simple environments [28].

Overall, wayfinding constitutes a dynamic decision-making process during which people have to make decisions on the spot. Temporal constraints depend on the mode of travel; for example, pedestrians usually have more time during spatio-temporal decision situations than car drivers. There has still been little research about how mobile, location-based decision-making is different from other types of decision-making. General decision theory covers a wide range of models with different foci such as describing how decisions could or should be made or specifying the decisions that are made [9]. In the cognitive literature, behavioral decision theory has been emphasized because human decision-making is not optimized in a strictly mathematical and economical sense [29].

Mobile, location-based decision-making involves spatio-temporal constraints that relate not only to people's behavior in large-scale space [17], but also to their interaction with mobile devices and the environment, and perceptual, cognitive, and social processes. This involves multiple psychologies of space [22] and different time scales [6]. Special tools have been developed for studying the interaction between individuals, environments, and mobile devices [19].

Mobile devices have the general challenge of presenting information to people on the move. Despite their technological limitations (e.g., a small screen size), users can reduce the complexity of a spatio-temporal decision situation by off-loading what would otherwise be cognitive work (e.g., [3]). "Cognitive work" in this context refers to the effortful processing that often accompanies explicit decision-making. People can offload cognitive work onto the environment during wayfinding by, for example, referring to a digital map. Accordingly, the cognitive load theory (CLT; [2]) offers a way of assessing and affecting some critical components during the design process of digital maps.

Adaptive location-based services (LBS) change the presentation of the map, or of the wayfinding instructions in general, depending on the current context, a user model, and a task model [25]. A large number of factors can be considered as context relevant for adaptive LBS, including position, time, speed, means of transportation, or weather information [27].

Cognitive off-loading depends on the interactions between each individual's cognitive abilities, the task at hand, and the immediate environment. During wayfinding, spatial abilities become especially critical [1]. Spatial abilities may vary according to age, gender, working memory capacity, reasoning strategies, preferred learning styles, attitudinal differences, and so forth [34]. One way of predicting wayfinding performance, specifically, is through a participant's self-reported sense of direction. For example, Hegarty and colleagues [12] found that participants' scores on the Santa Barbara Sense of Direction Scale (SBSODS) were more related to tasks that required updating over self-motion than those that required learning spatial information second-hand (e.g., as from a physical map).

In this paper, we propose a model for the complexity of pedestrian wayfinding decision situations in street networks. Our model describes the complexity of a decision situation with three elements: an *environmental model, a user model*, and an *instruction model*. We argue that a combination of these three elements is better suited for describing the complexity of a wayfinding decision situation than any single element or any combination of two of them. Three models are evaluated in terms of the above-mentioned factors. This evaluation demonstrated that models incorporating these factors are able to capture the complexity of a wayfinding decision situation better than a simple model using only the number of branches. Our dependent measures included the duration of making wayfinding decisions, the number of head movements, the number of gaze switches from the environment to the map, and the total time spent on the map. These measures can serve as an indication of cognitive load.

A context-aware pedestrian wayfinding assistant could use our model to minimize the complexity of the decision situations its user will be facing along the route. The route-planning algorithm would consider the complexity of each node in the street network for the given user, instead of choosing a user-independent route that is only optimized by environmental factors. The wayfinding assistant could also consider several possible route instructions for each decision point and choose the least complex one for the given user.

In section 2, we define the term *wayfinding decision situation* and introduce a conceptual model to describe its complexity. Section 3 introduces the wayfinding study used to evaluate three operational models. In section 4, we present the results of this evaluation, and in section 5, we discuss their implications for future research and LBS design.

2   Wayfinding Decision Situation

Wayfinders utilize environmental information, instructions (e.g., verbal or pictorial) and their spatial and cognitive abilities in order to make wayfinding decisions [23]. The complexity of these decisions is characterized by the structure of the given environment, the goals and task of the wayfinder, as well as her own characteristics. Thus, taking only environmental aspects into account, such as the number of branches at a decision point (as in the ICD model), is rather limited. For instance, a decision point with six branches can be less complex for a wayfinder than one with four branches if the given instructions for the former decision point are less complex. It is even possible that the same decision point is less complex for one wayfinder than for another because of their individual differences and spatial abilities. We propose a model that incorporates environmental, instruction, and user factors in order to characterize the complexity of wayfinding decisions and define it as wayfinding decision situation:

"A wayfinding decision situation occurs when a specific wayfinder has to make a wayfinding decision in a certain environment with a certain instruction."

In the following, we provide a conceptual model that describes the complexity of wayfinding decision situations and then evaluate three operational models.

2.1   Conceptual Model

The conceptual model is composed of environmental, instruction and user factors (see Figure 1) and aims at describing the factors that influence the complexity of wayfinding decision situations. The proposed conceptual model integrates several factors that can have an impact on the complexity of wayfinding decision situations but raises no claim to completeness.

Environmental Model. The environmental information that is available to a wayfinder, such as the geometry of a decision point, is crucial for making wayfinding decisions. The number of branches at a decision point is often used as a criterion for complexity [24]. Obviously, as the number of wayfinding options increases, the complexity of a decision point also increases. Landmarks are an important factor of the environmental context and are often used in navigation instructions [26]. Architectural differentiation [31], the availability of objects in the environment identifiable as landmarks, the unambiguity and saliency of landmarks, and their advance visibility [33] can have an impact on the complexity of a decision point. Even a decision point with only three branches may become extremely complex if the environmental cues cannot optimally be utilized by the user. Environmental factors that have an impact on the complexity of wayfinding decision situations can be classified into two categories. The first category contains all factors that contribute to complexity independent of the instructions (e.g., the number of branches, the geometry of an intersection). The factors that become (mostly) important through their use in an instruction, such as environmental landmarks, constitute the second category. The set $E = \{(c_1, f_1), (c_1, f_2), ..., (c_2, f_n)\}$ is comprised of all environmental factors that can influence a wayfinding decision situation as well as their corresponding category c_i. More than one element of this set can coexist in a given wayfinding decision situation, thus having a weighted additive linking. Environmental complexity $c(e), e \subseteq E$, is computed based on the existing environmental factors at the given decision point and a weight function w_E defines the impact of each factor on complexity.

Instructions Model. To reach a goal (e.g., while walking from a starting point A to a destination B), wayfinders have to perform different activities and interact with the environment in order to make several wayfinding decisions. Wayfinders use aids (e.g.,

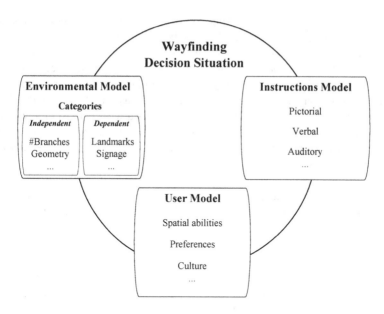

Fig. 1. The figure depicts a wayfinding decision situation. The environmental, instruction and user factors of the wayfinding decision situation model are used to assess its complexity.

maps; verbal, auditory and pictorial instructions; knowledge provided by other humans; [30]) to make wayfinding decision situations less complex. These aids help us fulfill our tasks when cruising in unfamiliar environments or when looking for a hospital. The complexity of an instruction is strongly related to the represented environmental factors. For instance, the complexity of landmark based instructions is related to the saliency and the advance visibility [33] of the incorporated landmarks (among other factors). An instruction, apart from being a wayfinding aid, can also have a negative effect on the complexity of a decision situation if its complexity is high. Thus, instructions are an important factor to be considered in wayfinding decision situations. The set $I = \{t_1, t_2, ..., t_n\}$ contains all the different instruction types. In contrast to the environmental model, in a given decision situation, only one instruction type can be active. Combinations of instruction types, such as the combination of verbal and pictorial instructions, form an additional instruction type. The complexity of the instructions $c(t_i), t_i \in I$, strongly depends on the instruction type (e.g., landmark based instructions); therefore, it is necessary that measures for assessing the complexity are type-specific (e.g., landmark based measures). A weight function w_I defines the impact of the type-specific measures on the instruction complexity.

User Model. A wayfinding decision situation differs for every wayfinder. Individuals' spatial abilities, preferences, interests, general knowledge, and cultural background have an impact on decision making during wayfinding [10]. It is more likely that a wayfinder with high spatial abilities will be able to process the environmental information and decrypt instructions faster than a wayfinder with low spatial abilities.

A wayfinder with better problem solving abilities is able to process environmental information more easily. For example, she may be able to incorporate the slope of the branches at an intersection as a criterion when it comes to finding the way to an orthopedics by making the inference that a place like that would be easily accessible and not on the top of a hill. The set $U = \{f_1, f_2, ..., f_n\}$ contains all factors representing the user characteristics that can have an impact on complexity and the function $f(U)$ represents the link between these factors. The user factors always coexist in a wayfinding decision situation and a weight function w_U defines the importance of the given factors.

Our proposed conceptual model takes into account the factors mentioned above and can be summarized as $c(e, t_i, U) = c(e) \oplus c(t_i) \oplus f(U)$, where $c(e), e \subseteq E$ is the resulting environmental complexity, $c(t_i), t_i \in I$ the complexity of the instructions, and $f(U)$ are the user factors that can account for more or less complexity of the wayfinding decision situation. The operator $\oplus$ represents a linking between the factors.

2.2 Operational Model

A model that describes the complexity of a wayfinding decision situation will have a significant impact on several aspects of wayfinding assistance. As a first step towards an operational model, we use a subset of the factors introduced in the conceptual model to construct three models. We then compare them to a widely used model that incorporates only the number of branches at a decision point [24].

$$Branches\ Model = number\ of\ branches \tag{0}$$

Each model introduced below is a step-wise extension of the previous, starting with the simple model (0) that uses only the number of branches as a complexity measure. The conceptual model $c(e, t_i, U)$ allows for incorporating a whole range of factors (as any context model's instantiations will necessarily always be incomplete). We will test the models against data collected in a human participants study. Therefore, we only incorporate factors in the operational models that correspond to data provided by the experiment. For this reason, the instruction type t_i used for the operational models is equal to pictorial landmark-based instructions, the user factors U are limited to the values obtained through the Santa Barbara Sense of Direction Scale (SBSODS) [12], and the environmental factors $e \subseteq E$ are limited to the number of branches at a decision point.

Incorporating Environmental and Instruction Factors. In a first step, we incorporate only the environmental factor, namely the number of branches $\subseteq E$ and the pictorial landmark-based instructions $\in I$:

$$c(e, t_i) = c(e) \oplus c(t_i) \implies$$

$$c(e, t_i) = (1 - w_1) * \#br + w_1 * (\beta * adv_{vis} + (1 - \beta) * lm) \tag{1}$$

#br: number of branches, adv_{vis}: advance visibility and lm: $landmark_{matching}$

$c(e, t_i) \in [0, 1], e \subseteq E, t_i \in I$

The first part of the model describes the environmental complexity as the number of branches at the decision point where a wayfinding decision situation occurs. The second part of the model defines the complexity of the instructions and is computed as the weighted addition of the advance visibility adv_{vis} [33] of the landmarks used in the given instruction and the landmark matching $landmark_{matching}$ value. The landmark matching value is represented as the ease with which the pictorial representations of landmarks can be matched with images of the corresponding real landmark. The advance visibility measure was introduced by Winter [33] and classifies landmarks based on how salient they are and how early they are visible on a path segment towards a decision point. The values for landmark matching were retrieved through an experiment described in section 3.2.

Incorporating the User. In a second step, we extend the model by incorporating user characteristics. We use the SBSODS score as a value for the weight w_1 introduced in the previous model (1). The underlying assumption for this step is that users with higher spatial abilities would be affected more by the complexity of the instructions, rather than by the complexity of the environment.

$$c(e, t_i, U) = c(e) \oplus c(t_i) \oplus f(U) \implies$$

$$c(e, t_i, U) = (1 - sa) * \#br + sa * (\beta * adv_{vis} + (1 - \beta) * lm) \tag{2}$$

sa: SBSODS, #br: number of branches, adv_{vis}: advance visibility and lm: $landmark_{matching}$
$$c(e, t_i, U) \in [0, 1], e \subseteq E, t_i \in I$$

We also introduce a third model that incorporates the user factors using an additive linking.

$$c(e, t_i, U) = c(e) \oplus c(t_i) \oplus f(U) \implies$$

$$c(e, t_i, U) = w_1 * \#br + w_2 * (\beta * adv_{vis} + (1 - \beta) * lm) + w_3 * sa \tag{3}$$

sa: SBSODS, #br: number of branches, adv_{vis}: advance visibility and lm: $landmark_{matching}$
$$c(e, t_i, U) \in [0, 1], e \subseteq E, t_i \in I$$

The weights w_1, w_2, and w_3 are constrained to sum up to one. The weight β, as well as the values for adv_{vis}, lm and #br are within 0 and 1.

In the following, all three models will be evaluated with regard to how well they fit the data collected during an outdoor wayfinding study. All the weights of the introduced models were estimated using a genetic algorithm that is discussed in section 4. The factors used in the models were normalized using the maximum values obtained in two experiments (discussed in section 3). The normalized values from the SBSODS were inverted, with a higher value denoting lower spatial abilities (since a higher score of the model denotes higher complexity).

3 Experiments

In the following we report on two experiments that were conducted in order to collect the data necessary for the evaluation of the operational models.

3.1 Outdoor Wayfinding Experiment

An outdoor wayfinding experiment was conducted in the city of Zurich and constituted one task of a larger study [15]. The data collected from this experiment were used to fit the operational models introduced in section 2.2.

Participants. We recruited 14 participants for the wayfinding experiment. Each participant was provided a small monetary compensation for his/her participation. All participants were recruited through collaboration with a nearby hostel and were unfamiliar with the city of Zurich. Due to errors in the recording software, three data sets were lost. The remaining 11 participants (seven females) had an average age of 26.8 years (min 21, max 50, SD 8.3). They had different cultural backgrounds, none of them was a geographer or cartographer, and none of them was using maps in their profession.

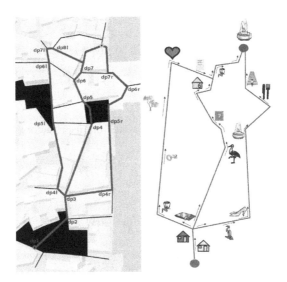

Fig. 2. The left side of the figure illustrates the area of the experiment and the decision points of the three routes. The pictorial map on the right was given to the participants.

Experiment Set-up and Procedure. The experiment took place on the streets of the old town part of Zurich (see Figure 2, left), where no cars are allowed. At the starting position, participants were given the task on a 28 × 28 cm paper print ("On this map you can see three possible routes that lead from your current position (green point at the bottom) to the next goal (red point at the top). Please make your way to the goal").

They had to reach a destination marked on the map with a red dot (see Figure 2, right) printed on the back side of the paper. This abstract map illustrated three routes that could be chosen in order to reach the destination as well as icons representing landmarks in the environment (i.e., buildings, signs) as a wayfinding aid.

The participants were equipped with a mobile eye tracking system[1] and had to carry a backpack ($\sim$ 2 kg) with the accompanying eye tracking hardware. They were not allowed to interact with other people or with the experimenters. The experiment ended either when the participants reached and correctly identified the destination or when they gave up.

During wayfinding, we tracked the eye movements of each participant and their field of view as recorded through the front camera of the eye tracker. We used these data in order to extract additional measures: number of head movements, number of gaze switches from the environment to the map, total time of map usage, and time spent in a wayfinding decision situation.

Data Post Processing. It was necessary to validate the captured eye tracking data because of possible distortions due to changing light conditions. We manually analyzed each frame of the captured eye movements in order to validate pupil detection and manually corrected frames where the pupil was not correctly detected. The validation and correction procedure can be manually achieved using the software[2] provided by the eye tracking vendor.

Extraction of Measures. Two human raters qualitatively analyzed the captured video frames (field camera) as well as the eye movements in order to perform a segmentation of each wayfinding trial and define the start and end point of every wayfinding decision situation. A wayfinding decision situation started immediately after the end of a previous one and ended when the participant had decided and was heading towards one of the available branches of the decision point. Overall, 75 decision situations were identified.

For each of these segments, we registered its duration (time to make a decision) as well as the number of head movements (change of the field of view), based on a manual analysis of the video frames of the field camera. Moreover, we used the captured eye movements to register the gaze switches from the environment to the map as well as to compute the total duration of map usage. These measures were separately used in the evaluation to estimate the fit of the operational models. Monocular eye trackers, such as the one used in our study, suffer from the parallax error [13]. They can be calibrated only for one distance at a time. Due to varying distances between the participant and the objects in the environment, we could not use the gazes in the environment, for example, to extract measures based on the gazes towards landmarks.

The advance visibility used to assess the complexity of the instructions was computed based on the values gathered from an analysis of the experiment area. We used a 3D model of the area in a GIS software[3] and computed the isovists for every landmark used in the instructions as well as the intersection of each isovist with the corresponding route segment towards the decision point.

---

[1] Dikablis - www.ergoneers.com
[2] Ergoneers - DLab Analysis.
[3] ArcGIS 10.1.

3.2 Web Experiment for Landmark Identification

We performed a web experiment for the evaluation of the selected map icons that served as landmark representations. The collected data were used to score the map icons based on how well they represented the real landmarks in the environment. These data were used for the computation of instruction complexity ($landmark_{matching}$).

Experiment Set-up and Procedure. The web experiment was implemented using JavaScript. The first page contained a task description and an example illustrating the task. When the participants started the actual experiment, they were directed to a website displaying one image of the real environment and the corresponding map icon. Participants were then instructed to click as fast as possible on the position of the real-world image (i.e., where they thought the corresponding landmark was located). After each click was performed, the next image and map icon were shown.

In total, seventy-two participants around the world took part in the experiment. The images and their corresponding map icons were randomly ordered for each participant. We registered the time needed to decide and click on the image, as well as whether the map icon was matched correctly.

The average time needed to perform a correct match was used for ranking the 16 map icons. A linear regression revealed a significant positive correlation of the ranking with the total number of errors that occurred for each map icon ($R^2 = .475, p < .010$).

4 Results

The data collected from the experiments were normalized using their maximum values and used to estimate the parameters (weights) of the models. The best-fitting parameter for the 1-parameter model (2) was determined through a brute-force search. The best-fitting parameters for the two- (1) and four-parameter models (3) were determined through a genetic algorithm.

4.1 Parameter Estimation Algorithm

A custom-written genetic algorithm was used in order to estimate the values of all parameters. Using this algorithm, we attempted to find the minimum summed and squared error (SSE) between the observed values of each dependent variable (i.e., decision time, time on map, map switches, and head movements, separately) and the values predicted by two- (1) and four-parameter models (3). Observed values were not aggregated over decision point or participant; thus, the genetic algorithm was used to fit 75 values. The algorithm started with 1000 randomly generated combinations of parameter values (i.e., "organisms"). The starting values for all parameters were constrained to fall between 0 and 1. Each iteration of the genetic algorithm consisted of three steps: selection, reproduction, and mutation. During selection, the best-fitting of every eight organisms was chosen for reproduction (i.e., "tournament selection"; [8]). During reproduction, the organisms were randomly paired and converted to bits, a random crossover point was determined, and every pair of organisms exchanged bits below that crossover point.

During mutation, every bit of every organism had a 0.5% chance of changing from a zero to a one or vice versa. Each parameter was represented by 17 bits, corresponding to a precision of approximately ±0.0001. The best-fitting organism over 100 iterations was maintained and ultimately used to evaluate each model. In order to compare models with different numbers of freely varying parameters, SSE for each model was converted to Bayes' information criterion (BIC; [18]).

The reliability of the genetic algorithm was validated by estimating known parameter values for both two-parameter (1) and four-parameter models (3). These initial parameter values were randomly generated using each model's constants (i.e., number of branches, advance visibility, landmark matching, and SBSODS score). Each parameter was constrained to fall between 0 and 1. For the four-parameter model (3), w_1, w_2, and w_3 were constrained to sum to one. Standardized decision times were generated using these initial parameter values and the constants for each combination of decision point and participant. The genetic algorithm was then used to estimate the initial, randomly generated, parameter estimates. This validation procedure was repeated 100 times. Variability and skew in the distribution of the differences between estimated and initial parameter values were used to evaluate the genetic algorithm's performance for each model.

4.2   Validation Results

In general, the validation procedure suggested that the genetic algorithm performed excellently for both models. For the two-parameter model (1), variance in the distributions of the differences between estimated and initial parameter values was 0.0000001 and 0.0000091 for w_1 and β, respectively; skew in the distributions of the differences between estimated and initial parameter values was -0.13 and -6.37 for w_1 and β, respectively. Because of the extreme precision with which β was estimated, this amount of skew is negligible (though perhaps notable). For the four-parameter model (3), variance in the distributions of the differences between estimated and initial parameter values was 0.0103, 0.0084, 0.0388, and 0.0701 for w_1, w_2, w_3, and β, respectively; skew in the distributions of the differences between estimated and initial parameter values was 0.41, -0.18, 0.36, and -0.62 for w_1, w_2, w_3, and β, respectively.

4.3   Estimated Parameters and Fit of the Models

The parameter estimates and overall fit for each model are illustrated in Table 1. The models are ordered from least to most complex (in terms of the number of free parameters). Models (1), (2), and (3) were compared to model (0) in terms of BIC (i.e., the lowest BIC indicates the best-fitting model). For each dependent measure, model (1), (2) and (3) fit better than model (0). This indicates, that each additional free parameter increased the fit of the model being developed. Qualitative trends in the parameter estimates (across dependent measures) are clear in some respects and less clear in others. For example, the SBSODS did appear to contribute to the overall performance of model (3). In contrast, the parameter estimates for β indicate that advance visibility did not contribute to the fit of the largest model (3). However, advance visibility did appear to contribute to the fit of models (1) and (2). Possible reasons for these trends are

Table 1. The table depicts the estimated parameters of the models, the summed and squared error (SSE) as well as the Bayes' information criterion (BIC) transformation

dec dur: decision situation duration, head mov: number of head movements

map sw: number of map switches and map dur: map usage duration

	#Parameters	w_1	w_2	w_3	β	SSE	BIC	Model
	0	-	-	-	-	19.6649	-100.39	(0)
dec dur	1	-	-	-	1	7.2884	-170.5228	(2)
	2	0.151	-	-	0.737	4.4185	-203.7416	(1)
	4	0.5565	0.0326	0.4109	0	3.2591	-217.9328	(3)
	0	-	-	-	-	25.7513	-80.1750	(0)
head mov	1	-	-	-	1	8.7034	-157.2155	(2)
	2	0	-	-	0.788	2.1307	-258.4428	(1)
	4	0.9009	0.0991	0	0	1.0166	-305.3069	(3)
	0	-	-	-	-	19.8388	-99.7385	(0)
map sw	1	-	-	-	1	5.7005	-188.9526	(2)
	2	0.082	-	-	0.752	1.4792	-285.8140	(1)
	4	0.4386	0.0795	0.4819	0	0.8382	-319.7790	(3)
	0	-	-	-	-	15.0279	-120.5680	(0)
map dur	1	-	-	-	1	4.5029	-206.6400	(2)
	2	0.152	-	-	0.644	2.3217	-252.0042	(1)
	4	0.3249	0.2453	0.4298	0.038	1.997	-254.6682	(3)

briefly explored in section 5. We estimated parameters that would fit the models based on the time that was spent in a wayfinding decision situation, which is a commonly used measure for wayfinding complexity. Moreover, we estimated parameters in order to find a fit for the models based on the number of head movements, the number of gaze switches from the environment to the map as well as the total time spent on the map. These are measures that can serve as an indication for cognitive load. The results demonstrate that, for all measures, each additional parameter increased the overall fit of the operational model.

5   Discussion and Outlook

According to the evaluation results of the three linear operational models, a combination of environmental, instruction, and user factors is better suited for describing the complexity of a wayfinding decision situation than any single element or any combination of two of them. Advance visibility did not contribute to the fit of the largest model (3), which probably can be explained by the small differences in the obtained values.

The conceptual model of a wayfinding decision situation (as discussed in section 2.1) is more general than the operational model used for the evaluation. A number of factors possibly impacting the complexity of a decision situation have not been considered in our evaluation, such as signage at the decision points, the user's cultural background, or different types of instructions. As future work, studies investigating the influence of these factors on the complexity of a decision situation are required. The resulting enriched models will hopefully lead to significant and strong correlations with the study data.

Also, an analysis of the interrelation of these factors in a combined model will help to complement previous findings from the wayfinding literature about each single factor.

Another logical next step is the implementation of a pedestrian wayfinding assistant which recommends the route with the least complexity, based on the wayfinding decision situation model. One challenge in this context consists of defining a user model. Some properties of the user, such as spatial abilities, may not be available when a user starts the wayfinding assistant for the first time. One possible solution for this could be to learn parts of the user model during wayfinding (e.g., from wayfinding behavior or from interaction with the system).

Acknowledgments. We would like to thank Jördis Graf and Mikko Schmitter for their help with the data pre-processing and validation. Thank you also Cristina Iosifescu for co-supervising our experiment sessions. The City Backpacker Hotel Biber, Zurich kindly supported our recruitment of participants. We also thank the participants for taking part in our experiments.

References

1. Allen, G.: Spatial abilities, cognitive maps, and wayfinding: Bases for individual differences in spatial cognition and behavior. In: Golledge, R.G. (ed.) Wayfinding Behavior - Cognitive Mapping and Other Spatial Processes, pp. 46–80. Johns Hopkins University Press, Baltimore (1999)
2. Bunch, R.L., Lloyd, R.E.: The cognitive load of geographic information. The Professional Geographer 58(2), 209–220 (2006)
3. Clark, A.: Being There: Putting Brain, Body, and World Together Again. MIT Press, Cambridge (1997)
4. Denis, M.: The Description of Routes: A Cognitive Approach to the Production of Spatial Discourse. Cahiers Psychologie Cognitive 16(4), 409–458 (1997)
5. Dogu, U., Erkip, F.: Spatial factors affecting wayfinding and orientation: A case study in a shopping mall. Environment and Behavior 32(6), 731–755 (2000)
6. Frank, A.U.: Different types of "times" in GIS. In: Egenhofer, M.J., Golledge, R.G. (eds.) Spatial and Temporal Reasoning in GIS, pp. 40–62. Oxford University Press, Oxford (1998)
7. Gärling, T., Lindberg, E., Mantyla, T.: Orientation in buildings: Effects of familiarity, visual access and orientation aids. Journal of Applied Psychology 68(1), 177–186 (1983)
8. Goldberg, D.E., Deb, K.: A comparative analysis of selection schemes used in genetic algorithms. Foundations of Genetic Algorithms, pp. 69–93 (1991)
9. Golledge, R.G.: Spatial behavior: A geographic perspective. Guilford Press (1997)
10. Golledge, R.G.: Wayfinding Behavior: Cognitive Mapping and Other Spatial Processes. Wayfinding Behavior. Johns Hopkins University Press (1999)
11. Haque, S., Kulik, L., Klippel, A.: Algorithms for reliable navigation and wayfinding. In: Barkowsky, T., Knauff, M., Ligozat, G., Montello, D.R. (eds.) Spatial Cognition 2007. LNCS (LNAI), vol. 4387, pp. 308–326. Springer, Heidelberg (2007)
12. Hegarty, M., Richardson, A.E., Montello, D.R., Lovelace, K., Subbiah, I.: Development of a self-report measure of environmental spatial ability. Intelligence 30(5), 425–447 (2002)
13. Holmqvist, K., Nyström, M., Andersson, R., Dewhurst, R., Jarodzka, H., Van de Weijer, J.: Eye tracking: A comprehensive guide to methods and measures. Oxford University Press (2011)

14. Hölscher, C., Meilinger, T., Vrachliotis, G., Brösamle, M., Knauff, M.: Up the down staircase: Wayfinding strategies in multi-level buildings. Journal of Environmental Psychology 26(4), 284–299 (2006)
15. Kiefer, P., Giannopoulos, I., Raubal, M.: Where am I? Investigating map matching during self-localization with mobile eye tracking in an urban environment. Transactions in GIS (2013)
16. Klippel, A.: Wayfinding choremes. In: Kuhn, W., Worboys, M.F., Timpf, S. (eds.) COSIT 2003. LNCS, vol. 2825, pp. 301–315. Springer, Heidelberg (2003)
17. Kuipers, B.J., Levitt, T.S.: Navigation and mapping in large scale space. AI Magazine 9(2), 25 (1988)
18. Lewandowsky, S., Farrell, S.: Computational modeling in cognition: Principles and practice. Sage Publications (2010)
19. Li, C., Longley, P.: A test environment for location-based services applications. Transactions in GIS 10(1), 43–61 (2006)
20. Lovelace, K.L., Hegarty, M., Montello, D.R.: Elements of good route directions in familiar and unfamiliar environments. In: Freksa, C., Mark, D.M. (eds.) COSIT 1999. LNCS, vol. 1661, pp. 65–82. Springer, Heidelberg (1999)
21. Mark, D.M.: Automated route selection for navigation. IEEE Aerospace and Electronic Systems Magazine 1(9), 2–5 (1986)
22. Montello, D.R.: Scale and multiple psychologies of space. In: Campari, I., Frank, A.U. (eds.) COSIT 1993. LNCS, vol. 716, pp. 312–321. Springer, Heidelberg (1993)
23. Montello, D.R.: Navigation. In: Shah, P., Miyake, A. (eds.) The Cambridge Handbook of Visuospatial Thinking, pp. 257–294. Cambridge University Press (2005)
24. O'Neill, M.J.: Evaluation of a conceptual model of architectural legibility. Environment and Behavior 23(3), 259–284 (1991)
25. Raubal, M., Panov, I.: A formal model for mobile map adaptation. In: Location Based Services and TeleCartography II, pp. 11–34. Springer (2009)
26. Raubal, M., Winter, S.: Enriching wayfinding instructions with local landmarks. In: Egenhofer, M., Mark, D.M. (eds.) GIScience 2002. LNCS, vol. 2478, pp. 243–259. Springer, Heidelberg (2002)
27. Reichenbacher, T.: Adaptation in mobile and ubiquitous cartography. In: Multimedia Cartography, pp. 383–397. Springer (2007)
28. Schneider, L.F., Taylor, H.A.: How do you get there from here? Mental representations of route descriptions. Applied Cognitive Psychology 13(5), 415–441 (1999)
29. Simon, H.A.: A behavioral model of rational choice. The Quarterly Journal of Economics, 99–118 (1955)
30. Weiser, P., Frank, A.U.: Cognitive transactions–a communication model. In: Tenbrink, T., Stell, J., Galton, A., Wood, Z. (eds.) COSIT 2013. LNCS, vol. 8116, pp. 129–148. Springer, Heidelberg (2013)
31. Weisman, J.: Evaluating architectural legibility way-finding in the built environment. Environment and Behavior 13(2), 189–204 (1981)
32. Werner, S., Long, P.: Cognition meets Le Corbusier - cognitive principles of architectural design. In: Freksa, C., Brauer, W., Habel, C., Wender, K.F. (eds.) Spatial Cognition III. LNCS (LNAI), vol. 2685, pp. 112–126. Springer, Heidelberg (2003)
33. Winter, S.: Route adaptive selection of salient features. In: Kuhn, W., Worboys, M.F., Timpf, S. (eds.) COSIT 2003. LNCS, vol. 2825, pp. 349–361. Springer, Heidelberg (2003)
34. Wolbers, T., Hegarty, M.: What determines our navigational abilities? Trends in cognitive sciences 14(3), 138–146 (2010)

Understanding Information Requirements in "Text Only" Pedestrian Wayfinding Systems

William Mackaness, Phil Bartie, and Candela Sanchez-Rodilla Espeso

School of GeoSciences, The University of Edinburgh, Edinburgh, UK

Abstract. Information that enables an urban pedestrian to get from A to B can come in many forms though maps are generally preferred. However, given the cognitive load associated with map reading, and the desire to make discrete use of mobile technologies, there is increasing interest in systems that deliver wayfinding information solely by means of georeferenced spoken utterances that essentially leave the user "technology free." As a critical prior step, this paper examines the optimal delivery of such georeferenced text based instructions in anticipation of their spoken utterance. We identify the factors governing the content, location of instruction and frequency of delivery of text instructions such that a pedestrian can confidently follow a prescribed route, without reference to a map. We report on street level experiments in which pedestrians followed a sequence of text instructions delivered at key points along a set of routes. In examining instructions that are easy to follow, we compare landmark based instructions with street name based instructions. Results show that a landmark based approach is preferred because it is easier to assimilate (not because it is faster). Analysis also revealed that some degree of redundancy in the instructions is required in order to bring "comfort" to the user's progress. There still remains the challenge of modeling the saliency of landmarks, knowing what is the most efficient set of instructions, and how to vary the frequency of instruction according to the complexity of the route. The paper concludes by identifying a set of design heuristics useful in the design of text based instructions for wayfinding.

1 Urban Pedestrian Wayfinding

The smartphone has become a conduit by which we access many different services [1–3], instantly, at any time, anywhere ("Where am I?", "Where is my nearest … ?", "What's on where?", "How do I get to … ?"). It has become a sophisticated device for both capturing and sharing information (deliberately and unwittingly). The predominant mode of interaction is through hand and eye [4–6] –but this often distracts us from other street level tasks. But Weiser and Brown [7] spoke of concealed, 'calm' technology—technology that "informs but doesn't demand our focus or attention," technology that provides only what is necessary, rather than a firehose of information. "What is necessary" (and how best to deliver that information [8, 9] is governed by the decisions being made at any given instant (perhaps one of several tasks), the ambient conditions and the context in which the user finds themselves. The Space-Book project is an EU funded project that examines these ideas in the context of the

M. Duckham et al. (Eds.): GIScience 2014, LNCS 8728, pp. 235–252, 2014.

ambling city tourist, exploring the city whilst remaining inconspicuous [10]. Intentionally it uses only dialogue based interaction to govern the flow of information [11] – an idea whose origin stems from work by Bartie and Mackaness [12]. The ambition is that the device remains essentially concealed at all times.

Among various tasks that a tourist performs as they explore the city, is one of wayfinding [13]. Wayfinding is known to be a complex process [14, 15]. In the absence of a map, and only having spoken instructions by which to guide the user, a number of issues arise: 1) what is the most efficient instruction set? 2) How do the instructions "link" the tourist to the city? 3) what type of instruction is preferred? 4) what level of redundancy is required to re-assure the pedestrian that they are on the right route? 5) when should an instruction be issued? These questions were explored through street level experiments, where text based instructions were provided to the pedestrian, in anticipation of being presented in spoken form via dialogue based interaction. Through close observation, analysis of wait times, the conditions by which pedestrians became lost, and through questionnaires it was possible to identify the conditions governing the timing and richness of route following descriptions.

1.1   Previous Work

Various research has explored the optimal form of wayfinding instructions [16–18], and compared instructions based on street names versus landmarks [19, 20]. It has been suggested that landmark based instructions are easier to follow because a street sign can be difficult to find, whereas "walk towards the castle" is easy to interpret. The instruction "walk towards the castle" is dependent on the castle being unique, and it being visible. Indeed for any given section of a route, there will be a number of candidate landmarks by which to guide the pedestrian, the choice of which will depend on their salience [21]. The salience of something is very dynamic [22], can depend on the direction by which it is approached from, how much it consumes the field of view, and its form (architecture, texture, color), and building type [23]. Its relevance to the task will depend on its location relative to a decision point: does it lie at a key junction, or is it a confirmatory cue to ratify that the pedestrian is on the right path? From these observations we discern that we need 1) a 3D model of the city in order to calculate the visibility of a feature, 2) a database recording various attributes of the candidate landmarks, 3) a way of determining the location of the pedestrian, 4) a way of constructing easy to follow descriptions that reflect the topological and morphological form of the city (rather than say, descriptions based on angles and distances which are known to be more difficult to follow).

In addition to the issue of landmark selection, there remains the challenge of deciding what is the optimum level of instructions and the timing of their delivery [24–26]. Systems that are verbose may become annoying, particularly when the route is obvious and simple. But in the absence of other information, a terse set may lead to wasted time searching for difficult to identify landmarks, or indeed getting lost. High cognitive load arises when information is either insufficient or excessive – resulting in poor situational awareness [27] with associated risk of becoming lost (Fig 1). Better that there be some redundancy in the instruction (grey dashed line Fig 1), in order to reduce the chance of getting lost. Redundancy can also be in the form of confirmatory cues that build confidence in the user (e.g., "You should now be passing the church on your left"). From this discussion it is possible to summarizes the key design criteria of such a system (Table 1).

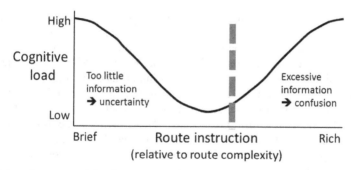

Fig. 1. Uncertainty or confusion? Erring on information redundancy to reduce chances of becoming lost

Table 1. Design criteria

Descriptive efficiency	The least number of instructions for the user to follow, in their simplest form
Effective	Instantaneous recognition (mapping between object description and real world object)
Salient	Unambiguous referent/ no confusion over what is being referred to
Robust	Minimize the chance of becoming lost by including confirmatory cues

2   A City Model to Underpin Landmark Based Instructions

An experiment was conducted that enabled comparison between street based instructions and landmark based instructions. A variety of automated database systems exist for text based instructions that use street names (e.g. TomTom, Google). We used Google maps, which ostensibly uses distances and street names to direct the user. A human, walking the route, can readily devise a landmark based solution that would, invariably, be better than street based instructions. Such an approach cannot be automated, and is not generalizable [28]. For these experiments, we built a city model sufficiently rich to automatically generate a set of landmark based instructions [10]. The city model contained a 2.5D digital terrain model and a digital surface model derived using LiDAR data, linked to a variety of open source data, data scraped from gazetteers, such as the "Gazetteer for Scotland" and National Mapping Agency Data. It also contained a topological network of pedestrian pathways, and an algorithm for calculating shortest path. Whilst OpenStreetMap was attribute rich and topologically more complete, it lacked geometric precision. This made tracking of the position of the pedestrian difficult. Thus we used Ordnance Survey's MasterMap data to manage the tracking of the subject, using GNSS technology on the smartphone to determine their location.

The main name of any building was determined using Point of Interest data. Various authors have identified factors governing a landmark's saliency [21, 29] – the degree of saliency provides a basis for selecting the most appropriate landmark by which to guide the user. The city model enables calculation of relative bearings between observer, landmark and junction, the façade, distance to the observer. Additionally the visibility engine (using the 3D modeling) enables calculation of vertical exposure (the degree to which a landmark sticks out above the foreground objects). We also examined the tags associated with images to identify how many times a landmark had been uniquely photographed (as an indication of how photo-worthy it was); additionally we used a count of FourSquare checkins to indicate frequency of visitation – again an indication of a landmark's importance. We then combined these variables to give a measure of the semantic and structural saliency of the feature. This was used to generate a list of salient landmarks along each segment (leg) of the route. This formed the basis of "confirmatory cues." These included the direction and distance to the object, and if available a description of the object (e.g. it has a tall clock tower at the front). In addition the visibility of salient objects near route turning points was generated. For instructions given a long way ahead, the salient object may be either in front or behind the turning point, but for instructions given closer to the junction a salient object was selected which was near the junction and if possible in the direction of the turn that the subject was intended to make. Where no salient objects could be found then road names and directions were used.

Junctions were automatically analyzed based on the travel direction to determine if they formed a well-known junction shape (e.g. T, Y, X). If a junction was found from the approach direction to form a known type then the description was customized (e.g. "take right fork for a Y junction," or "turn right at the T junction"). For more complex junctions an exit numbering system was used, similar to UK roundabouts, whereby exits are numbered in a clockwise direction starting from the first on the left. In addition a description based on the turn angle was added to assist the user in these more complex junctions, such as "take the 3rd exit heading straight on." Thus in summary, the city model was used to determine 1) the visibility of landmarks, and to measure their saliency, 2) generate rich descriptions of the landmark (e.g. "Old College has a green dome with a gold statue on it"). Because the city contained a detailed 2.5D model, it was possible to describe the slope of the features (e.g. head downhill/uphill, walk up/down the steps). We also calculated the "bendy-ness" of a road. This provided a rich and varied way to describe the landmark in relation to the subject's location (Table 2).

It enabled us to variously construct phrases that could: 1) orientate the user (so they set off in the right direction), 2) get the user to turn/ select the right path at junctions, 3) convey the distance required to be travelled, 4) assure the user (by providing confirmatory cues), 5) convey road qualities such as gradient, or bends in the road, 6) enable the user to move between streets, open spaces, and stairs. Table 3 is an example taken from one of the experiments, for the center of Edinburgh city in Scotland, comparing the two sets of descriptions.

Table 2. Adjectives, prepositions, verbs, adverbs, nouns and proper nouns

Adjectives	left, right, sharp, straight,
Prepositions	towards
Verbs	Turn, walk, carry on,
Adverbs	After, Before, Downhill, Uphill, immediately
Nouns	meters, minutes, steps, bend, distance,
Proper nouns	Streets and Landmarks

Table 3. Route descriptions from SND and LMD for the same route

Street based instructions	Landmark based instructions
1. Head west on Crichton St toward Charles St - go 45 meters	1. Stand with Informatics Forum on your right and Appleton Tower on your left.
2. Turn right onto Charles St About 1 min - go 94 meters	2. Walk about 50 meters towards George Square.
3. Turn left toward Teviot Pl About 2 mins - go 120 meters	3. Turn right before George Square at the cross roads.
	4. Walk about 100 meters (with Informatics forum on your right).
4. Turn left onto Teviot Pl About 2 mins - go 110 meters	5. Turn left to cross Bristo Square walking slightly uphill towards McEwan Hall (large building with a dome).
5. Continue onto 1/Lauriston Pl Continue to follow Lauriston Pl	6. Turn left on to Teviot Place, McEwan Hall on your left.
	7. Walk along Teviot Place continuing straight for about 100 meters. You should pass Royal Bank of Scotland on your right.
6. Destination will be on the right About 6 mins - go 500 meters	8. Carry on straight at the junction on to Lauriston Place. Walk for about 500 meters. You should pass George Heriot's School on your right.
* Your destination is: Edinburgh College Of Art	9. After the slight bend in the road, you will go downhill.
	10. Turn off right on to Lady Lawson Street, and walk for 40 meters. Your destination will be on your right, opposite the Novotel.
	* Your destination is: Edinburgh College Of Art

2.1   Factors Governing the Issuing of Instructions

The issuing of instructions for street based instructions is defined by the points at which there is a change of street name, or where the subject meets a junction (a node in the network). This is very different from landmark based systems – where there may be a large (or just a few) number of landmarks in the field of view - each of varying size and distance from the subject. Such a system risks overloading the subject (high redundancy in the instruction) though this may be required for routes that are complex (junctions and decision points being close to one another). Though not explored in these experiments, the richness of instruction is also influenced by the subject's familiarity with the city, their preferred type of landmark, level of situational awareness, and their capacity to interpret written/spoken instructions. The following rules (devised by empirical observation of subjects [30]) were therefore used to govern the selection of landmarks:

If a segment (distance between two decision points) is greater than 100m then
include a "you will be passing <feature>"
If the junction is complex (it has 3 or more exits) then
include a recognizable landmark on the exit road.

3   Experimental Design

To facilitate comparison between the street based instructions and landmark based instructions, the instructions were not spoken but delivered as text. This simplified approach meant the data could be stored on the phone, and issues of latency from a distributed system could be avoided. The "get directions" function in Google maps (maps.google.co.uk) was used to generate the route and associated street based instructions. Each street name and landmark instruction was geofenced so as to appear on the smartphone just 30 m prior to the subject arriving at the decision point. This distance was empirically derived from pilot experiments, and catered for the variability of the GNSS positioning and avoided the risk of the subject skirting around the trigger point and thus receiving no instruction. The phone vibrated just prior to an instruction being given – thus obviating the need for the subject to stare continuously at the screen waiting for the next instruction. By delivering instructions just at the point where they were needed we avoided the cognitive load associated with trying to memories a sequence of instructions for the whole route.

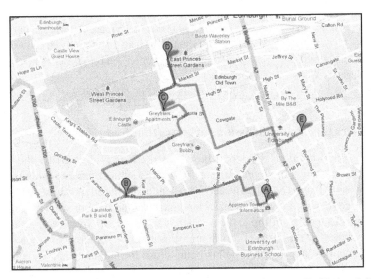

Fig. 2. : A: Informatics Forum, B: College of Art, C: Camera Obscura, D: National Gallery of Scotland, E: Institute of Geography

Delivering one instruction at a time also prevented the subject "short circuiting" the descriptions in cases where they might know the location of the destination, and therefore not need to interpret the route directions. For a set of four routes, we determined a set of instructions based on landmarks, and a set based on street names (Fig 2).

The routes varied in 1) the type of junction encountered, 2) the views offered across the city (both restricted and open vista), 3) the mix of landmark types and 4) a mix of pedestrian walkways, steps, footpaths and streets. For logistical purposes the routes were placed "end to end" and brought the observer and the subject back to near the start of the experiment. It was not possible to see one route from another (something that might have influenced the outcome). The length of each leg was such that it took about 1.5 hours to complete the experiment. A subject might stop for a variety of reasons (because they are lost, to search for a street name or landmark, for traffic lights). So an observer followed the subject.

The app was programmed in Java and ran on Android OS. The instructions were stored on the SD card in CSV format. The App used Google Fused Location Provider to control positioning information. The phone itself recorded: 1) a user input rating of each instruction in terms of its ease of comprehension, and speed of interpretation using a simple 1–5 star scale, 2) the time taken for the subject to complete each route; 3) accelerometer and GNSS based trajectory data, from which velocities, and dwell points could be determined, and 4) whether the subject was feeling lost at any time along the route. When a new instruction was given, the previous instruction was placed at the top of the screen, to help provide continuity should the user become lost and wish to reconsider the previous instruction. The user rated the quality of the instruction by using a simple star rating (Fig 3).

Fig. 3. (a) initial orientating instruction, (b) instruction at a decision point, (c) conformation instruction (the red indicates an instruction not correctly triggered due to GNSS positioning error/ canyon effect)

3.1   Subject Selection and Questionnaire Design

The call for subjects was promoted across several Facebook community groups, subjects were paid, and ethics and safety instructions were completed prior to commencement. The population comprised 15 males and 15 females; the average age of the population was just over 26 years, and none of the subjects had physical disabilities. The time spent living in Edinburgh ranged from 2 months to 62 years. About half of the subjects used Smartphones on a regular basis. Each subject did two of the legs using landmark instructions and two using street based instructions. The order and sequence in which subjects used these different instruction sets was managed by the phoneApp and the subjects were not aware of which system they were using during each leg.

At the end of the experiment, subjects were asked to complete a questionnaire that helped 1) to compare the two approaches, 2) to assess the ease of interpretation, 3) to determine the sufficiency of the instructions (non-ambiguous, sufficient to avoid getting lost), 4) to comment generally about their experience, 5) to comment on their familiarity with the area. Beyond this comparison, the primary goal was to develop a richer understanding of the "what and when" to provide landmark based instructions.

4   Observations

Analysis and observation came from two sources: analysis of the accelerometer and GNSS based trajectory data recorded on the smartphone, and observation of the subject and their responses to a questionnaire. The first step was to determine, and account for where subject's slowed or stopped though care was required in the interpretation of such data, since subjects were observed to have stopped for a variety of reasons. External factors such as congested pavements and traffic lights can slow or stop the pedestrian. Pedestrians slowed or stopped when hunting for the next visual clue, or when suspecting that they might be lost.

4.1   Analysis of Trajectories

Fig 4 shows the paths taken by subjects in response to the instructions (revealing a few who became lost for sections of the route). The location of each cultured dot is the point at which the text was displayed on the mobile device, and its color is the average level of satisfaction (from 5 – yellow – very effective instruction, to 2 - red – hard to interpret instruction). Three observations are made: 1) there are more instructions in the landmark based experiment (Fig 4b) since it was possible to take advantage of many more landmarks than there are streets, 2) subjects were far happier with landmark based instructions than street names (lighter cultured dots), 3) there is some correlation between poorer ratings being given to both forms of instructions and the complexity of the route at that point.

Various attempts have been made at modeling complexity (for example interior spaces [31] and in multi modal transportation systems [32]). Within the city model we modeled: visibility of roads (affected by changing gradient and bendy-ness of the road), the shape and degree node of the junction, and proximity from one decision point to another (junction density). A more complete model would include urban morphology (orthogonal networks are easier to navigate than mixed orientations and lengths), and the direction in which the route is being traversed (ease of instruction following is partly dependent upon direction in which the route is taken).

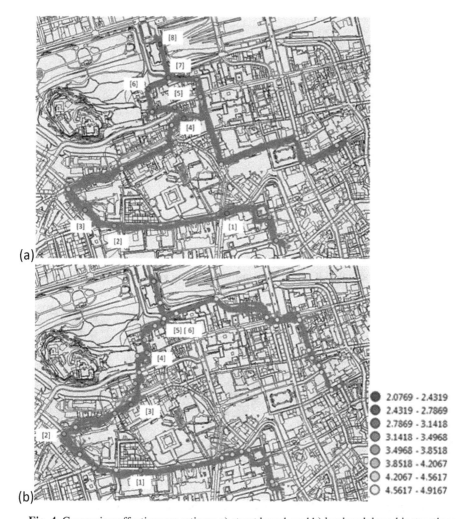

Fig. 4. Comparing effectiveness ratings: a) street based, and b) landmark based instructions

The distance it took between reading the instruction, and the user's comprehension of it was recorded (Fig 5 – blue for landmark instructions, red for street based). The red line reveals that the landmark based instruction was consistently understood sooner than street name instructions – though not by a significant amount of time. For example by 20m, 64% had rated the street based instruction, whereas 77% had rated the landmarked instruction. Because there were more landmark based instructions, the total elapsed time over the duration of the experiment was about the same. Though street names were difficult (or impossible) to find (Fig 6), it was not taking subjects much longer to find street names than it was to identify the landmark.

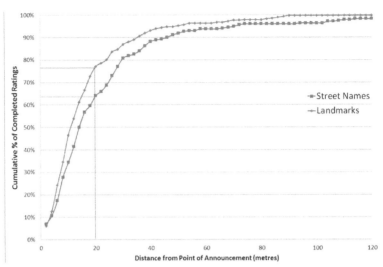

Fig. 5. Average distance interval between the instruction being given, and the user interpreting the instruction

<table>
<tr><td>(a)</td><td>(b)</td></tr>
</table>

Fig. 6. The Novotel on Lady Lawson street and the absence of any street sign

Fig 7a and 7b show the trajectories for all subjects, plotting distance against time for street based instructions (green), and landmark based (red). Gradient reflects speed; where the line rises steeply, the pedestrian is stood still. Negative gradients are possible where the subject was moving, but heading away from the destination. By applying kernel density estimation over the point clouds of Fig 7, it was possible to identify places along the route where subjects typically got lost or slowed during the process of searching for landmarks or street names. Table 4 is the average for all candidates, showing how the time was divided between going in the right direction, wrong direction or making no progress along the route (either standing still or moving but not in the direction of the destination). The distance to destination was calculated as the distance along the prescribed route from their current location. In effect this table is a summary of the trajectories shown in table 7a and 7b. Analysis revealed landmark following was marginally faster than street based, with greater consistency in the time taken (i.e. it was more variable among those following the street based instructions).

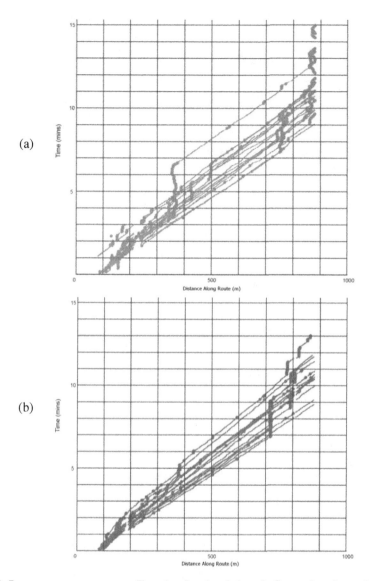

(a)

(b)

Fig. 7. 7a – green = streetname, 7b red = Landmark based. Comparing time taken overall against progress along route (larger blobs indicate low progression along the route).

Table 4. Progression along the route

	Landmark	Street based
Right direction	86.2%	82.4%
Wrong direction	5.7%	9.5%
No progress	8.1%	8.1%

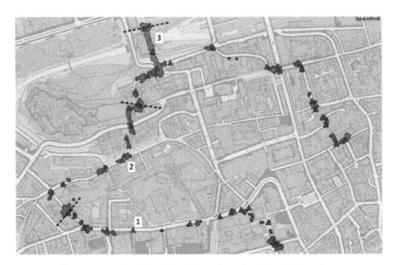

Fig. 8a. Results of landmark based analysis for all four legs

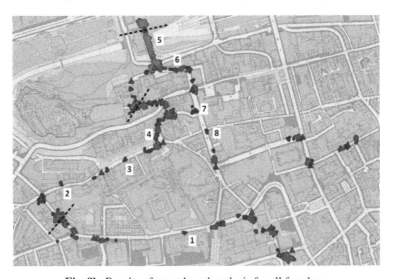

Fig. 8b. Results of street based analysis for all four legs

Fig 8a shows results from the landmark based experiment, combining announcement points and dwell points. Fig 8b is the same but for the street based directions. The dashed black line separates the legs; the triangles represent announcement points (purple indicating poor comprehension of the instruction and blue indicating good comprehension of instruction). The circles indicate dwell points (blue indicating that a few stopped, red indicates where many stopped). In Fig 8a [1] is an example of a poorly understood instruction, but where the pedestrian did not dwell. [2] indicates a slowing close to a poorly understood instruction – perhaps indicating uncertainty. [3] slowing where no instruction was given (or required), but where due to congestion, physical exertion climbing stairs or simply taking in the view meant that subjects slowed.

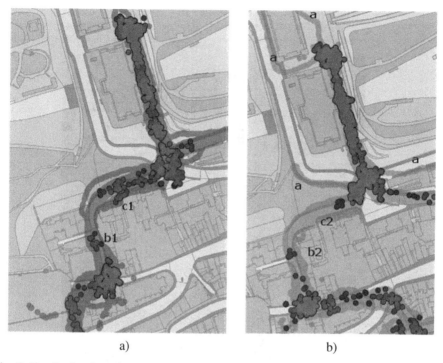

a)   b)

Fig. 9. Detail of trajectories comparing a) landmark based instructions with b) street based instructions

In Fig 8b, it is unclear why people paused at [1–6] and [7] indicates where people paused to cross a busy junction. Fig 9a is a detail of trajectories from 8a and Fig 9b is a detail from 8b. (a) indicates where subjects became lost under streetname guidance, (b1) and (b2) compares where people were tending to stop, as does (c1) and (c2). This might be because subjects are at junctions where they might expect an instruction, searching for a street sign, or where they are searching for landmarks that are in the field of view (see Section 4.2).

4.2   Urban Morphology, Junction Complexity, Interpretation of Instructions

The experiments revealed the subtle nature by which an instruction might be interpreted. Figure 5 shows the Novotel where users were given the instruction "take right fork by Novotel." For some, this led to uncertainty because they felt they should be standing by the main entrance (which would require them to cross the road), rather than looking across at it. This confusion was partially alleviated by the fact that Lady Lawson descends down a hill (Figure 5a) (something modeled in the database). Thus the full instruction was "take the right fork by Novotel, downhill for 40 meters." Downhill instructions were only included where the ratio of the difference in height against the length of a road segment was greater than 0.5 (its inclusion in this instruction is therefore fortuitous). The example illustrates the value of additional cues being built into the instruction.

(a)

(b)

Fig. 10. (a) The Black Watch monument – not so distinct against the trees, (b) The Playfair steps leading down to the Galleries - obscured by gathering crowds (blue arrow). Fig 10a is the view from c1,c2 in Fig. 9.

Fig 10 shows the crossing adjacent to a monument called "The Black Watch." The roads at this junction are of different gradients, and the angles at which they meet are "scissor-ed" rather than a simple "X." The Black Watch Monument is set to one side, not as clear to see given the relative height of surrounding tree vegetation (Fig 10 a). This is the location where quite a few subjects stopped (c1,c2 Fig 8). It is worth noting that from a different vantage point, and with a different background, the landmark becomes far more apparent. The ambition is get the pedestrian to cross the road so they can descend a set of stairs, also located at this road junction. The task is not made easy by the fact that the stairs are obscured by the crowds at the top of the stairs (Fig 10 b). The instructions, coming in close succession are: "at the crossroads go straight, towards the Black Watch monument," and "Immediately turn left and head down the Playfair steps." Though the instruction is topologically correct, the road shaping, angles and gradients coupled with the crowd obscured stairs meant that users sometimes struggled to follow the route at this point.

5 Analysis of Questionnaires

5.1 A Preference for Paper over Digital

From analysis of the questionnaire responses, a question on the acceptance of technology revealed a preference for non-digital forms of media (in particular paper maps). This was particularly the case for those following the street based instructions where all of the subjects expressed a desire to use an additional form of media. This may have reflected something of a frustration with street name based instructions, but there is no disguising the relative infallibility of the map:

"I am really not interested in electronic/GPS navigation – I would always just have a (paper) map. I much prefer to be more in control of where I am going and see where I am in relation with the rest of the city."

"It's fun to work out where you are on a map when you are exploring."

"I can fix a broken map with sticky tape myself but I can't fix a phone or any software myself so this reduces my feelings of independence."

5.2 Landmarks Easier to Interpret Than Street Names

The majority of subjects commented that landmarks were often easy to distinguish, whilst there were many negative comments that revealed the challenges of using a street name based approach:

"The lack of signage in Edinburgh meant that using only street names for directions wasn't particularly helpful."

"Reference to street names only very unhelpful, particularly in Edinburgh where there isn't a clear layout. Needs to be more descriptive."

"It didn't point out any landmarks or places of interest. Not done in a very engaging way."

This is contrasted to a landmark based approach:

"Really good and easy to follow instructions. Especially useful with easy to recognise places e.g. Doctors Pub or Novotel which let you know that you are definitely going in the right direction."

"Easy to use, very accurate and nice to have references (landmarks)."

5.3 Cardinality and Contextualization of Instructions

Subjects were asked what factors affected their capacity to interpret the instructions, and thus situate themselves. Their responses reflected a difficulty in gauging compass direction, and distances.

"Some of the instructions were unclear – i.e. 'head South' you would have to know which direction is South."

"I am not very good with depth perception so '100 m' etc made limited sense to me."

"Some of the instructions are not very clear however I do like that it gives you an estimation of how long it will take you to get to the next point as well as the distance."

5.4   Trust, Brevity, and Redundancy in Instruction

Some subjects desired more instructions when using the street based system as well as confirmatory cues:

"Instructions could have been more frequent to reassure me I had taken the right route."
"There was quite a long gap between instructions."
"They could have been clearer. Some description of the roads, including landmarks would have made things clearer. I found it difficult to know where I was going."

Some comments reflected the need to 'densify' instructions in areas of complex urban morphology:

"[...] They could have been clearer. For example, at the Mound there are a lot of different roads and the navigation instructions just said 'straight on' – they could have mentioned Playfair Steps as the sign is quite obvious.
The [landmark based] systems were more reassuring because they told you which buildings you should have walked past if you were going the right way."

6   Conclusion

That pedestrians prefer landmarks to other text based route instructions is well understood (though there are very few studies undertaken to analyze why). This study has tried to visualize and quantify the strength of those preferences. Key findings include:

1) where spoken/visual text is the sole means of conveying route following instructions, additional instructions are required as i) confirmatory cues, ii) at complex junctions, iii) where highly salient landmarks are few;

2) a city model is required that supports, among other things, functionality to i) detect complex junctions, ii) measure the visibility of landmarks and calculate landmark saliency, iii) construct rich descriptions of landmarks and the morphology of the city;

3) though subjects preferred landmarks to streets, the time taken was not significantly shorter.

The focus on a solely text based delivery system was in anticipation of delivery via dialogue based systems. Dialogue based systems are in pursuit of calm, concealed technology that takes account of the idea that pedestrians, at any given instance, are pursuing a basket of inter related tasks (crossing busy roads, talking with friends, avoiding other pedestrians, enjoying the city vista). We argue that the creation and form of delivery of wayfinding instructions needs to reflect this context. Dialogue interaction could have been used to resolve ambiguity and allow the user to request more detail should an instruction prove to be insufficient but given the complicating nature of dialogue based interaction, the experiment was deliberately simplified. The next phase of work will be to incorporate these ideas within a dialogue based context

— their correctness should ensure a minimum of interaction and a simplicity of interpretation that leaves the pedestrian to get on with the many other things they try and do as they race across the city!

Acknowledgments. The research leading to these results has received funding from the EC's 7th Framework Programme (FP7/2011-2014) under grant agreement no. 270019 (SpaceBook project).

References

1. Raper, J., et al.: Applications of Location Based Services: A Selected Review. Journal of Location Based Services 1(2), 89–111 (2007)
2. Bilandzic, M., Foth, M.: A review of locative media, mobile and embodied spatial interaction. Journal of Human Computer Studies 70(1), 66–71 (2012)
3. Jiang, B., Yao, X.: Location-Based Services and GIS in Perspective. Computers Environment and Urban Systems 30(6), 712–725 (2006)
4. Baldauf, M., et al.: Comparing Viewing and Filtering Techniques for Mobile Urban Exploration. Journal of Location Based Services 5(1), 38–57 (2011)
5. Kray, C., Baus, J., Cheverst, K.: A Survey of Map-Based Mobile Guides. In: Zipf, A. (ed.) Map Based Mobile Services - Theories Methods and Implementations, pp. 197–216. Springer, London (2005)
6. Kray, C., et al.: Presenting route instructions on Mobile Devices. In: Proceedings of the 8th Internatioanl Conference on Intelligent User Interfaces, pp. 117–124. ACM (2003)
7. Weisser, M., Brown, J.S.: Designing Calm Technology. In: Rutowski, T. (ed.) Annual of the American Center for Design, T, pp. 159–163. Watson-Guptill Publications, New York (1998)
8. Dunlop, M., Brewster, S.: The Challenge of Mobile Devices for Human Computer Interaction. Personal and Ubiquitous Computing 6(4), 235–236 (2002)
9. Frohlich, P., Simon, R., Baillie, L.: Mobile Spatial Interaction. Personal and Ubiquitous Computing 13(4), 251–253 (2009)
10. Mackaness, W.A., et al.: SpaceBook: Designing and Evaluating Spoken Dialogue Based Systems for Urban Exploration. In: GISRUK 2013. Liverpool (2013)
11. Janarthanam, S., et al.: A Multi-threaded Conversational Interface for Pedestrian Navigation and Question Answering. In: SIGDIAL 2013 (2013)
12. Bartie, P.J., Mackaness, W.A.: Development of a speech-based augmented reality system to support exploration of cityscape. Transactions in GIS 10, 63–86 (2006)
13. Cheverst, K., et al.: Developing a Context-aware Electronic Tourist Guide: Some Issues and Experiences. In: Proceedings of CHI 2000, Netherlands, pp. 17–24 (2000)
14. Gluck, M.: Making Sense of Human Wayfinidng: Review of Cogntive and Lingusitic Knowledge for Personal Navigation with a New Research Direction. In: Mark, D.M., Frank, A. (eds.) Cognitive and Linguistic Aspects of Geographic Space, pp. 117–135. Kluwer Academic Publishers, Dordrecht (1991)
15. Golledge, R.: Place Recognition and Wayfinding: Making Sense of Space. Geoforum 23(2), 199–214 (1992)
16. Caduff, D., Timpf, S.: On the Assessment of Landmark Salience for Human Navigation. Cognitive Processing 9, 249–267 (2008)
17. Millonig, A., Schechtner, K.: Developing Landmark Based Pedestrian Navigation Systems. In: 8th International IEEE Conference on Intelligent Transportation Systems (2005)

18. Raubal, M., Winter, S.: Enriching Wayfinding Instructions with Local Landmarks. In: Egenhofer, M., Mark, D.M. (eds.) GIScience 2002. LNCS, vol. 2478, pp. 243–259. Springer, Heidelberg (2002)

19. Tom, A., Denis, M.: Referring to Landmark or Street Information in Route Directions: What Difference Does it Make? In: Kuhn, W., Worboys, M.F., Timpf, S. (eds.) COSIT 2003. LNCS, vol. 2825, pp. 362–374. Springer, Heidelberg (2003)

20. Tom, A., Dennis, M.: Language and Spatial Cognition: Comparing the Roles of Landmarks and Street Names in Route Instructions. Applied Cognitive Psychology 18, 1213–1230 (2004)

21. Klippel, A., Winter, S.: Structural Salience of Landmarks for Route Directions. In: Cohn, A.G., Mark, D.M. (eds.) COSIT 2005. LNCS, vol. 3693, pp. 347–362. Springer, Heidelberg (2005)

22. Winter, S., et al.: Landmark Hierarchies in Context. Environment and Planning B: Planning and Design 35(3), 381–398 (2008)

23. Bartie, P.J., Mills, S., Kingham, S.: An Egocentric Urban Viewshed: A Method for Landmark Visibility Mapping for Pedestrian Location Based Services. In: Moore, A., Drecki, I. (eds.) Geospatial Vision. Springer, Heidelberg (2008)

24. May, A.J., et al.: Pedestrian Navigation Aids: Information Requirements and Design Principles. Personal and Ubiquitous Computing 7(6), 331–338 (2003)

25. Michon, P.-E., Denis, M.: When and Why are Visual Landmarks Used in Giving Directions. In: Montello, D.R. (ed.) COSIT 2001. LNCS, vol. 2205, pp. 292–305. Springer, Heidelberg (2001)

26. Tversky, B.: Pictorial and verbal encoding in short-term memory. Percept. Psychophysics 6, 225–233 (1969)

27. Endsley, M.R.: Towards a Theory of Situational Awareness in Dynamic Systems. Human Factors 37, 32–64 (1995)

28. Nothegger, C.: Automatic Selection of Landmarks. In: Institute of Geodesy and Geophysics. University of Technology, Vienna (2003)

29. Nothegger, C., Winter, S., Raubal, M.: Selection of Salient Features for Route Directions. Spatial Cognition and Computation 2, 113–136 (2004)

30. Schroder, C., Mackaness, W.A., Gittings, B.: Automating the Provision of Route Directions: The Requirements for Pedestrian Navigation Systems. Transactions in GIS 15(3), 419–438 (2011)

31. Raubal, M., Egenhofer, M.: Comparing the Complexity of Wayfinding Tasks in Built Environments. Environment and Planning B 25(6), 895–913 (1989)

32. Timpf, S., Heye, C.: Complexity of Routes in Multi Modal Wayfinding. In: GIScience 2002, Boulder, Colorado (2002)

Automatic Itinerary Reconstruction from Texts

Ludovic Moncla[1,2], Mauro Gaio[1], and Sébastien Mustière[3]

[1] LIUPPA, Avenue de l'Université Pau, France
[2] Computer Science and Systems Engineering Department, Universidad de Zaragoza, Spain
[3] Université Paris-Est, IGN, Laboratoire COGIT, Paris, 94160 Saint-Mandé, France

Abstract. This paper proposes an approach for the reconstruction of itineraries extracted from narrative texts. This approach is divided into two main tasks. The first extracts geographical information with natural language processing. Its outputs are annotations of so called expanded entities and expressions of displacement or perception from hiking descriptions. In order to reconstruct a plausible footprint of an itinerary described in the text, the second task uses the outputs of the first task to compute a minimum spanning tree.

1 Introduction

In the early nineties, Frank and Mark [1] wrote "It is conceivable that systems of the future might be able to assimilate and analyze explorer's journals such as Columbus' logs or the journals of Lewis and Clark, check them for consistency, and perhaps reach new inferences about the itineraries of their travels."

Since then, advances in automatic natural language processing (NLP), processing and representation of geographical information, but also the explosion of open geographical resources, have made developing such systems now possible.

In this paper, we propose a system for automatically reconstructing an itinerary from textual descriptions occurring in travelogues and guides.

The problem can be subdivided into two tasks. The first task entails annotating passages in the text that describe the various trips making up the itinerary. The second task entails creating a computational representation of the different descriptions, thus allowing the itinerary to be automatically reconstructed. Crucial to implementing such systems is the step known as toponym resolution. This essential step involves associating a non-ambiguous location with a place name[1].

Consider for example the following text from a true description of a hiking trail:

> "Cross Champagny-le-Haut and get around from the north of hamlet Friburge. You will see the Lac de la Plagne then walk to the refuge south of Lake Grattaleu." (1)

The proposed system proceeds as follows: the first task annotates the expanded spatial named entities *Champagny-le-Haut*, *hamlet Friburge*, *Lac de la Plagne* and *refuge south of Lake Grattaleu*. Some are associated with terms like *hamlet*, *lake*, *refuge* and/or spatial relations *from the north*, *south of*, allowing the ambiguity to be removed from

---

[1] Either a point, or a spatial footprint, in both cases expressed as geographic coordinates.

M. Duckham et al. (Eds.): GIScience 2014, LNCS 8728, pp. 253–267, 2014.

the nature of the geographic objects in question. This task will annotate some others spatial relations: *cross*, *get around*, *walk to*. Once all this information has been annotated, we apply a spatial analysis algorithm, guided by various clues obtained from the text, to reconstruct the itinerary. In this example, the expanded spatial named entities *Champagny-le-Haut*, *hamlet Friburge* and *refuge south of Lake Grattaleu* would be given priority while reconstructing the itinerary, as the spatial relations detected in the text imply the involvement of these entities in the itinerary to be taken. On the other hand, the spatial named entity *Lac de la Plagne* is not involved in the path, as the use of the verb *see* suggests.

This paper is set out as follows: Section 2 presents an overview of pertinent studies relating to the issue at hand; Section 3 describes our own contribution, proposal of a method of annotating expanded spatial entities, together with a method for automatic itinerary calculation; Section 4 describes our implementation and relates the early results of our experiments. Finally, in Section 5 we conclude the paper and propose future studies.

2   Relevant Work

2.1   Spatial Named Entity Recognition and Motion Expressions

Extraction and annotation of named entities is an important task in NLP, particularly in the case of automatic information extraction [2]. For named entity recognition and classification (NERC), two types of approaches have been proposed, those that use learning techniques and those based on ad-hoc rules. In the case of annotation of spatial entities in particular, there are also approaches that use external resources like gazetteers to search for and identify toponyms. These approaches can be used in a complementary manner in hybrid systems [3]. The ad-hoc approach relies on syntactic-semantic patterns developed manually with the help of experts. Amongst these rule-based approaches, several use transducers[2] with a finite number of states [2], which can also be used in cascade [4]. NERC methods automatically annotate different types of named entities: dates, people, organizations, themes, numeric values, as well as place names. There are a significant number of systems available, both proprietary and open source, such as OpenNLP[3] from Apache, OpenCalais[4] from Thomson Reuters, and CasEN [4]. More specific methods that are solely concerned with geographical data are known as geoparsing or toponym recognition [5]. The main difficulty in extracting geographical information is the ambiguity inherent in natural language. As stated in the introduction, there are actually several types of ambiguity involved in toponym resolution. In addition, a large number of spatial entity types exist: geopolitical entities (countries, administrative divisions), populated places (towns, addresses and postal codes), and natural geographical entities (parks, valleys, mountains, rivers, etc.), all of which can create ambiguities about the type of geographic object in question.

---

[2] Transducers are a type of finite-state machine that make insertions, replacements and deletions in a text.

[3] http://opennlp.apache.org/

[4] http://www.opencalais.com/

In itinerary analysis, it is not just spatial named entities that are important, but also their associated spatial relations. These enable the spatial named entity to be specified locally, as well as allowing the notion of movement between the different entities to be expressed. Many linguistic studies [6,7] deal with spatial relations with a view to describing the object to be located and the point of reference used. For French in particular, we could cite Vandeloise [8] with the term pair cible (target) and site (site), and Borillo [9] with the term pair entité concrète (concrete entity) and repère spatial (spatial reference). According to Talmy [10,11], a motion event is characterized by different conceptual components: a movement ("Motion"), a displaced object ("Figure"), a setting ("Ground"), a trajectory ("Path") and a "Manner". Syntactic parts of speech, in particular verbs, characterize a motion event. Many linguistic studies [12,13,14] have highlighted the importance of the use of motion verbs in language, especially in Romance languages. These studies suggest categorizing motion verbs according to their aspectual polarity. The three polarities are initial (e.g. to leave), median (e.g. to cross) and final (e.g. to arrive). The works also show the importance of the prepositions associated with these motion verbs. Without changing the intrinsic polarity of the verb, the preposition can change what it would be called the *focus*. More specifically, the association of a motion verb with a preposition of place (e.g. *from, in, at, to, by*, etc.) can change the focus of the displacement to take on the polarity of the preposition instead of the verb. Let us take the verb *to leave* for example. Alone or in association with the preposition *from*, the focus would be considered with initial polarity, but if used with the preposition for, the focus would then be considered as having final polarity. Undeniably, *leaving from Vienna* and *leaving for Vienna* have two radically opposite focus. If we consider the role played by the name, in one case, the place name is the origin of the displacement, and in the other case the place name is the destination. In the example *leaving for Vienna* it does not mean *to arrive in Vienna*, because we don't know if the destination is reached or not, but we know that we are leaving a place to go to *Vienna*. In terms of place name *Vienna* is the focus, so the polarity of the whole expression may be considered as final.

The use of contextual elements (other than toponyms), such as words that have a geographical denotation (downtown, valley, ridge, etc.), can be extremely important in toponym resolution and disambiguation [15]. In previous studies [16], we have put forward a method of marking non-toponymic terms associated with toponyms, especially those that have a topographical denotation (Wachau Valley, Lake Neusiedl, Saifnitzer Sattel, etc.).

2.2 Toponym Resolution

Toponym resolution [5] involves associating a non-ambiguous location with a place name. The use of resources like gazetteers is thus vital. In the last few years, we have seen a number of geographical resources emerge, such as Geonames[5], OpenGeoData[6],

---

[5] http://www.geonames.org/
[6] http://www.opengeodata.fr/

OpenStreetMap[7], Wikimapia[8], and BDNyme[9]. In an open data context, and with some benefiting from participative communities, these resources are expanding and being made more widely available through Web services. Some of these web-based geographic services are free, interoperable, and standardized, but the number and diversity of platforms makes using the data a complicated process. Before being able to use this mountain of data, first the most appropriate resources must be selected according to actual needs. Each resource can have different issues, for example the choice between a resource that covers a wide area but non-exhaustively and a more exhaustive resource covering a smaller area.

This resolution involves resolving the problem of ambiguities that toponyms may contain. Widely studied in recent years, the admittedly difficult task of toponym disambiguation remains a scientific problem today. According to [17] there are three main types of ambiguities: the same name is used for several places (referent ambiguity), the same place can have several names (reference ambiguity), the place name can be used in a non-geographical context, as in organizations or names of people (referent class ambiguity).

Using a corpus of hiking guides naturally reduce the number of ambiguities from referent class, as opposed to those used in a corpus of news articles for example. Then in this paper, we will focus on ambiguities resulting from the referent ambiguity class arose from the existence of homonyms (e.g. in a french formulation *Vienne* exists in Austria but *Vienne*, also exists in France) or arose from subtyping of toponyms [16]. Even once it has been identified that the reference is to the place named *Vienne* in France, ambiguities may remain concerning the geographic object that carries the name (*Vienne* the town, *Vienne* the county or *Vienne* the river). Another form of ambiguity arises from the presence of certain spatial expressions associated with the place name (e.g. *Paris-Nord railway station* is different from *the railway station in the north of Paris*).

A number of methods exist for disambiguating toponyms [18,19,20,21]. These methods can be classified into three categories [22]: map-based, knowledge-based, and data-driven or supervised. Many of these methods use toponyms that are geographically the closest to disambiguate the candidate toponyms. This can lead to poor results when important information is not included in the context, when the candidate toponym is not geographically close to non-ambiguous toponyms, or if it is not linked to a geopolitical entity [21]. Some studies use the notion of event to disambiguate toponyms. For example, Robert et al. [20] consider there to be three types of entities that participate in an event: people, organizations, and geopolitical entities. They use an ontology constructed from Geonames, and associate geopolitical entities with people and organizations using links from Wikipedia, but no other information or clues is used from the context. Knowledge-based methods use toponyms information extracted from gazetteers like importance, size, or population counts [23]. This kind of information is not the most suitable for a discriminating task in the case of documents describing hiking trails,

---

[7] http://www.openstreetmap.org/
[8] http://wikimapia.org/
[9] http://www.geoportail.gouv.fr/

because toponyms used in these documents are fine-grain toponyms or natural features such as mountains, lake, hamlet, and refuge [24].

These various methods are often applied to corpora of news articles [18,19,21] in which toponyms are associated with events, well-known figures or geopolitical entities and not with spatial relations. In this type of discourse, toponyms are not necessarily related to each other, and are not for example linked by motion events. Speriosu and Baldridge [21] show that toponym disambiguation methods that are based on the text (context extraction and interpretation of spatial relations) are more effective than methods based on metadata or heuristics that use distance calculations.

2.3 Wayfinding

With the rise of new needs and behaviors (e.g. route planning and tourist applications), the democratization of devices equipped with GPS and the wide availability of geographical information, the notion of itinerary is being studied more and more. Hao et al. [25] put forward a probabilistic model to identify place elements taken from travelogues. The aim of this work is to improve the tourist experience, providing them with information or recommendations about the places they are visiting. Zhang et al. [26] use these learning methods to extract the three elements they consider to be the most important in an itinerary: origin, destination, and the path taken (instructions). They work from a corpus of webpages where instructions giving directions can be found.

Other studies [27] have focused on ancient documents with the aim of finding and modeling historical roads that no longer exist. A large number of studies have looked into trajectories [28,29,30], with a focus on the movement of mobile objects (animal migrations, airplanes, ships, pedestrians, etc.). The concepts explored in these studies can be considered similar to those applicable to itineraries.

The notion of itinerary has already been a focus of research for our team. In previous studies [31], Loustau proposed a definition of the concept of itinerary, but more importantly contributed to the proposal of an initial approach to extracting constitutive information about itineraries.

3 Contribution

3.1 Problem Elucidation

Our aim is to identify geographical information in a text, as well as any textual clues that allow us to link some and exclude other information that should not be taken into consideration and then map the most likely route. In this study, we chose to test our approach on a french corpus of 1,295 descriptions of mountain bike and hiking trails in France.

As mentioned in the literature review (Section 2), spatial relations are an important factor in the disambiguation of toponyms. Particularly in forms of discourse, as can be found in our corpus and those of same categories where spatial relations exist on several levels of granularity and can be applied to the discourse at different scales. In this paper, we will examine two levels of granularity. The first involves local spatial relations that are part of a spatial named entity. To illustrate our discourse, let's take the next example (1) page 253. In the spatial named entity *south of Lake Grattaleu* the spatial relationship contained *south of* needs to be interpreted in order to solve the ambiguity of the referent. The second level involves spatial relations that associate various spatial named entities or a participant with a spatial named entity relative to another. In case of a hiking trail, the object under consideration is the participant in the motion event. In *Cross Champagny-le-Haut and get around from the north of hamlet Friburge* the spatial relations are *the north* and *around*, but also the motion verb *to cross* and *to get*. This is a description of a motion event relative to spatial named entities. Finally the toponym *Lac de la Plagne* is associated with a perception verb *to see*, which means that the toponym is not really part of the itinerary taken but a visual landmark. Moreover, the term *lake* serves to precisely identify the toponym, removing all ambiguity from the geographic object being referred to, i.e. it is most definitely a lake, and not a town for example.

3.2   Solution Adopted

The first step in our approach is a system whereby spatial expressions described in textual documents are automatically annotated. The method combines the notions of marking and extraction of named entities, through the use of local grammars[10] [33] or external resources (lexicons, gazetteers, etc.). The second step in our approach is a system capable of interpreting and linking information in order to automatically reconstruct an itinerary.

Expanded Spatial Named Entities. We define an *expanded spatial named entity* (*ESNE*) as an entity built from a proper name attributed to a place. This proper name can be associated with one or more ontological concepts with a geographical sense, and with one or more concepts relating to the expression of location in the language (spatial relations). For example *the north of hamlet Friburge* and *the refuge south of Lake Grattaleu* are two *ESNE* built from proper names (*Friburge* and *Grattaleu*), associated with ontological concept having a geographical sense (*hamlet*, *refuge* and *lake*), and spatial relations (*the north of* and *south of*).

We integrate spatial relations into a more generic concept that we have called *indirections* allowing a geographic object to be addressed indirectly. Indirections can be a part of the concrete entity, their role being to specify location, and grammatically they can belong to different word classes (prepositions, adverbs or adjectives). For example, in the *ESNE the north of hamlet Friburge*, as the toponymic name is the word *Friburge*, the concrete entity is *the north of hamlet*, which contains the indirection *the north*. Other parts of speech are annotated to reveal spatio-temporal sequences in the discourse, such as spatial adverbs of location [34]. These are prepositions of place, which occur frequently in hiking guides (e.g. here, there, near of, from the left, etc.). These prepositions structure the discourse by describing a spatial sequence (a step in a journey) and/or a temporal sequence (a succession of events).

Unlike traditional named entity annotation tools, we only annotate spatial named entities. We follow the proposal of Gaio et al. [35] for the recognition of spatial references

---

[10] These grammars are lexicalized graphs that make use of dictionaries and have the advantage of being able to be applied directly to texts. [32]

and spatial relations in language, as well as using a hybrid approach [36] combining the three main categories of spatial relations: topological relations [37], distance relations, and directional relations [38]. In order to establish the steps of an emerging route, itinerary is defined as being a special type of spatial relation. It is a spatio-temporal sequence of steps moving between different places. An itinerary could thus be thought of as a succession of spatial relations.

Expression of Motion. Based on preposition polarity and classification of verbs, we are able to establish simple linguistic rules in order to extract the source or target named entities in a motion event. We classified verbs into categories: motion verbs (*to go, to leave,* etc.), verbs of visual perception (*to glimpse, to see,* etc.), verbs we refer to as topographic (*to converge, to overhang,* etc.), topographic verbs are used when the narrator is describing a place using its topographical features. And finally location verbs (*to locate, to be,* etc.) [39]. The use of this last class of verbs, from a syntactic point of view, is very similar to that of motion verbs as previously described. They are often associated with prepositions of location or place names in hiking guides. They can be used, for example, to describe a step or stop in a journey. They also allow for better spatial representation and facilitate the location of the different events relative to each other. In order to formalist the relationship between expanded spatial named entities and verbs in travelogues, we use *VTo structures* [35,16]. *VTo structures* are formally defined as V, I, T, G: groups of classified verbs, indirections, geographical terms, and place name, respectively. $VTo = (v, t)$ where v is an instance of V and the t set is defined as $t = (te, i, g|t)$, in such a way that te is an instance of T, i is an instance of I and g is an instance of G. The symbol $g|t$ indicates that the third t group can be made up of either t (recursion) or g.

Here is a *VTo structure* from example (1):

{{get around,.v}{{from the north of,.i}{{hamlet,.te}{Friburge,.g},.t},.t},.VTo}

Itinerary Reconstruction. This final step entails using the information annotated in previous steps to reconstruct a plausible footprint of the itinerary.

Toponym Resolution. The information gathered from the marking process during the annotation step includes candidate spatial entities and candidate *VTo structures*. We use gazetteer-style external resources to verify the existence of toponyms and obtain their geographic coordinates. When toponyms exist in one of the resources they are validated. If the validated toponyms are part of a candidate expanded spatial entity, the entity is automatically validated. However, ambiguities still occur, for example when an entity is made up of several words and only one of those is validated. Toponyms may also be associated with several locations when the name occurs several times in the resources with different locations, e.g. *Pau* occurs nine times in BDNyme and three times in Geonames. In addition to those ambiguities, one may notice that every spatial named entity mentioned is not necessarily part of the itinerary. Some clues extracted from the text, such as negations, descriptions of places and scenic lookouts (e.g. with the use of perception verbs), allow us to conclude that some places should probably not be included in the representation of the itinerary, or only in a certain way.

Itinerary calculation. This step combines contextual information extracted from the text with geometric information extracted from gazetteers. This combined spatial and textual analysis aims at resolving some ambiguities and reconstructing the footprint of the itinerary. The approach proposed here is based on the idea that the most probable itinerary linking a set of places is the route linking all places and with a minimal length, to "optimize" the displacement. Finding this optimal itinerary should help to remove ambiguities or places appearing in the text but not actually crossed. This naturally leads to the notion of "minimal weighted spanning tree". The minimum weight spanning tree of a set of points is the tree connecting all the points together with the minimum weight, this weight being the sum of the weights of the edges linking points (e.g. equals to the distance between points). The implementation of the continuity detection with the notion of minimum spanning tree is not a new idea. It has already been developed for example by Zahn [40] for detecting clusters of points.

4   Implementation

As described in Section 3, we developed a two-step solution. The first step (Fig. 1a) entails annotating toponyms and the various spatial relations (indirections, expressions of motion). The second step (Fig. 1b) entails calculating an itinerary between the different toponyms previously annotated. We will now explain the implementation of our method in more detail.

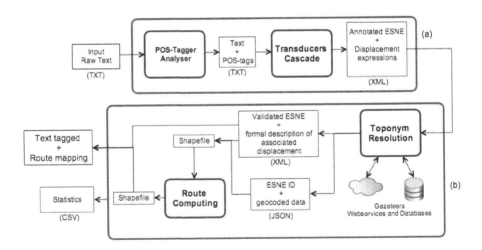

Fig. 1. Block diagram of our processing chain

4.1   Spatial Annotation for Itinerary Calculation

We developed an annotation processing chain[11] (Fig. 1(a)) that takes a raw text (written in natural language) as input. The main annotation module was designed using the finite-state transducer cascade creation program CaSys [4] available on the Unitex platform[12]. Transducers are represented by graphs on the Unitex platform, which simplifies both writing and maintenance. The cascade allows all or some of the elements labeled by preceding transducers to be used in those that follow. The cascade designed for annotation comprises six main transducers that mark the text in the following order: indirections, candidate toponyms, classified verbs, candidate terms with a geographical denotation, $ESNE$ and *VTo structures*.

Let us illustrate the execution of this cascade with an example. Take the sentence from example (1): "walk to the refuge south of Lake Grattaleu." The **indirection** transducer will be applied first. Indirections (*at, nearby, south*, etc.) are sought out with the help of lexicons and linguistic patterns (*the southern part, at the center of*, etc.) that make use of these lexicons. Our first main transducer gives the following output:
Walk to the refuge {{south,.**directional**} of,.**indirection**} Lake Grattaleu.

The second transducer of the cascade marks **candidate toponyms**. This transducer is designed to recognize the various complex forms of toponym construction. There even exist toponymic guidelines published by IGN[13]. For example, toponyms can be composed and formed of several distinct terms that may be accompanied by an article (e.g. *Champagny-le-Haut*). In our example the toponym is tagged like this: {Lake Grattaleu,.**toponymCandidat**}

The third transducer annotates **classified verbs**. It relies on subgraphs designed to label verbs according to different categories (motion verbs, position verbs, perception verbs, and topographic verbs). For motion verbs this transducer also specify the polarity of the verb (initial, median, or final). In our example the verb is tagged like this: {Walk,.**motionVerb+median**}. The fourth transducer annotates **common nouns** or common noun phrases. These will then be identified or not as candidate terms with a geographical denotation, thanks to *VTo structures*. Execution of this transducers tag common noun as follow: {the refuge,.**commonNoun**} The fifth transducer of the cascade annotates **candidate expanded spatial named entities**. The indirections, candidate toponyms, and candidate common nouns with a geographical denotation have already been annotated by the preceding transducers. This transducer uses these previously carried out annotations. Execution of this transducer finds this $ESNE$: {{the refuge,.commonNoun} {{south,.directional} of,.indirection} {Lake Grattaleu,.toponymCandidat},. **ESNE**}.

The final transducer in our cascade annotates **VTo structures**. It behaves in the same way as the previous transducer, in that it uses previously made annotations.

In our example, the observed structure is composed of a motion verb and an expanded spatial named entity. Execution of our cascade of transducers gives the following final output:

---
[11] Demonstration tool available online: http://erig.univ-pau.fr/PERDIDO/
[12] http://www-igm.univ-mlv.fr/~unitex/
[13] http://www.ign.fr/sites/all/files/charte_toponymie_ign.pdf

{{Walk,.motionVerb+median} to {{the refuge,.commonNoun} {{south,. directional}
of,.indirection} {Lake Grattaleu,.toponymCandidat} ,.ESNE},**.VTo**}.

The different annotations (toponyms, spatial named entities, *VTo structures*) are candidate annotations, i.e. they require a further step of validation. This is done next during
toponym resolution.

4.2   Experimental Results for the First Step

Our first experiment was to run our automatic tagging chain for french texts on a body
of reference in order to compare the results obtained automatically with those obtained
manually.

For the moment we have chosen the resources Geonames and BDNyme as they complement each other. BDNyme is the French benchmark toponymic database provided
by IGN. It lists 1,500,000 place names resulting from toponyms and georeferenced activities and points of interest. Geonames may only list around 135,000 names on French
soil, but it is especially useful for cross-border trail descriptions where part of the descriptions may refer to locations outside France.

To evaluate our method, we wanted to know the rate of correct recognition of spatial
named entities and if they are present or not in a *VTo structure*. The aim is also to
identify errors and causes in order to subsequently improve our processing chain.

Currently our body of reference is composed of 24 randomly selected hiking guides
manually annotated (ground truth) over 1295 available[14]. Each document of our corpus
has an average of:

  – 263.3 words (269.2 on the body of reference) with a standard distance of 188.5
    (242.9 for the body of reference) ;
  – 12.12 candidate toponyms (13.87 on the body of reference) with a standard distance
    of 9.88 (12.71 for the body of reference) ;
  – 4.72 candidate toponyms included in a *VTo structure* (5.46 on the body of reference)
    with a standard distance of 4.14 (4.31 for the body of reference).

The corpus has an average of 5.09% of candidate toponyms for 100 words (5.44%
on the body of reference). Furthermore 39.33% of candidate toponyms are in a *VTo
structure*. Figure 2 shows for each document of the body of reference, the number of
candidate toponyms that are included in a *VTo structure* over the total number of candidate toponym.

Of the 366 spatial expanded named entities present in the body of reference, 325
are correct recognitions and we get 49 false recognition which gives us a precision[15] of
85.37% and a recall[16] of 88.80%.

41 toponyms (11.20%) were not detected by the processing chain. Detection errors
(false recognition or non-detection) are due to several problems. The main problem is
the morpho-syntactic analyzer that tag proper names as common names. In this case if

---

[14] We are now developing a tool for a controlled manual tagging in order to easily enrich the
body of reference.

[15] Fraction of retrieved instances that are relevant.

[16] The fraction of relevant instances that are retrieved.

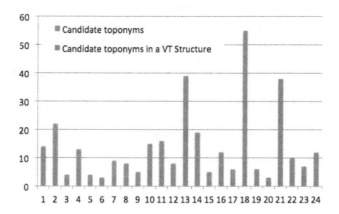

Fig. 2. Number of candidate toponyms in *VTo structure* over the total number of candidate toponyms per document

a proper name is considered as a common name, the annotation transducers of *ESNE* is not triggered. Some bad recognitions are also due to errors in the linguistic patterns described in the transducers, for example: missing rules for road names (like *GR55, A10, M25, etc*), this last kind of errors where corrected during the experiments.

49 toponyms (14.63%) are false detection, this problem is mainly a problem of ambiguities (e.g., toponyms recognized instead of names of people or organizations). 46.97% of the toponyms are in a *VTo structure*. These figures are partly explained by the fact that many cases are not listed as *VTo structures* (examples: *Continue to...*, *Take direction...*, etc). In addition, 59.33% of *VT structures* are not composed of toponyms. But they are composed of common nouns referencing toponyms, we call this *VTr structures*. This important amount of *VTr structures* can be explained by the type of corpus studied. Indeed a number of descriptions of hiking use less of toponyms and more of landscape features or spatial relationship (for example: "walk along the trail far as the church"). Finally, 69.93% of well localized toponyms are detected by at least one resource. Among these toponyms localized, 63.50% are located in the two resources, 19.00% are geo-coded only by BDNyme and 17.50% are geo-coded only by Geonames. The figures about the toponyms resolution take no account of ambiguities. These first results concerning the annotation of expanded spatial named entities in a corpus of real descriptions of hikes, are very encouraging. Following this first experiment our objective is to correct and enrich some transducers.

4.3   Itinerary Calculation

The second step (Fig. 1(b)) in our method entails reconstructing an itinerary from the elements annotated in the first step. It can be divided into two tasks. The first task is toponym resolution, where candidate toponyms are validated. The second task is the true itinerary calculation step.

For toponym resolution, we developed a module that consults external resources to validate the existence of toponyms and obtains their geographic coordinates. This module takes as input the output of our transducer cascade. This step further improves the content of our annotated document (using XML markup), and thanks to unique identifiers links it to the validated and geocoded spatial named entities (in a separate document in JSON format). The toponym validation program also generates a file in shape format containing information on the spatial entities, such as the name, geographic coordinates, associated classified verb and its polarity. In this way, all additional information can be easily added to one of the three documents according to its purpose.

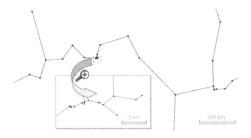

Fig. 3. Disambiguation based on the length of edges of the tree

As suggested before, we compute a minimum spanning tree algorithm in order to link the various located spatial entities. A first approach is to directly weight edges of the tree with the Euclidean distance between places. As exemplified in Figure 3, this tree can be use to disambiguate toponyms: in the actual area of interest (zoomed area in the figure), actual places are close to each other, and a simple selection of places linked by short edges in the tree is useful to focus on the area of interest and reject other toponyms.

Once this focus on the area of interest is done, the tree may be used to reconstruct an approximation of the actual itinerary. Figure 4 illustrates this idea applied to the description of hiking route, associated with a "ground truth" of the itinerary collected by GPS. Reconstructing the itinerary by simply ordering places as they appear in the text is of course inefficient because the discourse is rarely so linear (Fig. 4(b)). The result of the automatic creation of a minimum spanning tree on these locations is more efficient: the longest line of this tree is a first approximation of the itinerary (Fig. 4(c)). However, the built itinerary goes through places seen but not passed through. If one computes the tree minimizing the length weighted by information automatically extracted from the text, the result can be improved: one may under-weight edges linking places associated to a displacement verb, and over-weight edges linking places associated to a perception verb (Fig. 4(d)). In this example, the built itinerary is close to the actual one.

Those first experiments illustrate that neither language analysis alone, nor spatial analysis alone, may be sufficient. However a combination of those analyses is promising and should be explored more in depth to build itineraries or other spatial configurations from texts. Such a method could be enriched to consider other information issued from the text analysis, such as the polarity of verbs, negative forms, and so on.

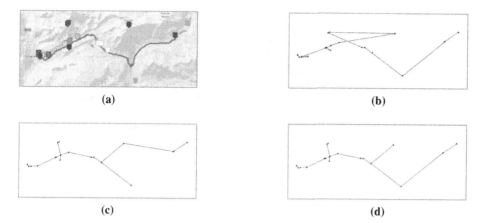

Fig. 4. Actual displacement recorded by GPS and spatial entities detected in the associated text (up), and itineraries built with different strategies (b,c,d)

5   Discussion and Outlook

This paper proposes an approach to reconstruct a plausible footprint of an itinerary extracted from narrative texts. This approach was divided into two main tasks. The first annotates expanded spatial named entities and expression of displacement or perception from textual hiking descriptions. The second task is a method to reconstruct itineraries as from the elements marked in the first stage. This method computes a minimum spanning tree weighted using information extracted from the text.

According to preliminary tests, the results are encouraging. The use of evidence extracted from the text (e.g. verbs of displacement, verbs of perception, spatial relationship etc) associated with the names of places improves the results of the minimum spanning tree. Furthermore our first observations show that disambiguation of names can be partly solved by the route calculation in the context of a body of descriptions of hiking. However, the first evaluation of our tagging method has identified some errors. A first short-term perspective is to increase the accuracy of the first task by refining transducers.

But our main goal is to improve the second task. Unlike many studies that use only metadata from geographic resources (e.g. number of inhabitants, area, etc.), we believe that information extracted from texts allows a better interpretation. As a first experimental approach we used the Euclidean distance to compute distances between places but it could be more efficient to use other methods like travel effort or a combination of information like the affordance. In addition to the concept of continuity already used in our calculation method of route, we plan to introduce in the method the two others laws of the gestalt theory (the similarity/complementarity and the proximity). Furthermore, the possibilities of interaction that a human could have with a named spatial object change the interest addressed to that object. This is why, like [41], we plan to integrate in our method the notion of affordance. According to the authors the affordance of a place consists of the following aspects: physical features, actions, narrative, symbolic representations, or socioeconomic and cultural factoring typologies. All these aspects are strongly present in the itineraries descriptions.

References

1. Frank, A.U., Mark, D.M.: Language issues for GIS. In: David, J., Maguire, M.F.G., Rhind, D.W. (eds.) Geographical Information Systems: Principles and Applications, pp. 147–163. Longman Scientific & Technical, Essex (1991)
2. Poibeau, T.: Extraction automatique d'information: du texte brut au web sémantique. In: Extraction Automatique D'information: Du Texte brut au web Sémantique. Hermès Lavoisier (2003)
3. Béchet, F., Sagot, B., Stern, R.: Coopération de méthodes statistiques et symboliques pour l'adaptation non-supervisée d'un système d'étiquetage en entités nommées. In: TALN 2011, Montpellier, France (2011)
4. Maurel, D., Friburger, N.: Finite-state transducer cascades to extract named entities in texts. Theoretical Computer Science 313, 93–104 (2004)
5. Leidner, J.L.: Toponym Resolution in Text: Annotation, Evaluation and Applications of Spatial Grounding of Place Names. Universal-Publishers (January 2008)
6. O'Keefe, J.: The spatial prepositions in english, vector grammar, and the cognitive map theory. In: Language and Space, pp. 277–316 (1996)
7. Bloom, P.: Language and space. MIT press (1999)
8. Vandeloise, C.: L'espace en français. Seuil, Paris (1986)
9. Borillo, A.: L'espace et son expression en français, l'essentiel. In: L'espace et son Expression en Français, L'essentiel. Orphrys (1998)
10. Talmy, L.: Lexicalization patterns: Semantic structure in lexical forms. In: Shopen, T. (ed.) Language Typology and Syntactic Description. Grammatical categories and the lexicon, vol. 3, pp. 57–149. Cambridge University Press, Cambridge (1985)
11. Talmy, L.: Toward a cognitive semantics. In: Toward a Cognitive Semantics. The MIT Press (2000)
12. Boons, J.P.: La notion sémantique de déplacement dans une classification syntaxique des verbes locatifs. Langue Française (76), 5–40 (1987)
13. Slobin, D.I.: Two ways to travel: Verbs of motion in english and spanish. In: Grammatical Constructions: Their Form and Meaning, pp. 195–219 (1996)
14. Aurnague, M.: How motion verbs are spatial: The spatial foundations of intransitive motion verbs in french. Lingvisticae Investigationes 34(1), 1–34 (2011)
15. Hollenstein, L., Purves, R.: Exploring place through user-generated content: Using flickr todescribe city cores. Journal of Spatial Information Science (1) (2010)
16. Nguyen, V.T., Gaio, M., Moncla, L.: Topographic subtyping of place named entities: A linguistic approach. In: The 15th AGILE International Conference on Geographic Information Science. Louvain, Belgique (2013)
17. Smith, D.A., Mann, G.S.: Bootstrapping toponym classifiers. In: Proceedings of the HLT-NAACL 2003 Workshop on Analysis of geographic References, vol. 1, pp. 45–49. Association for Computational Linguistics, Stroudsburg (2003)
18. Garbin, E., Mani, I.: Disambiguating toponyms in news. In: Proceedings of the Conference on Human Language Technology and Empirical Methods in Natural Language Processing, HLT 2005, pp. 363–370. Association for Computational Linguistics, Stroudsburg (2005)
19. Buscaldi, D., Magnini, B.: Grounding toponyms in an italian local news corpus. In: Proceedings of the 6th Workshop on Geographic Information Retrieval, GIR 2010, pp. 15:1–15:5. ACM, New York (2010)
20. Roberts, K., Adrian Bejan, C., Harabagiu, S.: Toponym disambiguation using events. In: Proceedings of the Twenty-Third International Florida Artificial Intelligence Research Society Conference (FLAIRS 2010), pp. 271–276 (2010)

21. Speriosu, M., Baldridge, J.: Text-driven toponym resolution using indirect supervision. In: Proc. 51st Annual Meeting of the Association for Computational Linguistics (ACL), pp. 1466–1476. Sofia, Bulgaria (2013)

22. Buscaldi, D.: Approaches to disambiguating toponyms. SIGSPATIAL Special 3(2), 16–19 (2011)

23. Overell, S., Rüger, S.: Using co-occurrence models for placename disambiguation. International Journal of Geographical Information Science 22(3), 265–287

24. Derungs, C., Purves, R.S.: From text to landscape: locating, identifying and mapping the use of landscape features in a swiss alpine corpus. International Journal of Geographical Information Science 1–22

25. Hao, Q., Cai, R., Wang, C., Xiao, R., Yang, J.M., Pang, Y., Zhang, L.: Equip tourists with knowledge mined from travelogues. In: Proceedings of the 19th International Conference on World Wide Web, WWW 2010, pp. 401–410. ACM, New York (2010)

26. Zhang, X., Mitra, P., Klippel, A., MacEachren, A.: Automatic extraction of destinations, origins and route parts from human generated route directions. In: Fabrikant, S.I., Reichenbacher, T., van Kreveld, M., Schlieder, C. (eds.) GIScience 2010. LNCS, vol. 6292, pp. 279–294. Springer, Heidelberg (2010)

27. Breier, M.: The way is the Goal–Modelling of historical roads. In: 26th International Cartographic Conference (August 2013)

28. Lee, J.G., Han, J., Li, X.: Trajectory outlier detection: A partition-and-detect framework. In: Alonso, G., Blakeley, J.A., Chen, A.L.P. (eds.) ICDE, pp. 140–149. IEEE (2008)

29. Kim, J., Sridhara, V., Bohacek, S.: Realistic mobility simulation of urban mesh networks. Ad. Hoc. Networks 7(2), 411–430 (2009)

30. Yuan, Y., Raubal, M.: Extracting dynamic urban mobility patterns from mobile phone data. In: Xiao, N., Kwan, M.-P., Goodchild, M.F., Shekhar, S. (eds.) GIScience 2012. LNCS, vol. 7478, pp. 354–367. Springer, Heidelberg (2012)

31. Loustau, P., Nodenot, T., Gaio, M.: Spatial decision support in the pedagogical area: Processing travel stories to discover itineraries hidden beneath the surface. In: AGILE Conf.

32. Constant, M.: Grammaires locales pour l'analyse automatique de textes: méthodes de construction et outils de gestion. PhD thesis, Université Paris-Est (2003)

33. Gross, M.: The Construction of Local Grammars. In: Schabès, E.R.Y. (ed.) Finite-State Language Processing, pp. 329–354. MIT Press (1997)

34. Borillo, A.: Quand les adverbiaux de localisation spatiale constituent des facteurs d'enchaînement spatio-temporel dans le discours. In: Information Temporelle, Procédures Et Ordre Discursif, Genève, pp. 123–138 (2004)

35. Gaio, M., Sallaberry, C., Nguyen, V.T.: Typage de noms toponymiques à des fins d'indexation géographique. TAL 53, 1–35 (2012)

36. Gaio, M., Sallaberry, C., Etcheverry, P., Marquesuzaà, C., Lesbegueries, J.: A global process to access documents' contents from a geographical point of view. Journal of Visual Languages & Computing 19(1), 3–23 (2008)

37. Egenhofer, M., Franzosa, R.: Point-set topological spatial relations. International Journal for Geographical Information Systems 5(2), 161–174 (1991)

38. Frank, A.U.: Qualitative reasoning about distances and directions in geographic space. Journal of Visual Languages and Computing 3(4), 343–371 (1992)

39. Borillo, A.: A propos de la localisation spatiale. Langue Française 86(1), 75–84 (1990)

40. Zahn, C.T.: Graph-theoretical methods for detecting and describing gestalt clusters. IEEE Transactions on Computers 100(1), 68–86 (1971)

41. Abdalla, A., Frank, A.U.: Combining trip and task planning: How to get from a to passport. In: Xiao, N., Kwan, M.-P., Goodchild, M.F., Shekhar, S. (eds.) GIScience 2012. LNCS, vol. 7478, pp. 1–14. Springer, Heidelberg (2012)

Integrating Sensing and Routing
for Indoor Evacuation

Jing Wang[1,2], Stephan Winter[2], Daniel Langerenken[3,2], and Haifeng Zhao[2]

[1] School of Mathematics and Computer Science, Wuhan Polytechnic University, China
[2] Department of Infrastructure Engineering, The University of Melbourne, Australia
[3] Cognitive Systems Group, University of Bremen, Germany

Abstract. Indoor evacuation systems are needed for rescue and safety management. A particular challenge is real-time evacuation route planning for the trapped people. In this paper, an integrated model is proposed for indoor evacuation used on mobile phones. With the purpose of employing real-time sensor data as references for evacuation route calculation, this paper makes an attempt to convert sensor systems to sensor graphs and associate these sensor graphs with route graph. Based on the integration of sensing and routing, sensor tracking and risk aware evacuation routes are generated dynamically for evacuees. Experiments of the proposed model are illustrated in the paper. The benefit of the integrated model could extend to hastily and secure indoor evacuation and it potentially presents an approach to correlate environmental information to geospatial information for indoor application.

1 Introduction

Applying geographic information systems (GIS) for indoor environments has gained attention in way-finding contexts and in building facility management. The challenges for indoor GIS include the representation of indoor space for routing tasks and the integration of relevant building facilities. There is no standard representation for routing networks in indoor space until now because indoor space has multiple functional purposes which lead to different proper representations for the same indoor environment. Furthermore, people move more freely in indoor space than vehicles move in outdoor space [1]. Many data structures realize in a unique way the representation of the navigable space in an indoor environment, with varying advantages and disadvantages [2–5].

The prior research on indoor space representation especially focused on utilization of the routing capabilities of GIS in response to indoor disasters, such as fire, gas leakage or terror attacks. This research was also used for simulation of navigation in evacuation scenarios considering building structure and emergency propagation characters [6–12]. Generally, evacuation routes in these systems are built on route graphs which are originally derived from building floor plans and then dynamically updated in references to the predicted result of emergency simulation models or risk assessment.

However, in modern buildings different kinds of sensors obtain environmental parameters continuously, such that prediction can be replaced by real-time sensing. Some other evacuation applications take already advantage of wired or wireless building sensor systems to monitor a current environmental state for dynamic evacuation [13–15].

M. Duckham et al. (Eds.): GIScience 2014, LNCS 8728, pp. 268–283, 2014.

In contrast to open (outdoor) environments the detecting area of a sensor inside a building is limited by the building structure. For example, near sensors can be independent because of a wall in between. Thus, linking sensor readings to routes requires complex analysis. Additionally, wireless building sensor systems (WSN) rely on the ad-hoc communication network of the sensor vertices, and special routing protocols [16] are required to guarantee a near-real-time updating of evacuation routes.

The above review implies the need to identify an integrated data model that not only represents the routing properties of indoor environments, but also the observations of building sensor systems in their impact on routing. This data model should combine geospatial information such as route graphs and sensor graphs, and environmental information such as sensor readings to achieve real-time risk-aware evacuation planning in emergency situations. In this study we introduce such a data model for indoor evacuation. The suggested data model, implemented in a personalized evacuation system, will allow generating an evacuation route dynamically for each evacuee, depending on current sensor readings, and taking risk assessments into account. The integrated model enables indoor evacuation applications valuable for both accurate emergency localization and reliable evacuation guidance. The potential benefits of the model can also extend to offer an approach to correlate route network to sensor network for a wide range of indoor applications.

The paper is structured as follows. Section 2 presents related work. Section 3.1 introduces the method of formulating route graphs, and Section 3.2 propose an algorithm to create sensor graphs with completely coverage of all area in building floor plans. Section 3.3 explains the integration of route graphs and sensor graphs. Section 4 provides a sensor tracking and risk aware evacuation plan using this novel data model. Experiments of the model are discussed in Section 5. Section 6 concludes the current work and explores future work.

2 Related Work

Since many outdoor concepts of routing networks are not suitable for indoor environments, indoor routing models have been presented, such as a representation in *Indoor Geography Markup Language* (IndoorGML) [2], and the *Node-Relation-Structure* (NRS) [3]. In [6] a 3D geometric network model based on NRS is introduced to represent an indoor environment, before an ant colony optimization algorithm is applied to find the shortest path during evacuation. Pre-simulations deploying the *Fire Dynamic Simulation* (FDS) provided feedback of smoke spread in the system.

Han et al. have proposed an integrated real-time evacuation route planning method for high-rise building fires. The sensor data acquisition is also pre-simulated in FDS, and the data transmission is based on a wireless network. An interpolation method is applied to quantitatively assess a *casual risk* for each evacuee. Casual risk is based on the distance to the nearest exits [9].

Nguyen et al. proposed a fire evacuation model under the assumption that evacuees follow the boundaries of obstacles or walls to find the exits when their visibility is limited by smoke. In their agent-based model, they do not discuss the method to generate an evacuation route on account of smoke [17].

In [11] a *multi-node hierarchical data model* is proposed to extract route graphs by topological relationships among indoor units of a building. A hierarchical route planning algorithm dynamically schedules evacuation routes considering distance, risk level and waiting time for evacuees. In their approach route planning is risk aware, but no environmental features are involved.

In [14] a WSN-based safe route identification algorithm has been introduced which depends on a dense sensor network and wireless communication technologies, and which routes along the sensor network. A location-based routing protocol is applied to dynamically generate an evacuation route in which the building structure and risk prediction are ignored.

Stahl et al. developed YAMAMOTO (*Yet Another Map Modeling Toolkit*) to generate route graphs automatically from floor plans for multi-level buildings by adding route vertices at key points on the basis of the outlines of spatial regions [18–21]. The toolkit provides an approach for finding navigable routes that also allows for modeling environmental monitoring devices, such as sensors or actuators. In this paper, the integration of sensing and routing, and then also the simulated evacuation experiments are built on top of YAMAMOTO.

3   The Proposed Model

The model consists of three main parts, presenting a route graph for the indoor environment, sensor graphs for sensors in this environment, and the integration of the route graph with all sensor graphs.

3.1   Route Graph

The main purpose of route graph is to associate evacuation routes with the building structure to ensure the feasibility of the generated evacuation route. While in principle any of the suggested route graphs of Section 2 can be applied, the following is specifically based on the route graph of YAMAMOTO [21], which is specified formally here for the first time. For this formal specifications, some definitions are introduced before generating a route graph.

Definition 1. *A region R is a planar and accessible area with a unique identifier in the model, denoted as R_i.*

Thus a region in this paper refers to a two-dimensional bounded area with at least one entrance that could provide access for pedestrians, such as a room, corridor, or lobby. Entrances form virtual parts of the boundary of a region. Every two regions R_1 and R_2 must not overlap, but may neighbor, in which case they share at least one boundary line. Overlapping areas P and Q are separated into regions, either by $\{P, (P \cup Q) \cap \overline{P}\}$, or by $\{Q, (P \cup Q) \cap \overline{Q}\}$.

Every region is represented by a polygon. If a part of the boundaries of a region is curved, this curved part is substituted by a polyline, for example, using the successive bisection algorithm of [22] adapting to a chosen level of accuracy. For convex regions, each pair of vertices inside or on a boundary of the region can be linked together by a straight line edge without piercing any boundary. These edges are *passable*. However, for concave regions the links between some pairs of vertices are not passable. For those pairs of vertices there exists always a set of passable edges between them, including a sequence of concave boundary vertices [23].

There are three types of vertices in a route graph, called *inner vertex*, *access vertex* and *connecting vertex*.

Definition 2. *An* inner vertex *is a shifted concave boundary vertex of concave regions.*

Inner vertices are used to form a passable edges of route graphs. They are placed one meter away from walls, boundaries or corners in order to keep a natural distance for pedestrians.

Definition 3. Access vertices *are placed pair-wise on both sides of the virtual boundary of an entrance.*

The edge between a pair of access vertices represents the accessibility of two adjoining regions in a route graph. The two vertices belong to different regions respectively. Also access vertices are placed one meter away from the virtual boundary of the entrance in order to keep a natural distance for pedestrians.

Definition 4. Connecting vertices *are placed at the beginning, end, and turning points of staircases.*

The edges between a sequence of connecting vertices are route graph elements connecting two different floors.

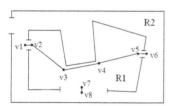
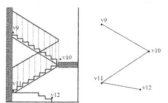

(a) Inner vertices and access vertices	(b) Connecting vertices

Fig. 1. Three types of vertices in route graphs

In Figure 1(a), there are three pairs of access vertices between R_1 and R_2 among which v_2, v_5, and v_7 belong to R_1. A straight link between v_2 to v_5 is not passable in the concave region R_1, but one of the passable routes is $v_2 - v_3 - v_4 - v_5$ with v_3 and v_4 being inner vertices. Figure 1(b) shows four connecting vertices; v_{10} and v_{11} are placed at turning points of a staircase. Elevators, while connecting floors as well, are not represented in the route graph because they cannot be used for evacuation.

Definition 5. *A route graph $G = (V, E)$ consists of a set V of inner vertices, access vertices, and connecting vertices, together with a set E of edges. The set E consists of the edges between each pair of access vertices, the edges between every pair of vertices in the same region if the edge is passable, and the edges between every two consecutive connecting vertices along the same staircase.*

As an example, Figure 2 shows a floor plan of a building. The five regions are named as R_1 to R_5, and R_5 is a concave region. The corresponding route graph G is shown in Figure 3.

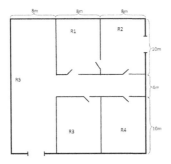

Fig. 2. Sample floor plan

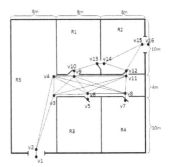

Fig. 3. Route graph

3.2 Sensor Graphs

In contrast to route graphs, the objective of sensor graphs is to give timely feedback of environmental factors in order to predict the impact on evacuation plans, including a risk assessment. The definition of a sensor graph is also based on some concepts that need to be defined first.

Definition 6. *A sensor detecting area S_i of sensor s_i is the collection of all points where sensor data can be obtained by sensor s_i. S_i is represented by a polygon with a clockwise ordered set of corner vertices $S_i := \{p_1, \cdots, p_m\}$, $m \geq 3$. Each vertex p_k, $1 \leq k \leq m$, is denoted by its position (x_k, y_k, z_k).*

The typical detection range of most types of sensors is a circular area of radius r_s of which the center is the point of installation of the sensor. However, there are significant exceptions. For example, walls shield the reception of sensors [24]. Hence, the actual detection area of each sensor is the intersection between the uninhibited sensor detection area and the boundary of the region in which the sensor is installed. For example, in Figure 4 S_3 is the sensor detecting area of s_3 in R_1, thus, $S_3 := \{p_1, p_2, p_3, p_4, p_5, p_6, p_7, p_8\}$.

Definition 7. *S_i of sensor s_i and S_j of sensor s_j is considered intersecting if (a) s_i and s_j are in the same region and $S_i \cap S_j \neq \emptyset$, or (b) s_i and s_j are in different regions and $S_i \cap S_j$ covers at least one entrance between the regions, or (c) s_i and s_j are at the opposite entrances of a staircase (i.e., on neighboring floors).*

Definition 8. *A sensor graph $G' = (V', E')$ consists of a set V' of sensor vertices s_i representing the location of the sensors, together with a set E' of edges if two sensors' detection areas are intersecting.*

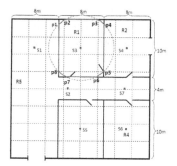

Fig. 4. A sensor detection area

When forming a sensor graph, there is a complete coverage problem of sensor vertices. According to the objective of sensor graphs, the sensor network should cover all areas of a floor plan, or otherwise events may be not detected, and also the event simulation becomes incoherent. Therefore, a sensor graph according to Definition 8 is completed by adopting an algorithm suggested originally in [25].

All together, the following steps describe the sensor graph generation.

Step 1: A floor plan is partitioned by a grid T, with equally spaced tiles $t_{p,q}$ (p is the row number and q is the column number). For those tiles that are divided by borders between regions, each part of the tile will be treated as separated tiles in different regions.

Step 2: The sensor vertices S are added to G' for each actual sensor in the environment.

Step 3: For a region R_i, a matrix $A(R_i) = [a_{p,q}]_{n \times m}$ ($1 \leq p \leq n, 1 \leq q \leq m$) is formed corresponding to the $n \times m$ tiles that just cover R_i and initialized by 0s. If there are some tiles in this matrix that are not part of the region the corresponding entries are set to infinite.

Step 4: For each tile $t_{p,q}$, if $t_{p,q} \cap S_j \neq \emptyset$ ($1 \leq j \leq |V'|$) in R_i the entry $a_{p,q}$ in $A(R_i)$ will be set to 1.

Step 5: If $A(R_i)$ contains one or more zero elements, i.e., tiles inside the region that are not observed by a sensor, a virtual sensor is placed at the center of the first zero tile encountered in row order, and added to V'. Then go to Step 4. Otherwise, go to Step 6.

Step 6: If all regions in the floor plan have been operated go to Step 7. Otherwise, go to Step 3.

Step 7, connecting floors: For each staircase connecting two adjacent floors, two virtual sensor vertices are added to V', which are placed in the middle of the entrances of the same staircase.

Step 8: For all pairs s_i, s_j, if S_i and S_j are intersecting add an edge between them to E'.

Step 9: If all floor plans in the building have been observed, stop. Otherwise, go to Step 1.

In Figure 5, the floor plan is firstly partitioned into tiles with an equal spacing, here of $r_s/2$ ($r_s = 6m$). The origin of the grid is at the top left corner. There are originally seven smoke sensors s_1 to s_7 in the sensor system, denoted by red squares.

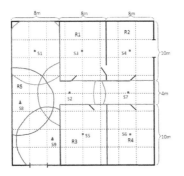

Fig. 5. Sensor detection areas in R_5

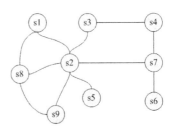

Fig. 6. Sensor graph

For R_1, the matrix $A(R_1)$ is 4×4 although some of the tiles in R_1 are only partly inside. Since s_3 is the only actual sensor in R_1 and all grids in R_1 intersect with S_3 (Figure 4), all entries in $A(R_1)$ equal to 1 and no virtual sensor vertex need to be added in R_1. R_2, R_3, and R_4 are similar, and S_4, S_5 and S_6 are added to V'.

For R_5, however, Figure 5 shows some gaps in the sensor coverage. There are three actual sensors s_1, s_2 and s_7 whose detection areas are denoted with lines of different colors. Corresponding $A(R_5)$ is

$$
A(R_5) = \begin{bmatrix}
1 & 1 & 1 & \infty & \infty & \infty & \infty & \infty \\
1 & 1 & 1 & \infty & \infty & \infty & \infty & \infty \\
1 & 1 & 1 & \infty & \infty & \infty & \infty & \infty \\
1 & 1 & 1 & 1 & 1 & 1 & 1 & 1 \\
0 & 1 & 1 & 1 & 1 & 1 & 1 & 1 \\
0 & 1 & 1 & \infty & \infty & \infty & \infty & \infty \\
0 & 0 & 0 & \infty & \infty & \infty & \infty & \infty \\
0 & 0 & 0 & \infty & \infty & \infty & \infty & \infty
\end{bmatrix}
$$

containing some zero values. Since the first zero value (in row order) refers to $t_{5,1}$ a virtual sensor s_8 is placed at its center, denoted by a red triangle, and S_8 is added, denoted by a purple line. After another iteration,

$$
A(R_5) = \begin{bmatrix}
1 & 1 & 1 & \infty & \infty & \infty & \infty & \infty \\
1 & 1 & 1 & \infty & \infty & \infty & \infty & \infty \\
1 & 1 & 1 & \infty & \infty & \infty & \infty & \infty \\
1 & 1 & 1 & 1 & 1 & 1 & 1 & 1 \\
1 & 1 & 1 & 1 & 1 & 1 & 1 & 1 \\
1 & 1 & 1 & \infty & \infty & \infty & \infty & \infty \\
1 & 1 & 0 & \infty & \infty & \infty & \infty & \infty \\
0 & 0 & 0 & \infty & \infty & \infty & \infty & \infty
\end{bmatrix}
$$

still contain some zero values. $t_{7,3}$ is the first zero value in search order, and a virtual sensor s_9 is added to V'. In the next iteration, $A(R_5)$ does no longer contain a zero value. The algorithm proceeds to Step 8 to generate edges E'. The final result of G' is shown in Figure 6.

This sensor graph generation algorithm is easy to program, but not optimal because the number of virtual sensors is not the minimum. For the purpose of this paper the algorithm is sufficient. In the future, another algorithm can be plugged in covering the whole area with a minimal number of virtual sensors.

Generally, when a building has different types of sensors simultaneously, separate sensor graphs are formed in the model.

3.3   Integration of Sensor Graph and Route Graph

The integration of sensor graph and route graph is actually a mapping from sensor network observations to route network elements. The integration facilitates the representation of *affectedness* of route graph vertices and edges, for passive and active sensors as well as for sensors at risk according to some prediction model. The following method integrates *one* sensor graph with a route graph. With other sensor graphs the same method can be applied.

If a vertex v_i in route graph G is inside S_k of some sensor s_k, vertex v_i is one of the affected vertices of s_k. Also, if an edge e_j in G intersects with S_k, edge e_j is one of the affected edges of s_k. The latter provides a mechanism to represent that an edge is blocked although the endpoints may not. The relationship between route graph vertices and sensors is $n : m$ since route graph vertices can be inside none, one or multiple S_i, and there may be zero, one or more route graph vertices in S_i. The same applies for relationships between route graph edges and sensors because a straight edge may go through more than one S_i and the range of a sensor may intersect with more than one route graph edge.

The integration of the sample floor plan is shown in Table 1. There are no affected route graph vertices or edges in S_1 which suggests that S_1 will not be crossed during evacuation. However, when s_8 is activated in an event, two route graph edges will be blocked (but no vertices are affected).

4   Sensor Tracking and Risk Aware Evacuation Planning

This section introduces evacuation planning based on the integrated route and sensor graph data model. It considers sensor states of being activated by an event, or at risk according to some prediction model. In emergency, evacuees who are familiar with the building tend to choose their habitual routes for evacuation although it may lead them to blocked or risky areas. Other evacuees who are unacquainted with the building may follow the crowd or the signed evacuation routes. Also these strategies may be in conflict with the event itself. Evacuation based on outdated information is *blind* evacuation. Recently, mobile application systems have emerged able to provide assistance for evacuees. Some of these applications are also blind, based on by the event outdated data. Their evacuation route planning is *static*, it does not adapt dynamically to the situation.

Table 1. Integration of sensor graph and route graph

Sensor	Affected vertices	Affected edges
s_1	$\emptyset$	$\emptyset$
s_2	v_3, v_4, v_6, v_9	all edges start or end on v_3, v_4, v_6, v_9
s_3	v_{10}, v_{13}	all edges start or end on v_{10}, v_{13}
s_4	v_{12}, v_{14}, v_{15}	all edges start or end on v_{12}, v_{14}, v_{15}
s_5	v_5	all edges start or end on v_5
s_6	v_7	all edges start or end on v_7
s_7	v_8, v_{11}	all edges start or end on v_8, v_{11}
s_8	$\emptyset$	$v_2\text{-}v_4, v_2\text{-}v_3$
s_9	v_2	all edges start or end on v_2

Even if these applications receive the initial location of the emergency (updating their data initially) they can not generate dynamically escape routes.

Few indoor evacuation systems are tracking sensors and risk aware. A *sensor tracking evacuation* is obtaining timely the sensor data from building sensor systems to generate real-time evacuation routes for evacuees. In addition, a *risk aware evacuation* also avoids or reduces risks after adopting some strategies to predict near-future states of the event.

The proposed model in this paper can be applied to realize sensor tracking and risk aware evacuation planning. There are two reasons. Firstly, sensor data is continuously read out by a building sensor system. When unusual situations are reported at a time (i.e., some sensors get activated), the identifiers of these sensors localize the event. If these identifiers are forwarded to an evacuation application, the affected route graph vertices and edges are directly accessible in the integrated model, and are marked as blocked when generating a real-time evacuation routes for individuals. Consequently, this application meets the requirements of sensor tracking evacuation planning. The application is also capable of risk awareness. When building sensor systems report an event (i.e., some sensors are activated), the connectivity in the sensor network can be used to predict the near-future spread of the event. Reporting risky sensors to the evacuation application allows to mark the affected route graph vertices and edges being at risk, e.g., by a risk rank value, which may be considered by some weighting in the computation of individual evacuation routes. Table 2 illustrates the differences among these three evacuation plans in which ST&RA refers to applying a sensor tracking and risk aware evacuation plan, and the symbol $\sqrt{}$ means the corresponding leftmost item is aware by the evacuation plan.

Both sensor tracking and risk awareness requires a standing communication between building sensor systems and evacuation planning applications. However, even only occasional updates and iterative evacuation planning must be superior to blind or static evacuation planning. This is tested in the following section.

Table 2. Difference among evacuation plans

	ST&RA	Static	Blind
Emergency location	✓	✓	×
Original route graph	✓	✓	Evacuee's experience
Real-time sensor reading	✓	×	×
Risk analysis	✓	×	×
Dynamic evacuation route generation	✓	×	×

5 Experiments

This section demonstrates a mobile evacuation system application, implementing the proposed model, in an office building. The objective of the experiments is to compare sensor tracking and risk aware evacuation plans with blind and static evacuation plans. In this experiment a fire scenario is assumed.

5.1 System Implementation

The system implementation is built on a framework shown in Figure 7. There are six separately modules in the system. The building sensor system gathers continuously the environment data, and reports an alarm when it detects some unusually situation. The sensor graphs, the route graph, and their integration have been pre-calculated and stored in a server. On the user side, four modules cooperate with each other to generate a safe and short evacuation route for each evacuee. The location module gives the current position of each user. The risk assessment module evaluates the emergency situation and renews the risk rank value of affected vertices and edges in route graph. Path generation module sets an optimal evacuation path dynamically by considering both the distance and the risk value. The building model provides a platform to the user to follow the evacuation route and related instructions. The thick lines in Figure 7 denote the dynamic data and the filaments denote the pre-generated information.

The entire evacuation system is developed by expanding YAMAMOTO, in C#. Sensor graphs, route graph and the integration are all produced in a suited XML format, from planar floor plans. The current location of each evacuee is obtained by QR-code scanning which obviously can be substituted by any other positioning mechanism. The risk assessment module is accomplished by a series of risk assessment strategies synthesized from Dow's Fire and Explosion Index [26] and Fire Safety Evaluation System [27]. In particular, the risk assessment mechanism is an independent part of the entire application, and could also be easily replaced by any other risk assessment method. The path generation module is that of YAMAMOTO, but now using dynamic route graph of the proposed model. The user interface is built by Open Graphics Library for Embedded System. Figure 8 shows section of the office building model together with sensor detection areas, and Figure 9 gives the interface of the system.

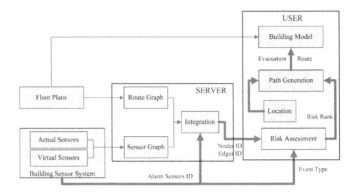

Fig. 7. System framework

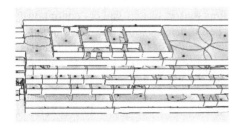

Fig. 8. Section of building model with sensor layers

Fig. 9. System implementation

5.2   Simulation and Result Analysis

The office building used in this experiment has five floors and three blocks including office rooms, lecture theaters and laboratories. Each block with an exit is connected to an adjacent block by a staircase. Blind, static and sensor tracking and risk aware evacuation plans according to the explanation of Section 4 are simulated in Repast for the same twenty fire scenes to keep the results comparable. The speed of user movement is assumed to be $1.5m$ per second. The initial position of 100 evacuees are randomly given and recorded into an XML file which will be imported at the beginning of each scene simulation to ensure all evacuees escape from the same initial position.

Figure 10 shows a sensor tracking and risk aware evacuation scene with a part of the route graph. The blue cube indicates the evacuee, the red circles the blocked, and the yellow circles the risky vertices in the route graph. For readability the sensor graph is not shown. Only sensor vertices are indicated (red cube: blocked, yellow: risky, gray:

normal, dark: extinguished). The fire source is on the fourth floor, and most of the evacuees are near the staircase except E_1. The evacuation system generates evacuation routes, so most evacuees are escaping toward the staircase, and E_1 escapes to the other block as the only pathway to the staircase in this block is blocked.

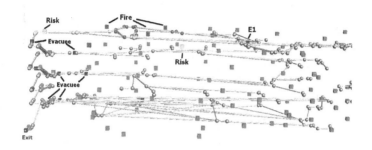

Fig. 10. Evacuation simulation

Figure 11(a) includes part of the data exported by Repast. The *success* columns of three evacuation plans are the number of successful evacuees in different scenes. From the data and the corresponding line chart in Figure 11(b), all evacuees in the twenty scenes of the sensor tracking and risk aware evacuation plan are successful, while there are some unfortunate people who fail to evacuate in static or blind evacuation plans. Therefore, the sensor tracking and risk aware evacuation plan is the safest for evacuees.

The *AveEvaTime* columns in Figure 11(a) and the corresponding chart in Figure 11(c) depict the average evacuation time in each scene of the three plans. Generally, the average evacuation time in sensor tracking and risk aware planning is the least but there are some exceptions. In Scene 16, the *AveEvaTime* of this mode is 103.5 seconds which is greater than the other two. There are two possible reasons. The first one is the average evacuation time is the quotient of the sum of evacuation time and the number of success evacuees. People who fail to evacuate in static or blind evacuation plans are not included in the calculation, but those people may need a little more time to evacuate with sensor tracking and risk aware plans which leads to the special case. Another reason is a blocked sensor vertex at T_1 may change to extinguished at T_2. The sensor tracking and risk aware planning will generate a longer but safer evacuation route at T_1. When evacuees follow this route it will take more time to escape. In blind evacuation, since no feedback from sensor and risk assessment evacuees may follow the shortest path at T_1 and at T_2 they are so lucky to pass the extinguished area which were blocked at T_1. The special cases are only 15% of the total scenes marked with gray background in Figure 11(a).

The *MaxTime* columns are the maximum evacuation time in each scene in different plans. The maximum evacuation time refers to the escape time of the last successful person. Regardless of the unsuccessful evacuees, the average maximum evacuation time of sensor tracking and risk aware plans are still the least which is 198.3 seconds.

Thus, the results of the simulation demonstrate that sensor tracking and risk aware evacuation plans are safer and more reliable than the other two evacuation plans.

Emergency Scenes	ST&RA			STATIC			BLIND		
	Success	AveEvaTime	MaxTime	Success	AveEvaTime	MaxTime	Success	AveEvaTime	MaxTime
1	100	80.68	177	100	80.68	177	96	98.42	200
2	100	83.38	177	100	83.38	177	89	99.52	199
3	100	65.74	193	95	60.95	177	95	61.48	177
4	100	63.08	177	100	63.08	177	100	63.32	177
5	100	63.20	177	100	65.08	177	100	66.80	189
6	100	81.03	216	86	78.75	373	89	78.55	375
7	100	64.15	194	95	62.14	194	95	61.53	194
8	100	65.27	293	98	67.31	155	98	67.36	155
9	100	63.15	177	100	63.15	177	100	63.15	177
10	100	63.08	177	100	63.08	177	100	64.76	193
11	100	63.08	177	100	63.08	177	100	63.08	177
12	100	69.55	177	97	71.24	177	97	71.92	177
13	100	62.49	276	99	62.18	155	99	62.04	155
14	100	63.08	177	100	63.08	177	100	63.08	177
15	100	65.97	177	100	72.95	314	100	73.44	320
16	100	103.50	239	97	99	224	81	92.81	230
17	100	54.87	188	97	61.19	155	99	61.93	155
18	100	73.39	177	100	72.39	193	93	61.47	182
19	100	67.99	242	99	67.99	302	99	71.00	312
20	100	58.03	177	95	60.19	177	95	60.31	177
Average	100	68.7	198.3	98	69.0	200.6	96	70.3	204.9

(a) Data table

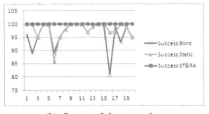

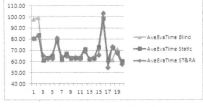

(b) Successful evacuation   (c) Average evacuation time

Fig. 11. Result analysis

5.3 Real-Time System Efficiency

Time is extraordinarily sensitive for all evacuation systems. The presented model in this study, however, is mainly based on pre-calculated networks and their integration, as described in Section 5.1, which has no effect on real-time system efficiency. Dynamic updates only apply to the presumably small areas of impact. These dynamic updates have the following complexity:

1. The cost to communicate affected sensor nodes to the route graph; this cost is linear because the integration is organized through a table.
2. The cost to renew the risk rank value for affected nodes and edges in the route graph. These costs depend on the chosen risk assessment model (the presented simple model is also linear), but for small areas of impact it will affect not many sensor nodes.
3. The cost of updating the route graphs on all mobile devices, which is a communication cost in a push or pull mode.
4. Positioning costs of the evacuees' mobile devices are constant, since each device locates itself.

5. The cost to calculate the shortest paths for each evacuee on the updated route graph happens in parallel on the mobile devices. We use an A* algorithm, which has a time complexity of $O(2^n)$ in the worst case, and $O(n \log(n))$ in the best [28, 29], n being the number of nodes in the route graph.

In brief, the eventual time efficiency of the system to a great extent relies on the size of the route graph.

6 Conclusion and Future Work

This study proposes an integrated data model for indoor evacuation on the basis of building sensor systems and building structure. With the purpose of employing real-time sensor data as references for evacuation route calculation, this paper converts monitoring sensors to sensor graphs, and integrates these sensor graphs to a route graph. With this integration, sensor tracking and risk aware evacuation routes may be generated dynamically for evacuees, which have been shown to be superior to static or blind evacuation plans.

Future work may optimize the sensor graph generation algorithm, and aim for automatic indoor positioning. But fundamentally, this model lays foundations for refined integration of modern building facilities and indoor GIS applications, for example through a more detailed risk model. Any improved risk model would only improve further the results of the presented model. Currently, only the risk of nodes and edges in route graph to be affected next by the event is considered. Other risks could be introduced as well, such as event propagation models, or congestion risk. Congestion risk, however, which is a function of network centrality and distribution of evacuees at the time of the alarm, requires either a central planning or a coordinated planning; in the current implementation route planning happens individually on the mobile devices. Standardization of interfaces between building sensor systems and evacuation systems will help adopting mobile evacuation applications on a large scale.

Acknowledgments. Funding of the European Union via mSAFE under grant agreement No. FP7-PEOPLE-2011-IRSES 295269 is gratefully acknowledged (http://msafe.informatik.uni-bremen.de/). Access to YAMAMOTO was kindly provided by Christoph Stahl, Bremen.

References

1. Richter, K.F., Winter, S., Rüetschi, U.J.: Constructing hierarchical representations of indoor spaces. In: Tenth International Conference in Mobile Data Management, Workshop on Indoor Spatial Awareness, pp. 686–691. IEEE Press, Taipei (2009)
2. OGC: Indoor geography markup language introduction (2013),
   http://www.opengeospatial.org/projects/groups/indoorgmlswg
3. Lee, J., Kwan, M.: A combinatorial data model for representing topological relations among 3D geographical features in micro-spatial environments. International Journal of Geographical Information Science 19(10), 1039–1056 (2005)

4. Stoffel, E.-P., Lorenz, B., Ohlbach, H.J.: Towards a semantic spatial model for pedestrian indoor navigation. In: Hainaut, J.-L., et al. (eds.) ER Workshops 2007. LNCS, vol. 4802, pp. 328–337. Springer, Heidelberg (2007)
5. Becker, T., Nagel, C., Kolbe, T.H.: A multi-layered space-event model for navigation in indoor spaces. In: Lee, J., Zlatanova, S. (eds.) 3D Geo-Information Sciences. Lecture Notes in Geoinformation and Cartography, pp. 61–77. Springer, Berlin (2008)
6. Wu, C., Chen, L.: 3D spatial information for fire-fighting search and rescue route analysis within buildings. Fire Safety Journal 48, 21–29 (2012)
7. Liu, J., Lyons, K., Subramanian, K., Ribarsky, W.: Semi-automated processing and routing within indoor structures for emergency response applications. In: Buford, J.F., Tolone, W.J., Jakobson, G., Ribarsky, W., Erickson, J. (eds.) Cyber Security, Situation Management, and Impact Assessment II; and Visual Analytics for Homeland Defense and Security II, vol. 7709, p. 77090Z. The International Society for Optical Engineering (2010)
8. Kraus, L., Stanojevic, M., Tomasevic, N., Mijovic, V.: A decision support system for building evacuation based on the EMILI SITE environment. In: 20th IEEE International Workshops on Enabling Technologies: Infrastructure for Collaborative Enterprises (WETICE), p. 334 (2011)
9. Han, Z., Weng, W., Zhao, Q., Ma, X., Liu, Q., Huang, Q.: Investigation on an integrated evacuation route planning method based on real-time data acquisition for high-rise building fire. IEEE Transactions on Intelligent Transportation Systems 14(2), 782–795 (2013)
10. Jiyeong, L.: A three-dimensional navigable data model to support emergency response in microspatial built-environments. Annals of the Association of American Geographers 97(3), 512–529 (2007)
11. Zhang, L., Wang, Y., Shi, H., Zhang, L.: Modeling and analyzing 3D complex building interiors for effective evacuation simulations. Fire Safety Journal 53, 1–12 (2012)
12. Tang, F., Ren, A.: GIS-based 3D evacuation simulation for indoor fire. Building and Environment 49, 193–202 (2011)
13. Lei, Z., Gaofeng, W.: Design and implementation of automatic fire alarm system based on wireless sensor networks. In: Proceedings of the International Symposium on Information Processing, pp. 410–413 (2009)
14. Nauman, Z., Iqbal, S., Khan, M.I., Tahir, M.: WSN-based fire detection and escape system with multi-modal feedback. In: Dziech, A., Czyżewski, A. (eds.) MCSS 2011. CCIS, vol. 149, pp. 251–260. Springer, Heidelberg (2011)
15. Sha, K., Shi, W., Watkins, O.: Using wireless sensor networks for fire rescue applications: Requirements and challenges. In: IEEE International Conference on Electro/Information Technology, pp. 239–244. IEEE (2006)
16. Al-Karaki, J.N., Kamal, A.E.: Routing techniques in wireless sensor networks: A survey. IEEE Wireless Communications 11(6), 6–28 (2004)
17. Nguyen, M., Ho, T., Zucker, J.: Integration of smoke effect and blind evacuation strategy (SEBES) within fire evacuation simulation. Simulation Modelling Practice and Theory 36, 44–59 (2013)
18. Stahl, C., Schwartz, T.: Modeling and simulating assistive environments in 3D with the YAMAMOTO toolkit. In: International Conference on Indoor Positioning & Indoor Navigation (IPIN), p. 1 (2010)
19. Munzer, S., Stahl, C.: Learning routes from visualizations for indoor wayfinding: Presentation modes and individual differences. Spatial Cognition and Computation 11(4), 281–312 (2011)
20. Stahl, C.: New perspectives on built environment models for pedestrian navigation. In: Spatial Cognition 2008 Poster Proceedings. Universität Freiburg (September 2008)
21. Stahl, C.: Spatial modeling of activity and user assistance in instrumented environments. PhD thesis, DFKI, University of Saarland (2010)

22. Chandra, A.M.: Dividing a circular arc into equal numbers of division. IOSR Journal of Mathematics 4, 38–39 (2012)
23. Stanford University: Geometric algorithms design and analysis (1992), http://graphics.stanford.edu/courses/cs268-09-winter/notes/handout7.pdf
24. National Fire Protection Association: National Fire Alarm Code. 2007 edn. An International Codes and Standards Organization (2007)
25. Aziz, N.A.A., Aziz, K.A., Ismail, W.Z.W.: Coverage strategies for wireless sensor networks. In: Proceedings of the World Academy of Science, Engineering and Technology, vol. 50, pp. 145–150 (2009)
26. American Institute of Chemical Engineers, ed.: DOW's Fire & Explosion Index Hazard Classification Guide. 7th edn. Wiley (1994)
27. National Fire Protection Association: Guide on Alternative Approaches to Life Safety. 2013 edn. An International Codes and Standards Organization (2013)
28. Hart, P.E., Nilsson, N.J., Raphael, B.: A formal basis for the heuristic determination of minimum cost paths. IEEE Transactions on Systems Science and Cybernetics 4(2), 100–107 (1968)
29. Hart, P.E., Nilsson, N.J., Raphael, B.: Correction to a formal basis for the heuristic determination of minimum cost paths. ACM SIGART Bulletin (37), 28–29 (1972)

Significant Route Discovery:
A Summary of Results

Dev Oliver[1], Shashi Shekhar[1], Xun Zhou[1], Emre Eftelioglu[1], Michael R. Evans[1],
Qiaodi Zhuang[1], James M. Kang[2], Renee Laubscher[2],
and Christopher Farah[2]

[1] Department of Computer Science, University of Minnesota, USA
[2] National Geospatial-Intelligence Agency, USA

Abstract. Given a spatial network and a collection of activities (e.g., pedestrian fatality reports, crime reports), Significant Route Discovery (SRD) finds all shortest paths in the spatial network where the concentration of activities is unusually high (i.e., statistically significant). SRD is important for societal applications in transportation safety, public safety, or public health such as finding routes with significant concentrations of accidents, crimes, or diseases. SRD is challenging because 1) there are a potentially large number of candidate routes ($\sim 10^{16}$) in a given dataset with millions of activities or road network nodes and 2) significance testing does not obey the monotonicity property. Previous work focused on finding circular areas of concentration, limiting its usefulness for finding significant linear routes on a network. SaTScan may miss many significant routes since a large fraction of the area bounded by circles for activities on a path will be empty. This paper proposes a novel algorithm for discovering statistically significant routes. To improve performance, the proposed algorithm features algorithmic refinements that prune unlikely paths and speeds up Monte Carlo simulation. We present a case study comparing the proposed statistically significant network-based analysis (i.e., shortest paths) to a statistically significant geometry-based analysis (e.g., circles) on pedestrian fatality data. Experimental results on real data show that the proposed algorithm, with our algorithmic refinements, yields substantial computational savings without reducing result quality.

1   Introduction

Significant Route Discovery (SRD) has important societal applications in transportation safety, public safety, or public health such as finding routes with significant concentrations of accidents, crimes, or diseases. In transportation safety, domain experts attribute pedestrian fatalities largely to the design of streets, which have been engineered for speeding traffic with little or no provision for people on foot, in wheelchairs, or on bicycles [1]. In urban areas, more than 56% of the pedestrian fatalities in the US (2007-2008) occurred on arterial roads [1]. Figure 1(a) shows an example of a pedestrian at risk on a road without proper sidewalks. This lack of basic infrastructure can be lethal. Figure 1(b) shows a map of pedestrian fatalities that occurred on Orlando roads from 2000 - 2009. Transportation planners and engineers need tools to assist them in identifying which frequently used road segments/stretches pose significant risks for pedestrians and consequently should be redesigned.

M. Duckham et al. (Eds.): GIScience 2014, LNCS 8728, pp. 284–300, 2014.

Fig. 1. (a) Pedestrian at risk on a road without proper sidewalks [1] (b) Pedestrian fatalities occurring on arterials in Orlando, FL [2]. A large fraction of the bounding circles (e.g., C1, C2) of significant routes are empty.

Informally, the Significant Route Discovery (SRD) problem can be defined as follows: given a spatial network, a collection of activities (e.g., pedestrian fatality reports, crime reports), and a likelihood threshold θ, find all shortest paths in the spatial network where the concentration of activities is unusually high (i.e., statistically significant) and the likelihood exceeds θ. Depending on the domain, an activity may be the location of a pedestrian fatality, a carjacking, a train accident, etc. Figures 2(a) and 2(b) illustrate an input and output example of SRD, respectively. The input consists of seven nodes, six edges (with edge weights set to 1 for illustration purposes, shown as the second number on each edge), twenty activities (shown as the first number in red on each edge), and $\theta = 2$, indicating that we are interested in shortest paths whose likelihood exceeds $\theta = 2$. The output contains two shortest paths, $\langle N_1, N_2, N_3 \rangle$ and $\langle N_6, N_5, N_7 \rangle$ that are at least twice as likely to have pedestrian fatalities.

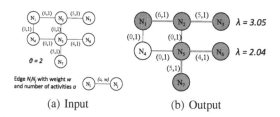

(a) Input   (b) Output

Fig. 2. Example of Significant Route Discovery

SRD is challenging due to the potentially large number of candidate routes ($\sim 10^{16}$) in a given dataset with millions of activities or road network nodes. For large roadmaps such as the 100 million road-segments in the US, this results in prohibitive shortest path computation times. Additionally, significance testing does not obey the monotonicity property, meaning that there is no ordering between the likelihood of a path and its super-paths, or vice-versa. In other metrics such as activity count, for example, a path will always have less than or equal to the number of activities of its super-paths, a property which may be exploited for computational speedup. However, this property does not hold for significance testing. Furthermore, depending on the method used to determine statistical significance, computation times may also be impacted (e.g., $m = 1000$ Monte Carlo simulations may be required to calculate statistical significance).

Related Work and their Limitations. Dividing spatial data into statistically significant groups is an important task in many domains (e.g., transportation planning, public health, epidemiology, climate science, etc.). Previous methods for this type of partitioning have generally been geometry-based [3–6] or network-based [7–10].

Geometry-based techniques [3, 4, 6] partition spatial data using geometric shapes (e.g., circles, rectangles). This is useful in domains such as public health, where finding spatial clusters with a higher density of disease is of interest for understanding the distribution and spread of diseases, outbreak detection, etc. Kulldorff, et al proposed a spatial scan statistics framework for disease outbreak detection [3]. The spatial scan statistic employs a likelihood ratio test where the null hypothesis is the probability that disease inside a region is the same as outside, and the alternate hypothesis is that there is a higher probability of disease inside than outside. All the spatial regions, represented by a circle or ellipsoid in the spatial framework, are enumerated and the one that maximizes the likelihood ratio is identified as a candidate. However, if we apply SaTScan to a road network, many significant routes may be missed since a large fraction of the area bounded by circles for activities on a path will be empty, as shown in Figure 1(b). Furthermore, geometry-based techniques may not be appropriate for modeling linear clusters, which are formed when the underlying generator of the phenomena is inherently linear (e.g., pedestrian fatalities, railroad accidents, etcetera).

Network-based techniques [7–10], on the other hand, leverage the underlying spatial network when partitioning spatial data. For example, Linear Intersecting Paths (LIP) [9] and Constrained Minimum Spanning Trees (CMST) [7] utilize a subgraph (e.g., a path or tree) to discover statistically significant groups.

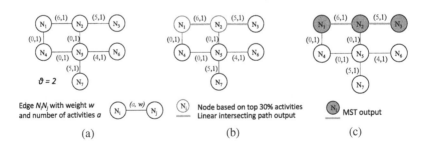

Fig. 3. Example (a) Input, (b) Output of Linear Intersecting Paths (LIP) [9], (c) Output of Constrained Minimum Spanning Trees (CMST) [7] (Best in color)

In LIP [9], one anomalous sub-component of a set of connected paths that intersect each other is discovered. The connected paths are based on locations in the spatial network with the highest percentage of activities, specified by the user. Hence the likelihood ratio is only evaluated on a portion of the graph specified by this percentage, not the entire spatial network. Figure 3 shows an example input and output of LIP. The user-specified percentage is 30%, which means all the candidates will have paths containing edge $\langle N_1, N_2 \rangle$ since this edge has six activities (out of a possible 20 activities). Examples of possible candidates are $\langle N_1, N_2, N_3 \rangle$, $\langle N_1, N_2, N_5 \rangle$, $\langle N_2, N_1, N_4 \rangle$, $\langle N_1, N_2, N_5, N_7 \rangle$, etc. The output is $\langle N_1, N_2, N_3 \rangle$, since it has the highest likelihood (Section 2 details how the likelihood ratio is calculated). However, in addition to returning only one statistically significant component, the results of this approach are sensitive to the percentage of the network selected. If the percentage is too high, the number of candidates may be highly restricted, which could result in not identifying statistically significant regions of interest. If the percentage is too low, LIP may be computationally prohibitive due to the large number of candidates.

CMST [7] finds one statistically significant tree in the spatial network. Figure 3(c) shows an example of this approach. Here the output is $\langle N_1, N_2, N_3 \rangle$, since this tree has the highest likelihood. However, in addition to returning only one statistically significant tree, the size of the tree is restricted, which could result in not identifying statistically significant regions of interest.

In contrast to previous methods, the proposed approach finds multiple statistically significant routes in the spatial network.

Contributions. Our contributions are summarized as follows:

– We introduce the problem of significant route discovery using shortest paths.
– We propose the Smart Significant Route Miner (SmartSRM) algorithm with algorithmic refinements that improve performance by pruning unlikely paths and speeding up Monte Carlo simulation. SmartSRM finds multiple significant routes in the spatial network.
– We present a case study comparing the proposed significant network-based analysis (i.e., shortest paths) to a significant geometry-based analysis (e.g., circles) on pedestrian fatality data.
– Experimental results on real data show that the proposed algorithm, with our refinements, yields substantial computational savings over a naïve approach without reducing result quality.

Scope and Outline of the Paper. This work focuses on finding significant discrete activity events (e.g., pedestrian fatalities, crime reports) associated with a point on a network. This does not imply that all activities must necessarily be associated with a point in a street. In addition, other network properties such as GPS trajectories and traffic densities of road networks [11] are not considered. In this work, it is assumed that the number of activities on the road network is fixed and does not change over time. A dynamically changing number of activities is presently beyond the scope of this research, as are techniques that do not explore statistical significance (e.g., DBScan [12], K-Means [13], KMR [14], and Maximum Subgraph Finding [15]). Furthermore, resolving activity hotspots to the sub-arc level requires a dynamic segmentation data model (currently not explored) that will introduce additional nodes and may create a computational bottleneck. Finally, modeling stochastic route choice (where one or several of the edge attributes are not deterministic [16]) also falls outside the scope of this paper.

The paper is organized as follows: Section 2 presents the basic concepts and problem statement of SRD. Section 3 presents both the Naïve and Smart Significant Route Miner algorithms to solve SRD. Section 4 presents a case study comparing the proposed significant network-based output (i.e., shortest paths) to a significant geometry-based output (e.g., circles) on pedestrian fatality data. The experimental evaluation is covered in Section 5. Section 6 presents a discussion. Section 7 concludes the paper and previews future work.

2   Basic Concepts and Problem Statement

This section introduces several key concepts in SRD and presents a formal problem statement.

2.1   Basic Concepts

We define our basic concepts as follows:

Definition 1. *A **spatial network** $G = (N, E)$ consists of a node set N and an edge set E, where each element u in N is associated with a pair of real numbers (x, y) representing the spatial location of the node in a Euclidean plane [17]. Edge set E is a subset of the cross product $N \times N$. Each element $e = (u, v)$ in E is an edge that joins node u to node v.*

Figure 2(a) shows an example of a spatial network where circles represent nodes and lines represent edges. A road network is an example of a spatial network where nodes represent street intersections and edges represent streets.

Definition 2. *An **activity set** A is a collection of activities. An **activity** $a \in A$ is an object of interest associated with only one edge $e \in E$.*

In transportation planning, an activity may be the location of a pedestrian fatality; in crime analysis, an activity may be the location of a theft. Each edge in Figure 2(a) is associated with a number of activities (e.g., edge $\langle N_1, N_2 \rangle$ has 6 activities).

Definition 3. *The **activity coverage inside a path**, a_p, is the number of activities on p. The **activity coverage outside** p is $|A| - a_p$, where $|A|$ is the total number of activities in the spatial network, G.*

For example, in Figure 2(a), the activity coverage *inside* path $\langle N_1, N_2, N_3 \rangle$ is 11 whereas the activity coverage *outside* $\langle N_1, N_2, N_3 \rangle$ is $20 - 11 = 9$.

Definition 4. *The **weight inside a path**, w_p, is the sum of weights of all edges in p. The **weight outside** p is $|W| - w_p$, where $|W|$ is sum of weights of all edges in G.*

In Figure 2(a), the weight *inside* $\langle N_1, N_2, N_3 \rangle$ is 2 whereas the weight *outside* $\langle N_1, N_2, N_3 \rangle$ is $7 - 2 = 5$.

Definition 5. *The **likelihood ratio of path** p,* $\lambda_p = \frac{a_p \div w_p}{(|A| - a_p) \div (|W| - w_p)}$ *[3,10].*

The likelihood ratio of path p, λ_p, is the ratio of the activity density *inside* path p to the activity density *outside* p. Activity density may be estimated in different ways across different domains. In transportation planning, activity density inside p may be estimated using $\frac{a_p}{VMT}$, where VMT is vehicle miles traveled (i.e., the total number of miles driven by all vehicles within a given time period and geographic area). Path weight may also be used to estimate activity density [10]. In Figure 2(a), $\lambda_{\langle N_1, N_2, N_3 \rangle} = \frac{11 \div 2}{9 \div 5} = 3.05$.

Definition 6. *An **active edge** is an edge $e \in E$ that has 1 or more activities. An **active node** is a node u joined by an active edge. An **inactive node** is a node that is not joined by any active edges.*

Edges $\langle N_1, N_2 \rangle$ and $\langle N_2, N_3 \rangle$ in Figure 2(a) are active edges because they each have at least one activity, and nodes N_1, N_2, N_3, N_5, N_6, and N_7 are all active nodes because they are all joined by active edges. By contrast, Node N_4 is an inactive node because it is not joined by any active edges.

Definition 7. *A **super-path** of path p is any path sp that contains p, where sp is a subset of G. A **sub-path** is a path making up part of the super-path.*

For example, in Figure 2(a), $\langle N_1, N_2, N_5, N_6 \rangle$ and $\langle N_1, N_2, N_5, N_7 \rangle$ are super-paths of $\langle N_1, N_2, N_5 \rangle$. Conversely, $\langle N_1, N_2, N_5 \rangle$ is a sub-path of $\langle N_1, N_2, N_5, N_6 \rangle$.

2.2 Problem Statement

The problem of Significant Route Discovery (SRD) can be expressed as follows:

Given.

1. A spatial network $G = (N, E)$ with activity count function $a(u, v) \geq 0$ and weight function $w(u, v) > 0$ for each edge $e = (u, v) \in E$ (e.g., network distance),
2. A likelihood ratio (λ) threshold, θ,
3. A p-value,
4. m, indicating the number of Monte Carlo simulations

Find. All routes $r \in R$ with $\lambda_r \geq \theta$ and a p-value significance level

Objective. Computational efficiency

Constraints.

1. Each route $r \in R$ is a shortest path between its end-nodes,
2. $r_i \in R$ is not a subset of any $r_j \in R \; \forall r_i, r_j \in R$ where $r_i \neq r_j$,
3. Each route $r \in R$ starts and ends with active nodes,
4. Correctness and completeness

The spatial network input for SRD is defined in Definition 1. The θ input is a threshold indicating the minimum desired likelihood ratio. The p-value input is the desired level of statistical significance and m indicates the number of Monte Carlo simulations for determining statistical significance. The output for SRD are all shortest paths meeting the desired likelihood ratio and level of statistical significance. The shortest paths returned are constrained so that they are not sub-paths of any other path in the output. This constraint aims to improve solution quality by reducing redundancy in the paths returned. The output is also constrained such that the shortest paths returned start and end with active nodes. This constraint also aims to improve solution quality by ignoring edges at the start and/or end of a path that do not have any activities.

Example. The network in Figure 2(a) can be viewed as a road network, composed of streets (edges) and intersections (nodes). The aim is to find significant shortest paths that meet the given likelihood threshold of 2. In other words, find shortest paths that are twice as likely to have pedestrian fatalities. In a transportation planning scenario, identifying such routes would guide street redesign efforts to reduce the risk of pedestrian fatalities (e.g., adding sidewalks, crosswalks, pedestrian refuges, street lighting, etcetera). In Figure 2(b), routes $\langle N_1, N_2, N_3 \rangle$ and $\langle N_6, N_5, N_7 \rangle$ are returned since they are shortest paths whose likelihood exceeds $\theta = 2$, they start and end with active nodes, and they are not sub-paths of any other path in the output.

In an alternative formulation of the problem, the spatial network may be modeled with an activity count function $a(u) \geq 0$ for each node. The idea is that activities may also occur at nodes in addition to being distributed within network edges. In this way, the current approach may be extended to capture activities at nodes (e.g., vehicle accidents). If activities are modeled as counts at each node, this may alter the computational structure. We plan to investigate this in future work.

3   Proposed Approach

First we describe a naïve version of our miner, Naïve Significant Route Miner (NaïveSRM). Then we present our proposed Smart Significant Route Miner (SmartSRM) with its two algorithmic refinements, Likelihood Pruning and Monte Carlo Speedup.

3.1   Naïve Significant Route Miner (NaïveSRM) Algorithm

Algorithm 1 presents the pseudocode for the NaïveSRM approach. The basic idea behind the algorithm is to find all statistically significant shortest paths in the spatial network whose likelihood exceeds θ, under the constraints that the shortest paths returned are a) not sub-paths of any other path in the output and b) both start and end with active nodes. Algorithm 1 proceeds by calculating all shortest paths, P, in the spatial network (Line 1). Line 2 evaluates each shortest path in P to determine if it meets the given likelihood threshold θ to form a *Candidates* set. In line 3, the statistical significance of each shortest path in *Candidates* is evaluated and the significant routes are stored in *SigRoutes*. In order to assess statistical significance, all shortest paths in each of the

Algorithm 1. Naïve Significant Route Miner (NaïveSRM) Algorithm

```
Input:
    1) A spatial network G   =   (N, E) with activity count function a(u, v)  ≥  0 and
    weight function w(u, v) > 0 for each edge e = (u, v) ∈ E (e.g., network distance),
    2) A likelihood ratio (λ) threshold, θ,
    3) A p-value threshold,
    4) m, indicating the number of Monte Carlo simulations
Output:
    All routes r ∈ R with λ_r ≥ θ and p-value significance level
Algorithm:
1: {step 1:} P ← calculate all-pairs shortest path in G
2: {step 2:} Candidates ← paths in P starting and ending with active nodes having
    λ ≥ θ
3: {step 3:} SigRoutes       ← significant paths in Candidates using m Monte Carlo
    simulations
4: {step 4:} return paths that are not sub-paths of any other path in SigRoutes
```

m simulated graphs are used to calculate the p-value. In line 4, all paths in $SigRoutes$ that are not sub-paths of any other path in $SigRoutes$ are returned, and the algorithm terminates. The purpose of returning significant routes that are not sub-paths of any other path is to improve solution quality. For example, if $\langle N_1, N_2 \rangle$ and $\langle N_1, N_2, N_3 \rangle$ are both found to be significant, only $\langle N_1, N_2, N_3 \rangle$ is returned.

NaïveSRM Example. Figure 4 shows an example execution trace of NaïveSRM. The spatial network has 7 nodes, 6 edges, and 20 activities, represented by the first number in red on each edge (e.g., edge $\langle N_1, N_2 \rangle$ has six activities). The given likelihood ratio threshold θ is set to 2 and the p-value is set to 0.05.

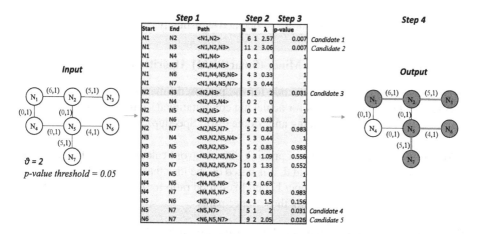

Start	End	Path	a	w	λ	p-value	
N1	N2	<N1,N2>	6	1	2.57	0.007	Candidate 1
N1	N3	<N1,N2,N3>	11	2	3.06	0.007	Candidate 2
N1	N4	<N1,N4>	0	1	0	1	
N1	N5	<N1,N4,N5>	0	2	0	1	
N1	N6	<N1,N4,N5,N6>	4	3	0.33	1	
N1	N7	<N1,N4,N5,N7>	5	3	0.44	1	
N2	N3	<N2,N3>	5	1	2	0.031	Candidate 3
N2	N4	<N2,N5,N4>	0	2	0	1	
N2	N5	<N2,N5>	0	1	0	1	
N2	N6	<N2,N5,N6>	4	2	0.63	1	
N2	N7	<N2,N5,N7>	5	2	0.83	0.983	
N3	N4	<N3,N2,N5,N4>	5	3	0.44	1	
N3	N5	<N3,N2,N5>	5	2	0.83	0.983	
N3	N6	<N3,N2,N5,N6>	9	3	1.09	0.556	
N3	N7	<N3,N2,N5,N7>	10	3	1.33	0.552	
N4	N5	<N4,N5>	0	1	0	1	
N4	N6	<N4,N5,N6>	4	2	0.63	1	
N4	N7	<N4,N5,N7>	5	2	0.83	0.983	
N5	N6	<N5,N6>	4	1	1.5	0.156	
N5	N7	<N5,N7>	5	1	2	0.031	Candidate 4
N6	N7	<N6,N5,N7>	9	2	2.05	0.026	Candidate 5

Step 1 — Step 2 — Step 3 — Step 4 headers span the table columns; Input and Output panels flank the table.

Input: $\vartheta = 2$, p-value threshold = 0.05

Fig. 4. Execution trace of Naïve Significant Route Miner (NaïveSRM). Circles represent nodes and lines represent edges (Best in color).

In step 1 of Figure 4, all shortest paths in the given spatial network are calculated. For example, the shortest path between nodes N_1 and N_3 is $\langle N_1, N_2, N_3 \rangle$. Next, in step 2, the likelihood ratio, λ, for each shortest path is determined (see Definition 5)

and those whose $\lambda \geq \theta$ are stored as candidate solutions. In the figure, the five high-lighted paths $\langle N_1, N_2 \rangle$, $\langle N_1, N_2, N_3 \rangle$, $\langle N_2, N_3 \rangle$, $\langle N_5, N_7 \rangle$, and $\langle N_6, N_5, N_7 \rangle$ are all candidates since their likelihood ratios meet or exceed the threshold of 2. In step 3, the statistical significance of each candidate is calculated using Monte Carlo simulations (discussed next). All five candidates meet the p-value threshold of 0.05. In step 4, the paths among significant paths that are not sub-paths of any other path are returned. In this example, paths $\langle N_1, N_2, N_3 \rangle$ and $\langle N_6, N_5, N_7 \rangle$ are returned. Paths $\langle N_1, N_2 \rangle$, $\langle N_2, N_3 \rangle$, and $\langle N_5, N_7 \rangle$ were not returned (even though they met and exceeded the likelihood and p-value thresholds) because they are each sub-paths of the two paths that were returned.

Finding Significant Paths. Each shortest path in SRM is evaluated for statistical sig-nificance using Monte Carlo simulations to determine whether or not it is truly anoma-lous. Here the null hypothesis states that the paths identified by SRM are random or by chance alone. The likelihood ratio is associated with a p-value to decide whether the null hypothesis should be rejected in the hypothesis test. The p-value is the probability of obtaining a value of a given likelihood ratio as equally or more extreme than that observed by chance alone.

In the Monte Carlo simulations, each activity in the original graph G is randomly associated with an edge so that the number of activities on each edge is shuffled, forming a new graph G_s. Note that all the activities in G are present in G_s, with no activities added or removed; the original activities in G are now shuffled so they may be on different edges in G_s. We then compare the highest likelihood threshold λ_{maxG_s} of randomized G_s with the highest λ_{maxG} of original G. If the original one is smaller (i.e., $\lambda_{maxG} < \lambda_{maxG_s}$), then $p = p + 1$. The above process repeats m times and after it terminates, the p-value is subsequently p/m. Paths whose p-values are less than or equal to the given p-value threshold are deemed statistically significant.

3.2   Smart Significant Route Miner (SmartSRM) Algorithm

Algorithm 2 presents the pseudocode for the proposed SmartSRM approach. The al-gorithm features two key ideas for achieving computational savings while maintaining result quality: Likelihood Pruning and Monte Carlo Speedup.

Likelihood Pruning: Likelihood pruning aims to avoid calculating all shortest paths in G based on the given threshold θ. It is based on the idea that for each shortest path p, it is possible to determine an upper bound likelihood ratio for the super-paths rooted at p's start node, without calculating those super-paths.

Definition 8. *The **upperbound likelihood ratio for path** p, $\hat{\lambda}_p = \frac{\hat{a}_p \div \hat{w}_p}{(|A| - \hat{a}_p) \div (|W| - \hat{w}_p)}$, where $\hat{a}_p = a_p + (|A| - a_t)$ (where a_t is the number of activities in the shortest path tree rooted at p's source node) and $\hat{w}_p$ is the weight of the shortest super-path of p, rooted at p's start node.*

The intuition behind the upper bound likelihood ratio for path p is that (1) the number of activities on all of p's super-paths rooted at p's start node are bounded by the number of activities in the spatial network minus the number of activities in the current shortest path tree rooted at the source node in p and (2) the weight of any super-path of p is at least the weight of the closest edge to p plus p's weight.

Algorithm 2. Smart Significant Route Miner (SmartSRM) Algorithm

---

```
Inputs and Outputs for SmartSRM are same as NaiveSRM
Algorithm:
   {Step 1: Likelihood Pruning}
 1: for each s ∈ active nodes in G do
 2:    Initialize D[v] ← inf; Pred[v] ← ∅; Λ̂[v] ← θ; a[v] ← 0; aₜ ← 0; D[s] ← 0; PQ ←
    N
 3:    while PQ ≠ ∅ do
 4:       u ← node in PQ with smallest distance in D[]; P ← shortest path (s, u) in
    Pred[]
 5:       aₜ ← aₜ+ number of activities on edge Pred[u]
 6:       if Λ̂[v] ≥ θ then
 7:          for each v adjacent to u do
 8:             sum ← D[u] + w(u, v)
 9:             if sum < D[v] then
10:                D[v] ← sum; update v's position in PQ based on sum; Pred[v] ← u
11:                a[v] ← a[u] + a(u, v); ŵ ← sum+ weight of closest neighbor w(u, v)
12:                Λ̂[v] ← calculate λₛᵥ based on a[v], aₜ and ŵ
13: {Step 2:} Candidates        ← paths in P starting and ending with active nodes
    having λ ≥ θ
   {Step 3: Monte Carlo Speedup}
14:    λₘₐₓG ← highest likelihood ratio in G
15:    for each simulation₁....simulationₘ do
16:       Gₛ ← assign activities in G to random edges
17:       λₘₐₓGₛᵢ ← 0
18:       for each shortest path p ∈ Gₛ do
19:          if λₚ > λₘₐₓGₛᵢ then
20:             λₘₐₓGₛᵢ ← λₚ; pₘₐₓᵣ ← pₘₐₓᵣ + 1
21:             if pₘₐₓᵣ/N ≤ p-value threshold then return ∅
22:             if λₚ > λₘₐₓG then break
23:       for each route r ∈ Candidates do
24:          if λₘₐₓGₛᵢ > λₘₐₓG then pᵣ ← pᵣ + 1
25:    for each route r ∈ Candidates do
26:       if pᵣ/N ≤ p-value threshold then SigRoutes ← r
27: {Step 4:} return paths that are not sub-paths of any other path in SigRoutes
```

Lines 1-12 of Algorithm 2 shows the pseudocode for likelihood pruning, which is similar to Dijkstra's algorithm [18] with a few exceptions: (1) the shortest paths from a single active node to all destinations are calculated for all active nodes in the spatial network, (2) if the upper bound likelihood ratio for path $\langle s...u \rangle$ is below the given likelihood threshold θ, u's neighbors are not visited (line 6), and (3) upperbound statistics are calculated and updated each time the weight from source s to a node v is updated (lines 9-12).

Likelihood Pruning Example. Figure 5(a) illustrates the basic idea behind likelihood pruning. In this example, we have set the likelihood threshold to $\theta = 5$, indicating that we are interested in paths that are five times as likely to have pedestrian fatalities. During the algorithm's execution, at some point the source node becomes N_1, and the shortest path between N_1 and every other active node in the spatial network is calculated.

When the shortest path between N_1 and N_5 is calculated, the upper bound likelihood ratio for path $\langle N_1, N_2, N_5 \rangle$ is determined to be 4, since based on Definition 8, the calculation would be $\frac{(6+(20-11))\div 3}{(20-((6+(20-11))\div(7-3)}$, where $\hat{a}_p = 6 + (20 - 11) = 15$ and $\hat{w}_p = 2+1 = 3$. We can, therefore, avoid calculating the shortest paths $\langle N_1, N_2, N_5, N_6 \rangle$ and $\langle N_1, N_2, N_5, N_7 \rangle$ for $\theta = 5$.

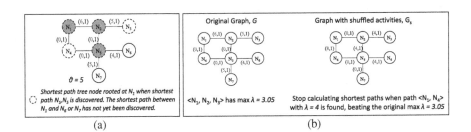

(a)   (b)

Fig. 5. (a) Example of Likelihood Pruning. Since we know the upper-bound likelihood for $\langle N_1, N_2, N_5 \rangle$ is 4, we can avoid calculating the shortest paths $\langle N_1, N_2, N_5, N_6 \rangle$ and $\langle N_1, N_2, N_5, N_7 \rangle$ for $\theta = 5$. (b) Example of Monte Carlo Speedup. (Best in color).

Monte Carlo Speedup: Monte Carlo speedup aims to calculate the p-value without considering all shortest paths in each simulated graph. The basic idea is that once a shortest path in the simulated graph is found to have a higher likelihood ratio than the maximum likelihood ratio in the original graph, the simulation immediately ends with the p-value being incremented. In other words, there is no reason to keep looking at all shortest paths in the simulated graph if we find one that already beats the maximum likelihood ratio in the original graph. Additionally, Monte Carlo speedup stops all simulations the moment p out of m simulations are found where the simulated likelihood ratio beats the original maximum likelihood ratio. In other words, there is no reason to execute all m simulations if we find that the p-value threshold will not be met. The pseudocode for Monte Carlo speedup is presented in Lines 14-26 of Algorithm 2.

Monte Carlo Speedup Example. Figure 5(b) illustrates one of the basic ideas behind Monte Carlo speedup. In this example, the graph on the left is the original graph G whereas the graph on the right, G_s, represents one simulation with the activities shuffled. In G_s, instead of looking at all 42 shortest paths, we can stop and increment p the moment a path that has a likelihood higher than the maximum likelihood in G is found. In this case, that path would be $\langle N_1, N_2 \rangle$ (on the right of the figure), with a likelihood ratio of 4.

SmartSRM uses filter and refine techniques (e.g., Likelihood Ratio pruning and Monte Carlo speedup) to achieve computational savings. Filter and refine techniques may not change worst case complexity but they can reduce runtime. Likelihood Ratio pruning creates a boundary via the upperbound likelihood ratio such that not all destinations are visited from each source node. Some of the destinations are pruned because the shortest paths to them will never meet the likelihood ratio threshold. Monte Carlo speedup avoids generating all shortest paths in cases where a shortest path in the simulated dataset has a higher likelihood ratio than the shortest paths in the original dataset.

Monte Carlo speedup also terminates early if the p-value threshold will not be met based on the number of times the maximum likelihood ratio in the simulated dataset beats the maximum likelihood ratio in the original dataset.

The computational costs of NaïveSRM and SmartSRM stem from 1) the cost of calculating all pair shortest paths and 2) the cost of assessing statistical significance for all shortest paths in the spatial network. For NaïveSRM, the total cost is $(N^2 logN \times C_{\lambda_p}) + (m \times N^2 logN \times C_{\lambda_p})$, where N is the set of nodes, $N^2 logN$ is the cost of calculating shortest paths in the spatial network, C_{λ_p} is the cost of calculating the likelihood ratio of path p, and m is the number of Monte Carlo simulations.

For SmartSRM, the total cost is $((N \times r_{\hat{\lambda}})^2 logN \times C_{\lambda_p}) + (m \times (N^2 logN \times r_m) \times C_{\lambda_p})$, where N is the set of nodes, $(N \times r_{\hat{\lambda}})^2 logN$ is the cost of calculating shortest paths for a set of shortest paths that is a superset of all paths in G with $\lambda_p \geq \theta$, $r_{\hat{\lambda}}$ (whose value is between 0 and 1) is the ratio of shortest paths with $\lambda_p \geq \theta$ to all shortest paths, C_{λ_p} is the cost of calculating the likelihood ratio of path p, m is the number of Monte Carlo simulations, and r_m (whose value is between 0 and 1) is the ratio of shortest paths calculated before finding a path whose likelihood beats the maximum likelihood in the original graph to all shortest paths.

In summary, SmartSRM may only consider a fraction of the paths considered by NaïveSRM, both in calculating all pair shortest paths and assessing statistical significance for all shortest paths in the spatial network.

4   Case Study

We conducted a qualitative evaluation of SmartSRM, comparing its analysis with the analysis of SaTScan [19] (continuous Poisson process) on a real pedestrian fatality data set [2], shown in Figure 6(a). As noted earlier, SaTScan discovers areas of significant activity that are represented as circles on the spatial network while SmartSRM discovers significant shortest paths. The input consisted of 43 pedestrian fatalities (represented as dots) in Orlando, Florida occurring between 2000 and 2009. For each edge (portion of road) in the network, fatality count was aggregated, yielding overall activity, and weight was the actual road network distance. The maps were prepared using QGIS' Open Layers plugin [20], and the road network was from the US Census Bureaus TIGER/Line Shapefiles [21].

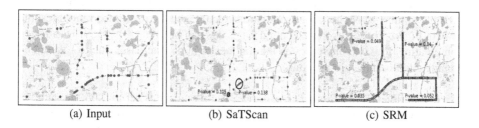

| (a) Input | (b) SaTScan | (c) SRM |

Fig. 6. Comparing SmartSRM and SaTScan's output for a p-value threshold of 0.15 and $\theta = 1.75$ on pedestrian fatality data from Orlando, FL [2] (Best in color)

When evaluating the techniques, we consider the outputs of circles vs. shortest paths. While p-value thresholds of 0.05 or lower are often desired, we used a p-value threshold of 0.15 because the circles chosen by SaTScan had high p-values for this dataset. As noted earlier, pedestrian fatalities usually occur on streets, particularly along arterial roadways [1]. Thus this activity can be said to have a linear generator. However, the results generated by SaTScan do not capture this. From Figure 6(b), it is clear that the circle-based output is meant for areas, not streets. In contrast, the shortest paths detected by SmartSRM fully capture the significant activities on the arterial roads (some of the paths in Figure 6(c) are overlapping). Furthermore, the paths in the figure make sense in this context due to the inherently linear nature of the activities.

5   Experimental Evaluation

The goal of our experiments was to evaluate the scalability of the proposed approach by varying and observing the effect of three workload parameters: nodes, likelihood ratio threshold θ, and p-value threshold. All experiments were performed on a Mac Pro with a 2 x Xeon Quad Core 2.26 GHz processor and 16 GB RAM. For each workload experiment we compared Naïve Significant Route Miner (NaïveSRM) and our Smart Significant Route Miner (SmartSRM).

5.1   Experiment Data Set

Our experiments were performed on real-world data obtained from the Fatality Analysis Reporting System (FARS) encyclopedia [2]. The dataset contained geospatial and temporal data describing 487 pedestrian fatalities in Orange County, FL (which includes Orlando), from 2001 to 2011. For each edge (portion of road) in the network, fatality count was aggregated, yielding overall activity, and weight was the actual road network distance. The road network was obtained from the US Census Bureau's TIGER/Line Shapefiles [21].

5.2   Experimental Results

Effect of the Number of Nodes. We varied the number of nodes from 500 to 2500, which is akin to varying the number of shortest paths (routes) from $250,000$ to $6,250,000$ (since there are $\binom{n}{2}$ shortest paths in the spatial network). We set the p-value threshold to 0.05, the number of Monte Carlo simulations to 100, and the likelihood ratio threshold θ to 20. Figure 7(a) gives the execution times. As can be seen, $SmartSRM$ is faster. Computational savings increases as the number of nodes increases due to Likelihood Pruning and Monte Carlo Speedup.

Effect of the Likelihood Ratio Threshold θ. The p-value was set to 0.05, the number of Monte Carlo simulations was set to 100, and the number of nodes was set to 1000. Figure 7(b) gives the execution times. Again, $SmartSRM$ beats the naïve algorithm. Computational savings increases as the likelihood ratio increases due to Likelihood Pruning and Monte Carlo Speedup.

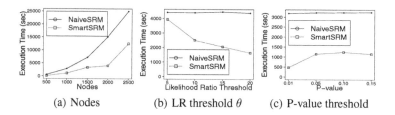

(a) Nodes   (b) LR threshold θ (c) P-value threshold

Fig. 7. Scalability of SRM with increasing (a) number of nodes, (b) likelihood ratio threshold θ, and (c) P-value Threshold

Effect of the P-value. The number of nodes was set to 1000, the likelihood ratio threshold θ was set to 20, and the number of Monte Carlo simulations was set to 100. Figure 7(c) gives the execution times. As can be seen, $SmartSRM$ is faster. Computational savings increases as the p-value increases due to Likelihood Pruning and Monte Carlo Speedup.

In summary, the experiments uniformly show that $SmartSRM$ is much better (2-3 times faster) than the naïve approach. This is because $SmartSRM$ prunes unlikely paths and speeds up Monte Carlo simulation.

6 Discussion

Non-Statistically Significant Techniques. Our work focuses on partitioning techniques that consider statistical significance. There are a myriad of other techniques that divide data into groups but that do not consider statistical significance including DBScan [12], K-Means [13], KMR [14], and Maximum Subgraph Finding [15]. However, statistical significance is important for determining the probability that an effect is not due to just chance alone. Post-processing the output of these techniques for statistical significance will not guarantee completeness as some of the clusters returned may not be statistically significant. For example, the algorithm from our previous work [14] on summarizing activities using routes may return routes that are not statistically significant. Figure 8(a) shows an example where DBScan [12] returns 7 chance clusters on a complete spatial randomness dataset.

Alternative network footprints. Summarizing significant network footprints of activities may be done using significant subgraphs, significant paths, significant shortest paths, etc. Each representation entails a tradeoff between fidelity and computational scalability. For example, subgraphs may offer accurate significant network footprints but their calculation may be computationally intensive due to their exponential number. As an initial step, we have selected shortest paths to summarize significant network footprints of activities. While shortest paths may lose some fidelity, they offer computational scalability because their number is bounded (i.e., $\binom{n}{2}$, where n is the number of nodes). The union of shortest paths may also be used to represent other network footprints.

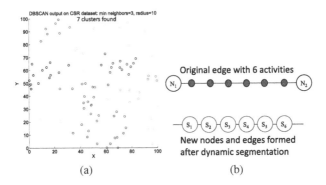

(a)   (b)

Fig. 8. (a) Colored dots are part of chance clusters identified by DBScan [12] on a complete spatial randomness dataset (b) Example of Dynamic Segmentation (Best in color)

Dynamic Segmentation. Resolving statistically significant routes to the sub-arc level requires a dynamic segmentation data model. In dynamic segmentation, the original nodes and edges in a statically segmented network (e.g., Figure 2(a)) are replaced by new nodes formed at the locations of activities, with new edges connecting these locations. Figure 8(b) shows an example where edge $\langle N_1, N_2 \rangle$ from Figure 2(a) has been dynamically segmented. As can be seen, the six activities on the original edge form six new nodes in the network, with new edges connecting these nodes. Dynamic segmentation has the potential to improve result quality in significant route discovery. This is because each segment in the dynamically segmented network structure corresponds to the locations of activities so the likelihood ratios of candidate routes are more precise. However, dynamic segmentation has the potential to introduce many new nodes in the spatial network, which could be computationally prohibitive for datasets with a large number of activities. Future research is needed to investigate this tradeoff.

7   Conclusion

This work explored the problem of significant route discovery in relation to important application domains such as preventing pedestrian fatalities and crime analysis. We proposed a Smart Significant Route Miner (SmartSRM) algorithm that discovers statistically significant shortest paths in a spatial network. SmartSRM uses Likelihood Pruning and Monte Carlo Speedup to enhance its performance and scalability. We presented a case study comparing SmartSRM with SaTScan on pedestrian fatality data. Experimental evaluation using real-world data indicated that the algorithmic refinements utilized by SmartSRM yielded substantial computational savings without sacrificing result quality.

In future work, we plan to explore other types of data that may not be associated with a point in a street (e.g., aggregated pedestrian fatality data at the zip code level). The present research is centered on finding high concentrations of activities whose counts and locations are deterministic. However, future work is needed to investigate attributes that may not be deterministic such as delay when moving between nodes, capacity constraints, etc. We will also generalize significant route discovery for all paths and explore additional spatial constraints (e.g., nearest neighbors). Finally, incorporating time and dynamic segmentation into SRD will be explored.

Acknowledgments. This material is based upon work supported by the National Science Foundation under Grants No. 1029711, IIS-1320580, 0940818 and IIS-1218168, the USDOD under Grants No. HM1582-08-1-0017, and HM0210-13-1-0005, and the Doctoral Dissertation Fellowship program at the University of Minnesota's graduate school. We would like to thank Kim Koffolt and the members of the University of Minnesota Spatial Computing Research Group for their comments.

References

1. Ernst, M., Lang, M., Davis, S.: Dangerous by design: Solving the epidemic of preventable pedestrian deaths. Transportation for America: Surface Transportation Policy Partnership, Washington, DC (2011)
2. National Highway Traffic Safety Administration (NHTSA): Fatality Analysis Reporting System (FARS) Encyclopedia, http://www.nhtsa.gov/FARS
3. Kulldorff, M.: A spatial scan statistic. Communications in Statistics-Theory and Methods 26(6), 1481–1496 (1997)
4. Neill, D.B., Moore, A.W.: Rapid detection of significant spatial clusters. In: Proceedings of the Tenth ACM SIGKDD International Conference on Knowledge Discovery and Data Mining, pp. 256–265. ACM (2004)
5. Kulldorff, M., Mostashari, F., Duczmal, L., Katherine Yih, W., Kleinman, K., Platt, R.: Multivariate scan statistics for disease surveillance. Statistics in Medicine 26(8), 1824–1833 (2007)
6. Kulldorff, M.: Spatial scan statistics: Models, calculations, and applications. In: Scan Statistics and Applications, pp. 303–322. Springer (1999)
7. Costa, M.A., Assunção, R.M., Kulldorff, M.: Constrained spanning tree algorithms for irregularly-shaped spatial clustering. Computational Statistics & Data Analysis 56(6), 1771–1783 (2012)
8. Duczmal, L., Assuncao, R.: A simulated annealing strategy for the detection of arbitrarily shaped spatial clusters. Computational Statistics & Data Analysis 45(2), 269–286 (2004)
9. Shi, L., Janeja, V.P.: Anomalous window discovery for linear intersecting paths. IEEE Transactions on Knowledge and Data Engineering 23(12), 1857–1871 (2011)
10. Janeja, V.P., Atluri, V.: Ls 3: A linear semantic scan statistic technique for detecting anomalous windows. In: Proceedings of the 2005 ACM Symposium on Applied Computing, pp. 493–497. ACM (2005)
11. Li, X., Han, J., Lee, J.-G., Gonzalez, H.: Traffic density-based discovery of hot routes in road networks. In: Papadias, D., Zhang, D., Kollios, G. (eds.) SSTD 2007. LNCS, vol. 4605, pp. 441–459. Springer, Heidelberg (2007)
12. Ester, M., Kriegel, H.P., Sander, J., Xu, X.: A density-based algorithm for discovering clusters in large spatial databases with noise. In: KDD, vol. 96, pp. 226–231 (1996)
13. MacQueen, J., et al.: Some methods for classification and analysis of multivariate observations. In: Proceedings of the fifth Berkeley Symposium on Mathematical Statistics and Probability, vol. 1, pp. 281–297 (1967)
14. Oliver, D., Bannur, A., Kang, J.M., Shekhar, S., Bousselaire, R.: A K-Main Routes Approach to Spatial Network Activity Summarization: A Summary of Results. In: IEEE International Conference on Data Mining Workshops (ICDMW), pp. 265–272 (2010)

15. Buchin, K., Cabello, S., Gudmundsson, J., Löffler, M., Luo, J., Rote, G., Silveira, R.I., Speckmann, B., Wolle, T.: Finding the most relevant fragments in networks. J. Graph Algorithms Appl. 14(2), 307–336 (2010)
16. Chawla, S., Roughgarden, T.: Single-source stochastic routing. In: Díaz, J., Jansen, K., Rolim, J.D.P., Zwick, U. (eds.) APPROX/RANDOM 2006. LNCS, vol. 4110, pp. 82–94. Springer, Heidelberg (2006)
17. Shekhar, S., Liu, D.: CCAM: A connectivity-clustered access method for networks and network computations. IEEE Transactions on Knowledge and Data Engineering 9(1), 102–119 (1997)
18. Cormen, T.: Introduction to algorithms. The MIT press (2001)
19. Kulldorff, M., Rand, K., Gherman, G., Williams, G., DeFrancesco, D.: SaTScan v 2.1: Software for the spatial and space-time scan statistics. National Cancer Institute, Bethesda (1998)
20. The QGIS Project: Quantum GIS OpenLayers Plugin, `http://plugins.qgis.org/plugins/openlayers_plugin/` (accessed: January 23, 2014)
21. US Census Bureau: Census TIGER/Line Shapefiles (2010), `http://www.census.gov/geo/maps-data/data/tiger-line.html` (accessed: January 23, 2014)

Location Oblivious Privacy Protection for Group Nearest Neighbor Queries

A.K.M. Mustafizur Rahman Khan[1], Tanzima Hashem[2], Egemen Tanin[1], and Lars Kulik[1]

[1] University of Melbourne, Victoria, Australia
[2] Bangladesh University of Engineering & Technology, Dhaka, Bangladesh

Abstract. Finding a convenient meeting point for a group is a common problem. For example, a group of users may want to meet at a restaurant that minimizes the group's total travel distance. Such queries are called Group Nearest Neighbor (GNN) queries. Up to now, users have had to rely on an external party, typically a location service provider (LSP), for computing an optimal meeting point. This implies that users have to trust the LSP with their private locations. Existing techniques for private GNN queries either cannot resist sophisticated attacks or are computationally too expensive to be implemented on the popular platform of mobile phones. This paper proposes an algorithm to efficiently process private GNN queries. To achieve high efficiency we propose an approach that approximates a GNN with a high accuracy and is robust to attacks. Unlike methods based on obfuscation, our method does not require a user to provide an imprecise location and is in fact location oblivious. Our approach is based on a distributed secure sum protocol which requires only light weight computation. Our experimental results show that we provide a readily deployable solution for real life applications which can also be deployed for other geo-spatial queries and applications.

1 Introduction

Internet accessibility through smartphones has fueled the increased popularity of Location Based Services (LBS). Finding a convenient meeting point for a group is a common problem. For example, a group of friends may want to know the restaurant which minimizes the total travel distance for them. Such queries are called Group Nearest Neighbor (GNN) queries. In a GNN query, the users issue a query to find out the point of interest (POI) which minimizes aggregate distance (AD) for the group. The AD could be the total distance of all users to the POI. To access GNN queries, users have to disclose their locations to a location service provider (LSP). The users risk their privacy by exposing locations while receiving such services. An LSP might use a user's location as a clue to derive sensitive and private information about her habits, health, social relations, etc. An alternative approach would be to compute ADs by sharing a user's location with other group members. However, users may want to hide their locations not only from an LSP but also from other group members. Thus, the major challenge for processing a privacy preserving GNN query is to compute ADs for POIs, stored on the LSP's database, without revealing user locations to others.

Obfuscation based techniques such as spatial cloaking, i.e., providing imprecise location, have been widely used to protect location privacy while accessing LBSs [1–4].

M. Duckham et al. (Eds.): GIScience 2014, LNCS 8728, pp. 301–317, 2014.

But, in many cases, this direction of research of blurring a point to a region is not enough to provide location privacy. For example, if a user frequently reveals her imprecise location, her trajectory can be derived. Obfuscation cannot protect a user from absence disclosure attack as well [5].

Most of the existing techniques are designed to preserve location privacy of an individual and cannot preserve location privacy for a group of users. Hashem et al. [3] first proposed a technique to preserve privacy for a group of users. Talouki et al. [6] and Huang et al. [7] followed them. In Sec. 2.1, we show that these techniques are either vulnerable to sophisticated attacks or computationally expensive for mobile phones.

In this paper, we also identify a new attack type that no previous work related to GNN queries has addressed. We show that if attackers can learn a group's ADs to a sufficient number of POIs, they can compute the locations of all users in the group without knowing which location corresponds to which user. We call it *Multi-point Aggregate Distance (MPAD) attack*.

In a privacy preserving GNN query, a user's location must be kept secret from other users as well as the LSP. If a user's distances to three POIs are known, the user's location can be calculated by 2D trilateration. Therefore, the group is required to compute ADs of POIs without revealing the users' locations or individual distances to POIs. In order to resist the MPAD attack, the number of revealed ADs of POIs must be limited.

In any LBS, a user expects a certain quality of that service. In our scenario, the quality depends on at least three factors: (i) the level of guaranteed privacy, (ii) the cost to the user, i.e., the communication and computation cost for a mobile device, and (iii) the accuracy of the results returned by the LSP. As there is no known algorithm that fulfills all three criteria, in this paper, we worked on addressing the problem in a way that could be readily deployed in the real world and could be used by real users. We have developed two new algorithms. Our new algorithms are LOOP and H-LOOP. Location Oblivious Private (LOOP) algorithm produces an accurate answer with very little cost. ADs of all POIs are calculated in a distributed privacy preserving manner, and the POI with the smallest AD is declared as the GNN. LOOP is based on a distributed secure sum protocol, proposed by Sheikh et al. [8], which does not use cryptography and thus is computationally inexpensive. Although LOOP provides better location privacy than existing methods, it still has two limitations: first, it cannot resist the MPAD attack from malicious users, second, LOOP requires the LSP to reveal locations of all POIs to the users. However in many cases an LSP might be unwilling to reveal all POIs' locations.

To overcome these limitations, we need to reduce the number of POIs whose ADs are calculated. In this paper, we propose H-LOOP, an improved version of LOOP, which fulfills all the requirements, i.e., ensures both high level of privacy and high accuracy of the query answer, and incurs less processing overhead. H-LOOP takes a heuristic approach to drastically narrow down the search space and returns an approximate answer with an extremely high level of accuracy. H-LOOP predicts the location of the GNN in the part of the search space which has the smallest minimum AD. Existing methods cannot maintain a high level of privacy, efficiency and accuracy simultaneously. H-LOOP is the first method which provides high levels of privacy in an efficient way while rarely sacrificing accuracy at a small degree.

Unlike methods based on imprecise location, our methods do not require a user to provide any information about her location and thus is location oblivious. A user divides her distance from a point/square into random parts and sends those to other users. Her distance remains undisclosed unless all other users collude and recover all parts. As a result, a user's location cannot be recovered or inferred by an attacker.

1.1 Attack Model

In our scenario, neither the LSP nor the users are trustworthy. The LSP may collude with malicious group members to organize an attack. Therefore, an honest user trusts neither the LSP nor the fellow group members. In this study, we assume that both users and the LSP act according to the protocol. We consider only passive attacks, where an attacker or a group of colluding attackers try to recover the locations of participating users using information gained through query processing.

Existing works related to privacy preserving GNN queries considered 2D trilateration based attacks. In these attacks, an attacker or a group of colluding attackers compute distances of an individual user to multiple POIs using gained information during the distance aggregation process. Then these distances are used in 2D trilateration to recover the exact location of that user. We, too, consider these attacks.

We identify a new attack type, called Multi-Point Aggregate Distance (MPAD) attack in this paper. In an MPAD attack, an attacker can use aggregate total distances of multiple POIs to calculate users' locations. Let the coordinate of a POI P_i be (P_{xi}, P_{yi}) and coordinate of a user U_j be (U_{xj}, U_{yj}). Aggregate total distance of P_i from n users is computed as

$$D_{Pi} = \sum_{j=1}^{n} \sqrt{(P_{xi} - U_{xj})^2 + (P_{yi} - U_{yj})^2}$$

Since the locations of users are unknown, there are $2n$ unknown variables in the above mentioned equation. Hence, if an attacker can learn aggregate total distances to $2n$ POIs, she can solve the equations. However, the equations have multiple solutions. If the attacker knows more than $2n$ ADs, she can use the extra equations to discard false solutions. Aggregate total distance functions are symmetric functions. Therefore, the set of solutions is invariant under any permutation of the unknown variables. An attacker can target only the whole group collectively and cannot distinguish individuals in the group. The attacker may associate a single point to a user using background knowledge or revealed imprecise location of that user.

2 Background

2.1 Finding an Optimal Meeting Point in a Privacy Preserving Manner

Papadias et al. introduced the GNN query [9]. Hashem et al. [3] first proposed a privacy preserving GNN query processing technique. The users provide their imprecise locations as regions instead of exact points to an LSP and thus preserve their location privacy. The LSP returns a set of candidate POIs with respect to the provided regions. Users update AD for every POI with respect to their actual locations such that each user's distance to POIs remain hidden in an aggregate form. After all users' updates, the POI with the smallest AD becomes the GNN. Talouki et al. [6] have shown that Hashem et al's [3] technique is vulnerable to a Partial Collusion (PC) attack and a user's location can be recovered. Moreover, as we mentioned in Sec. 1, imprecise location does not provide a high level of location privacy.

Talouki et al. took a cryptographic approach to solve the privacy preserving GNN query problem and proposed two solutions. In [6], a group of users calculate their centroid by averaging their coordinates using an anonymous veto network and homomorphic encryption thus hiding their locations. The LSP returns the nearest point to that centroid as the GNN. Unfortunately, the nearest point to the centroid is not necessarily the GNN. Suppose user A and B are located in $(0,0)$ and $(8,0)$ respectively and POI p and q are located in $(4,1)$ and $(8,0)$ respectively. The centroid of the users' locations is $(4,0)$. p is the nearest POI to the centroid and its aggregate total distance is $\sqrt{4^2 + 1^2} + \sqrt{4^2 + 1^2} = 8.24$. AD of q is $8 + 0 = 8$. Hence, p is not the GNN but q is. We can conclude that, their approach cannot guarantee a correct answer. In [10], Talouki et al. used imprecise location and calculated ADs of all candidate POIs using an anonymous veto network and homomorphic encryption. As mentioned in Sec. 1, we consider hiding location in a larger region is not enough to provide privacy.

Huang et al. also used a cryptographic technique to preserve privacy in GNN queries [7]. A single user creates a Garbled circuit, and another user evaluates it. All other users get their encrypted input bits from the circuit creator by Oblivious Transfers (OT) and send their input bits to circuit evaluator. Then the circuit evaluator evaluates the Garbled circuit to find out the GNN. Creating and evaluating a Garbled circuit when there were 10 users and 2000 POIs took 20 seconds in a centralized model and 4 seconds in a distributed model in their experiment. The number of OTs performed by a user is equal to the number of POIs. In [11], Huang et al. reported that Garbled circuit implementations on smartphones are about three orders of magnitude slower than desktop computers. Moreover, the circuit creator needs to send a circuit file of several GB to the circuit evaluator [12] and thus incur heavy communication cost. Hence, Garbled circuit based methods are not applicable in various real world mobile phone applications.

Existing privacy preserving techniques for GNN query [3, 7, 10] cannot bound the number of POIs whose ADs are calculated. As a result, they are open to MPAD attack as well. The number of candidate POIs in [3] and [10]'s technique depends on the size of cloaking rectangle. Larger rectangles lead to more candidate POIs. ADs to all candidate POIs are compared to find out the GNN. Huang et al.'s [7] technique requires calculation and comparison of ADs of all POIs.

2.2   Distributed Secure Sum

If a user's distances from three points are known, her position can be easily calculated by 2D trilateration. The group members have to calculate AD of a point, which is a function of all members' distances, without revealing their own distances. This problem is similar to secure multi-party computation problems, where a group of users compute a function which takes one input from each user. The function is public, but each user can keep her input secret. Sheikh et al. proposed a simple distributed secure sum protocol to compute the summation function [8]. We base our work on their technique in order to calculate aggregate total distance.

We illustrate their technique with a simple example. Suppose, there are four players A, B, C, and D and a coordinator. A, B, C, and D have 10, 21, 15, and 8 cards, respectively. They do not want to disclose their number of cards but want to calculate how many cards they have in total. The coordinator publishes the result and can help in computation if necessary. The solution is as follows.

Each player randomly divides her cards into four parts, keeps one part for herself and gives other three parts to the other three players. We assume that transactions between two players are not visible by others. Their transactions are shown in Fig. 1. Change of a player's stock is shown above/below her.

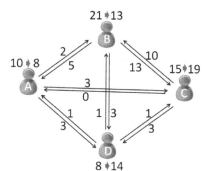

		Receives			Stock before
	A	B	C	D	transaction
A gives	2	2	3	3	10
B gives	5	0	13	3	21
C gives	0	10	2	3	15
D gives	1	1	1	5	8
Stock after transaction	8	13	19	14	54 (Total)

Fig. 1. Transactions between players

The players inform the coordinator about the number of cards they possess, i.e., 8, 13, 19, and 14. The coordinator adds these numbers and gets 54 as total. Since, the cards only change hands, the total numbers of cards before and after the transactions are equal. Even if the coordinator collude with dishonest players, they cannot calculate a player's initial stock if there is at least one honest player other than the targeted player.

3 Our Solution

In our model, all users can communicate with the LSP and the users can communicate with each other without the help of the LSP. While forming a group, group members exchange their identities (e.g., IP address, phone number) with other members. A member sends all group members' identities and the POI type to the LSP and inquires about the GNN. The LSP runs either LOOP or H-LOOP. During the course of execution, the LSP communicates with the users to direct the GNN search. The users communicate with each other to compute and compare ADs. Computation and comparison of ADs are done with two privacy preserving protocols, which we study in Sec. 3.1. Finally, the LSP broadcasts the answer to all users.

3.1 Privacy Preserving Protocols

In this section, we present two privacy preserving protocols. These protocols safeguard users' location privacy. We present our algorithms to process the GNN query using these protocols in Sec. 3.2.

Private Aggregate Distance (PAD) Protocol. We use PAD to calculate the AD of all group members to a point in a privacy preserving manner. As a result, users can calculate their ADs from any point without revealing their individual distances from the point. We assume that the attackers cannot observe communications between the users. This protocol is based on Sheikh et al. [8]'s work, described in Sec. 2.2. Let $\{U_1, U_2, \ldots, U_n\}$ be the group of users. Distance from a user U_i to a point or a rectangle is d_i, which is kept private. The AD has to be computed as

$$D = \sum_{i=1}^{n} d_i.$$

The protocol to solve this problem is as follows:

Each user U_i does:

1. Choose an integer random number α_i, where $0 \leq \alpha_i < n$.
2. Choose α_i users from the group, excluding U_i. Let $\{U_{i1}, U_{i2}, \ldots, U_{i\alpha_i}\}$ be the chosen subset of users.
3. Choose α_i random numbers d_{ij}.
4. Calculate

$$\Delta_i = d_i - \sum_{j=1}^{\alpha_i} d_{ij}$$

5. Send d_{ij} to U_{ij}, where $j = \{1, 2, \ldots, \alpha_i\}$.
6. Receive numbers from other users if sent. Let Δ_i' be the sum of the received numbers.
7. Calculate $d_i' = \Delta_i + \Delta_i'$ and send d_i' to the coordinator.

Finally, the coordinator calculates

$$D = \sum_{i=1}^{n} d_i'$$

A user discloses only d_{ij} to U_{ij} and d_i' to the coordinator in step 5 and 7 respectively. The user keeps all other information private.

An LSP or any user or any third party can act as a coordinator. The number of messages sent by a user U_i is α_i. Since an attacker is not aware about communications between users, even if a user chooses $\alpha_i = 0$ and does not send any number to any user, the attacker still cannot learn the value of α_i. Hence, privacy is assured even at a lower value of α. If α_i has a mean value of α, total number of messages is $n\alpha$. When calculating ADs to a set of points/rectangles, users exchange an array of values instead of just one value in each communication. PAD cannot protect a user if all other users collude. Therefore, the group size must be at least three.

Comparing with Minimum Disclosure (CWMD) Protocol. Let there be n users and each user U_i has a set of POIs S_i. Suppose a user knows ADs of all POIs of her set. The users want to find out the POI from $S_1 \cup S_2 \cup \ldots \cup S_n$, which has the minimum AD. An attacker can use ADs to multiple points in order to find out the group's locations. Therefore, our goal is to find out the POI with smallest AD with minimum disclosure. CWMD solves this problem. If CWMD is used, a user cannot learn the AD of any additional POI and an LSP cannot learn enough to perform an attack.

Let $Dmin_i$ be the smallest AD of a POI in S_i. The protocol is as follows:

1. Each user U_i sends the smallest AD $Dmin_i$ to the LSP.
2. Let $Dmin_j$ be the smallest in $\{Dmin_1, Dmin_2, \ldots Dmin_n\}$. The LSP sends a disclosure request to U_j.
3. Let the POI P_{min} has the minimum AD in S_j. U_j sends P_{min} to the LSP.
4. The LSP publishes P_{min}.

There are $2n$ unknown coordinate values of n user locations. Therefore, an LSP needs at least $2n$ POIs and their ADs to perform an MPAD attack. In CWMD, an LSP can learn n ADs and the ID of only one POI. It cannot learn the IDs of the remaining related POIs. Even if the LSP cheats it can learn ADs of at most n POIs. It cannot invoke the protocol multiple times because the protocol needs active participation of the users. As a result, the LSP cannot perform an MPAD attack.

3.2 Privacy Preserving Algorithms

We present two privacy preserving GNN query processing algorithms. These algorithms are named as Location Oblivious Private (LOOP) algorithm and Heuristic Location Oblivious Private (H-LOOP) algorithm. They are based on PAD and CWMD.

Location Oblivious Private (LOOP) Algorithm. Let there be n users in the group. The LSP divides all POIs into n subsets such that POI P_j is an element of set S_l, where $l = (j \bmod n) + 1$. Then the LSP assigns each user U_i as the coordinator of POIs in S_i, where $i = 1, 2, \ldots, n$. Next the LSP sends a list of all POIs along with their locations and assigned coordinators' IDs to the users. The coordinator users communicate with other users and calculate the ADs of their assigned set of POIs using PAD. In this way, ADs of all POIs are calculated. Then the coordinators use CWMD to find out the POI with the smallest AD, which is the GNN.

Figure 2 shows an example of the flow of LOOP. At first, the LSP distributes 8 POIs among 4 users. Then the users communicate among themselves according to PAD and calculate ADs of all POIs. Finally, the users perform CWMD to find out the POI with the minimum AD.

Let there be x users in the group who are colluding. They want to calculate locations of the remaining $n - x$ users. So they need at least $2(n - x)$ ADs to find out $2(n - x)$ unknown coordinate values. Hence, ADs of at least $2(n - x) - 1$ POIs can be calculated safely. If the total number of POI is N, the colluders can learn ADs of at most $x \lceil \frac{N}{n} \rceil$ points. Thus, in order to prevent MPAD attack $2(n - x)$ must be greater than $x \lceil \frac{N}{n} \rceil$.

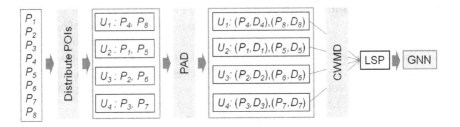

Fig. 2. Flow of LOOP

Hence, the maximum number of POIs can be handled by LOOP while resisting an MPAD attack from a group of x colluding users is

$$N_{max} = \begin{cases} n \times \lfloor \frac{2(n-x)-1}{x} \rfloor, & \text{if } 2(n-x)-1 \geq x. \\ 2(n-x)-1, & \text{otherwise.} \end{cases} \quad (1)$$

Table 1 shows N_{max} for a group of 10 users at different resistance levels. If the resistance level is y, MPAD attack from a group of y colluding users can be resisted.

Table 1. N_{max} at different resistance levels according to Equation 1

Number of colluders or resistance level	1	2	3	4	5,6	7	8
N_{max}	170	70	40	20	10	5	3

If there are a large number of POIs, i.e. larger than N_{max}, malicious users can learn sufficient number of ADs to perform MPAD attack. Since LOOP is a brute force method, users have to learn locations of all POIs and calculate their ADs. However, an LSP may not be willing to reveal locations of all POIs. In order to overcome above mentioned limitations, we need to reduce the number of POIs whose ADs are calculated. Next we propose a solution, H-LOOP, which prunes the search space and predicts the location of the GNN and searches in the surrounding area only.

Heuristic LOOP (H-LOOP) Algorithm. The base idea for H-LOOP is to locate the GNN in the part of the search space which has the smallest minimum AD and consider only the POIs located in that part as candidates. The LSP selects the smallest square which bounds all the POIs as the current search space and counts the number of POIs. If the number of POIs is larger than a threshold value β, then it divides the search space into γ equal sized non-overlapping squares. In H-LOOP, the users reach a consensus about the resistance level and then calculate N_{max} with (1). The users send N_{max} with the the query to the LSP. β is the minimum of N_{max} and the number of POI's location an LSP is willing to reveal. Lower value of β leads to higher resistance to MPAD attack. Then the LSP and users calculate aggregate minimum distances of these squares with PAD. Note that the minimum distance of a point from a region is defined as the distance of that point to its closest point in that region. If the concerned point is located inside the region, the minimum distance is zero. Next it selects δ squares with minimum ADs and narrows down the search space to those squares. Then it again counts the number of POIs in the new search space. The LSP divides each selected square into γ squares and narrows down the search space recursively until the number of POIs in the search space is less than β. Finally, LOOP is performed considering only those limited number of POIs in the search space to find out the GNN.

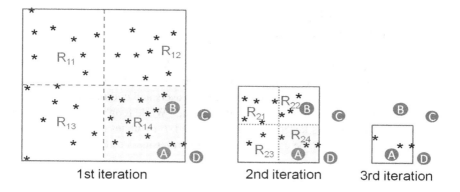

Fig. 3. Reduction of the search space in H-LOOP

We illustrate our approach with an example in Fig. 3. In this figure, asterisks are POIs and A, B, C and D are users. The square with solid boundary is the initial search space. Let $\gamma = 4, \delta = 1$ and $\beta = 4$. Since there are more than β POIs in the search space, the LSP divides it into four squares with dashed lines. These squares are R_{11}, R_{12}, R_{13} and R_{14}. Then the LSP calculates their centers and lengths and passes them to the users. Next the users calculate their minimum distances from these squares. Then they calculate their aggregate minimum distances from each of these squares using PAD and the LSP coordinates the calculation. The LSP selects R_{14} as the new search space since it has the minimum AD. The LSP divides the new search space again by dotted lines into four squares because it includes more than β POIs. It sends new squares R_{21}, R_{22}, R_{23} and R_{24} to the users. Again the users communicate with the LSP and calculate their aggregate minimum distances from each of these squares using PAD. The LSP selects R_{24} as the new search space since it has the minimum AD. Now the new search space has less than four POIs. Finally, LOOP is performed considering the POIs located inside R_{24} to find out the GNN.

An LSP runs the following algorithm for searching the GNN.

Input: A list of POIs with their locations, Division factor γ, Selection factor δ, Division threshold β
Output: The GNN

---

1    $SearchSpace \leftarrow$ bounding square of all POIs;
2   **While** number of POIs in $SearchSpace > \beta$
3        $sq \leftarrow$ divide($SearchSpace, \gamma$);
4   doPAD(sq);
5        $SearchSpace \leftarrow$ top δ squares with minimum ADs;
6   **end**
7   Perform LOOP considering POIs in $SearchSpace$;
8   **Return** the POI returned by LOOP;

The function 'divide($SearchSpace$, γ)' divides each square of $SearchSpace$ into γ non-overlapping equal sized squares. A square number should be used as γ. In the function 'doPAD()' the LSP communicates with the users and coordinate AD calculation according to PAD. The number of iteration in the "While loop" depends on γ and β. The search space shrinks faster if a larger γ is used. β regulates the number of candidate POIs, whose AD will be calculated. Thus the robustness of H-LOOP to MPAD attack depends on β.

Privacy Analysis. In order to recover the location of a single user, her individual distances to multiple points must be known. PAD enables a user to keep her distances private if there is at least another honest user. Therefore, in this case, a user's location cannot be recovered by traditional 2D trilateration.

As discussed in Sec. 3.2, if x users collude, they must learn at least $2(n - x)$ ADs to points to perform an MPAD attack. Since an LSP can learn only one AD to a POI, it cannot perform MPAD attack. Malicious users may perform MPAD attack in LOOP if the number of POIs is large. However, in H-LOOP, if the number of colluders is within the resistance level set by the users, the colluders cannot learn sufficient number of ADs to points and consequently cannot do MPAD attack. The users can choose any value that is less than $n - 1$ as resistance level. Although the LSP can learn aggregate minimum distances of multiple squares, this information cannot be used to solve (1) which requires ADs to points. An LSP can contribute less than a malicious user in MPAD attack because it knows AD of only one POI. Therefore, even if the LSP colludes with malicious users, their MPAD attack can be resisted.

Communication Cost. The communication cost in LOOP is low because both PAD and CWMD are performed only once. Though H-LOOP requires several rounds of communication like other privacy preserving methods, but in return H-LOOP can improve privacy level of users significantly compared to other methods. Hashem et al.'s [3] method needs anonymous communications between a user and an LSP. Talouki et al.'s [6, 10] methods setup an anonymous veto network at first. Users transfer data using Oblivious Transfer in Huang et al.'s [7] method. Anonymous communication and Oblivious Transfer are multi-step communication processes.

4   Experimental Setup

In this section we evaluate the performance of our proposed algorithms and compare them with related works through extensive experiments. We used both synthetic and real data in our experiments. The data space was normalized to an area of 1024×1024 square units. The data space can be considered as a map of a real city. We generated synthetic data sets using uniform and Zipfian distributions. We varied the size of datasets as 2K, 5K, 10K, 20K and 50K point locations. As real dataset we used all parks of California as POIs and locations of flats as user locations[1]. Since there are two major metropolitan areas in California, i.e., Los Angeles and San Francisco, we divided the dataset into two parts taking longitude -120 as a dividing line. California contains 6728 parks and 2700 flats. L.A. has 3407 parks and 1266 flats, where San Francisco has 3321 parks and 1434 flats. Flats were randomly selected as users' locations. As mentioned in Sec. 3.2, H-LOOP can be adjusted to prevent MPAD attack from a single user or a group of colluding multiple users. We varied the value of the division threshold β to deal with MPAD attacks by one to eight attackers. The value of both division factor γ and selection factor δ were set to four. Table 2 summarizes the values used for each parameter in our experiments and their default values. In our default scenario, we imagine 10 friends who want to select a park from nearby 2000 parks to have a picnic.

Table 2. Experimental setup

Parameter	Range	Default
Group size	3,5,10,15,20	10
POI distribution	Uniform, Zipfian	Uniform
User distribution	Uniform, Zipfian	Uniform
Number of POIs	2K, 5K, 10K, 20K, 50K	2K
Number of colluders	1,2,3,4,5,6,7,8	1

We considered 10000 private GNN queries for each set of experiments, evaluated the proposed algorithms for each of these GNN queries and determined the average experimental results. For each query we generated new set of POIs' locations and users' locations. We ran our experiments on a desktop computer with Intel(R) Core(TM) i7-2600 3.40GHz processor and 8GB RAM.

5   Results and Discussion

In this section, we present our results. We did a comparative analysis with [6], the most recent method. Very few techniques are currently available that consider privacy while

---

[1] http://www.cs.utah.edu/$\sim$lifeifei/SpatialDataset.htm

executing GNN queries. We do not consider [3, 10, 11] for comparison because of their limited privacy protection and extremely high cost as discussed in Sec. 2.1.

We measure *error rate* and *error size* to evaluate accuracy of the methods. Error rate means the percentage of wrong answers an algorithm produces. If POI x is the true GNN and we get POI y as answer using an algorithm then

$$\text{Error size} = \frac{\text{AD of } y}{\text{AD of } x} - 1$$

We report the error size in percent. If x and y are the same POI, then error size is 0%.

We measure computation time and number of rounds to evaluate cost. The number of rounds is the number of times the search space is reduced in H-LOOP.

5.1   Effect of User and POI Location Distribution

We studied the effect of user and POI location distributions by varying their distributions. Other parameters were set to the default values listed in Table 2.

Accuracy: Since H-LOOP and Talouki et al.'s work [6] are the only two approximate privacy preserving methods, we investigated their accuracy. Our experiments used four types of synthetic and three real datasets with default setup of Table 2. Figure 4 shows that H-LOOP outperformed Talouki et al.'s [6] work by three orders of magnitude in all cases. From Fig. 4, Talouki et al.'s [6] method has a probability of 0.45 to incur an error of size 1% or more, whereas in case of H-LOOP this probability is 0.0001. In 8% of the cases, Talaouki et al.'s [6] method incurred errors of size 10% or more, whereas the worst error size of H-LOOP was 7.31%. Mean and standard deviation of H-LOOP's error size were 5.013E-4% and 0.0425%, in contrast to 2.5547% and 4.3728% of Talouki et al.'s method. In practical applications, if the users are spread out over an area of 50 Km diameter, Talouki et al.'s [6] method will incur errors in the order of

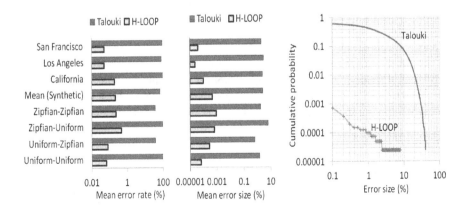

Fig. 4. Error in different datasets

Kms, whereas H-LOOP will incur errors in the order of meters. Even when the users were located in multiple cities of California H-LOOP performed well.

Cost: We randomly selected 2000 POI locations from each real dataset because the computation time depends on the number of POIs. Since LOOP takes into account all POIs, the distribution of user locations and POI locations have no effect on computation time. Overall average computation time per user in LOOP was 0.76ms. Table 3 shows the effect of location distribution on cost of H-LOOP. In H-LOOP, the average round count and computation time were higher when the POIs had Zipfian distribution. The average computation time per user was below one ms in all setups. Even if a smartphone is three orders of magnitude slower than a desktop computer [11], the computation of H-LOOP and LOOP will still take less than one second. This indicates that LOOP and H-LOOP are practical solutions. [In a compatible setup, Huang et al. [7] reported that their method took 4 seconds per user in a decentralized approach and 20 seconds in a centralized approach. These values translate to unrealistic times on today's smartphones. We did not investigate the computation cost of Talouki et al.'s [6] method because we felt that their method is not practically useful due to large errors.]

Table 3. Effect of Location Distribution on Cost of H-LOOP

User distribution	POI distribution	Average round count	Average computation time per user(ms)
Uniform	Uniform	3.00	0.53
Zipfian	Uniform	3.01	0.54
Uniform	Zipfian	3.99	0.68
Zipfian	Zipfian	4.02	0.70
San Francisco	San Francisco	3.35	0.60
Los Angeles	Los Angeles	3.04	0.55
California	California	3.33	0.59

5.2 Effect of Group Size

We studied the effect of group size by varying the group size. Other parameters were set to the default values listed in Table 2. Figure 5 shows that the error decreases with an increase of the group size in H-LOOP. There are two reasons for that. First, from (1) we can say that the number of POIs LOOP can handle while defending an MPAD attack increases rapidly with a group size increase. Thus, H-LOOP prunes the search space less when the group is larger. The number of rounds, as shown in the lower left part of Fig. 5, verifies this fact. A larger search space has a higher probability to include the GNN. Second, our heuristic of estimating a POI's AD with its bounding square's minimum AD works better for larger group sizes. The best error rate and error size of Talouki et al.'s [6] method were 94.1% and 0.68%, respectively.

Figure 5 also shows the effect of the group size on computation costs in H-LOOP. A user's computation cost can be mainly divided into two parts: calculating distances and running PAD. A user has to calculate distances to cells and POIs. The distances to cells are calculated in multiple rounds when H-LOOP is reducing the search space. As the number of rounds decreases with a larger group size, the time spent in calculating distances to cells also decreases. The distances to POIs are calculated in the final round. Since the number of candidate POIs increases, the time spent in calculating distances to POIs also increases. In our experimental setup, a user sends messages to all other users in PAD, which means that the number of messages is proportional to the group size, and the time spent in PAD is higher for a larger group. In case of LOOP, the computation time increased linearly up to 1.3 ms with the increase of group size.

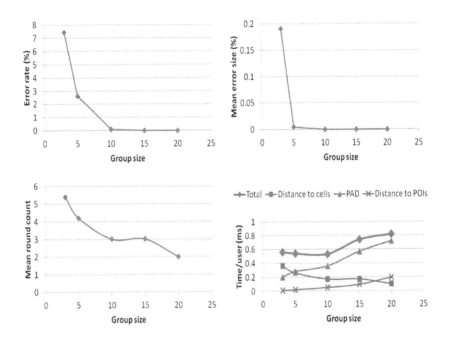

Fig. 5. Effect of group size on H-LOOP

We also studied the effect of the group size while keeping the percentage of colluders constant. In LOOP the POIs are divided into subsets and each subset is assigned to a group member for AD computation. When the the colluder percentage remains constant the subset size also remains constant. However, since there are more users in a larger group, the number of subsets are larger. Therefore, the number of POIs LOOP can handle while resisting an MPAD attack increases with a group size increase. As a result, both error rate and error size are smaller in H-LOOP for larger group sizes. Figure 6 shows the error rate and error size for different group sizes when 40% of the group members are colluders. The figure shows that error rate and error size decreases with the increase of group size.

5.3 Effect of the Number of POIs

We studied scalability by varying the number of POIs. Other parameters were set to default values listed in Table 2. The left part of Fig. 7 shows that the mean computation time per user in H-LOOP increased logarithmically with the increase of POI. H-LOOP reduces the search space exponentially in every round. So the number of rounds increases logarithmically. The communication and computation cost of users are both proportional to the number of rounds. Therefore, the communication and computation cost are logarithmic to the number of POIs. In case of LOOP, the computation time increased linearly up to 18 ms with the increase of POI. The right part of the figure shows that although the error rate increased linearly with the increase of POI numbers, it remained small. Even when the number of POIs is 50000, the error rate was only 1.4%. The mean error size did not show any trend and 1.37E-4% was the largest mean error size. Therefore, H-LOOP is scalable with respect to the number of POIs. The best error rate and error size of Talouki et al.'s [6] method were 98.5% and 1.22%, respectively.

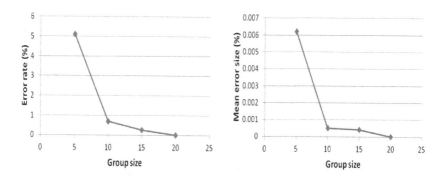

Fig. 6. Effect of group size on error when 40% of the group members are colluding

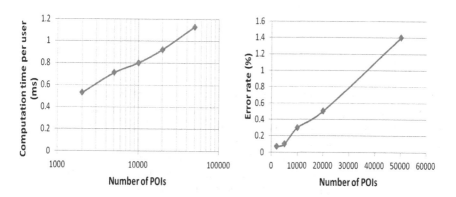

Fig. 7. Scalability of H-LOOP: number of POIs

5.4   Effect on H-LOOP for Different Collusion Group Sizes

We set the value of the division threshold β equal to the N_{max} listed in Table 1 and examined errors at different resistance levels. Other parameters were set to the default values listed in Table 2. Figure 8 shows our experimental results. Although the error increased with an increase of the robustness level, both the error rate and the mean error size remained very low. Even when 8 out of 10 users collude, their attack can be defended easily. In this case, the users travel on average 0.0026% more due to error.

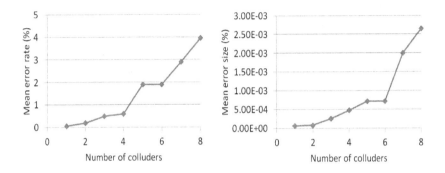

Fig. 8. Error in different resistance levels

Table 4. Practicality of privacy preserving GNN query methods

Algorithms	Resistance to attack	Accuracy	Cost
Hashem [3]	Low	Perfect	Low
Huang [7]	Medium	Perfect	High
Talouki [6]	High	Low	Low
Talouki [10]	Low	Perfect	Low
H-LOOP	High	High	Low

6   Conclusion

We proposed a framework for privacy preserving GNN queries. First, we developed two privacy preserving protocols, i.e., PAD and CWMD in order to privately compute and compare ADs respectively. Then we proposed a basic algorithm, LOOP, and a heuristic approach, H-LOOP, to search for the GNN.

A privacy preserving GNN query algorithm must be strong against attacks, highly accurate and fast to be applicable in real world. In Table 4, we compare H-LOOP, our best method, with existing works with respect to these three criteria based on our findings. We can observe that H-LOOP is a practically applicable solution.

We assumed that users and POIs are located in a 2D Euclidean space without any constraints on movements. In future, we intend to extend our work to road networks.

Acknowledgments. This research was partially supported under Australian Research Council's Discovery Projects funding scheme (project number DP110100757).

References

1. Gruteser, M., Grunwald, D.: Anonymous usage of location-based services through spatial and temporal cloaking. In: MobiSys, pp. 31–42 (2003)
2. Chow, C.Y., Mokbel, M.F., Liu, X.: A peer-to- peer spatial cloaking algorithm for anonymous location based service. In: ACMGIS, pp. 171–178 (2006)
3. Hashem, T., Kulik, L., Zhang, R.: Privacy preserving group nearest neighbor queries. In: EDBT, pp. 489–500 (2010)
4. Duckham, M., Kulik, L., Birtley, A.: A spatiotemporal model of strategies and counter strategies for location privacy protection. In: Raubal, M., Miller, H.J., Frank, A.U., Goodchild, M.F. (eds.) GIScience 2006. LNCS, vol. 4197, pp. 47–64. Springer, Heidelberg (2006)
5. Shokri, R., Freudiger, J., Hubaux, J.P.: A unified framework for location privacy. In: HotPETs (2010)
6. Ashouri-Talouki, M., Baraani-Dastjerdi, A., Aydın Selçuk, A.: GLP: A cryptographic approach for group location privacy. Computer Communications 35(12), 1527–1533 (2012)
7. Huang, Y., Vishwanathan, R.: Privacy preserving group nearest neighbour queries in location-based services using cryptographic techniques. In: IEEE GLOBECOM, pp. 1–5 (2010)
8. Sheikh, R., Kumar, B., Mishra, D.: A distributed k-secure sum protocol for secure multi-party computations. Journal of Computing 2(3), 68–72 (2010)
9. Papadias, D., Shen, Q., Tao, Y., Mouratidis, K.: Group nearest neighbor queries. In: ICDE, pp. 301–312 (2004)
10. Ashouri-Talouki, M., Baraani-Dastjerdi, A.: Homomorphic encryption to preserve location privacy. International Journal of Security and Its Applications 6(4), 183–189 (2012)
11. Huang, Y., Chapman, P., Evans, D.: Privacy preserving applications on smartphones. In: USENIX Workshop on Hot Topics in Security, p. 4 (2011)
12. Mood, B., Letaw, L., Butler, K.: Memory efficient garbled circuit generation for mobile devices. In: Keromytis, A.D. (ed.) FC 2012. LNCS, vol. 7397, pp. 254–268. Springer, Heidelberg (2012)

Practical Approaches to Partially Guarding a Polyhedral Terrain

Frank Kammer[1], Maarten Löffler[2], Paul Mutser[2], and Frank Staals[2]

[1] Institut für Informatik, Universität Augsburg, 86135 Augsburg, Germany
[2] Department of Information and Computing Sciences, Utrecht University

Abstract. We study the problem of placing guard towers on a terrain such that the terrain can be seen from at least one tower. This problem is important in many applications, and has an extensive history in the literature (known as, e.g., multiple observer siting). In this paper, we consider the problem on polyhedral terrains, and we allow the guards to see only a fixed fraction of the terrain, rather than everything. We experimentally evaluate how the number of required guards relates to the fraction of the terrain that can be covered. In addition, we introduce the concept of dominated guards, which can be used to preprocess the potential guard locations and speed up the subsequent computations.

1 Introduction

Terrains are a key concept in Geographic Information Science. They are the topic of interest in may different problems, ranging from determining how water flows along a terrain [16,22] to computing valleys and ridges [19]. We study the problem of guarding a terrain; that is, we wish to place a small number of guards such that they can together can see the terrain. The applications for this problem are numerous. Consider for example protecting the border of a country, placing watchtowers to protect against forest fires [6], or determining where to place base stations for a telecommunication network [8,24]. See also Floriani and Magillo [11] for an extensive treatment of the subject.

There are two standard representations for terrain data. We can store a terrain as a digital elevation model (DEM), which is a two-dimensional grid with height values, or as a *polyhedral terrain*: a planar subdivision—usually a triangulation—in which each vertex has an associated height. Heights are linearly interpolated in the interior of a face. This yields a polyhedral surface in $\mathbb{R}^3$.

For grid-based terrains, guarding is well-studied. Franklin et al. [12] present a greedy approach, Kim et al. [18] investigate heuristics for placing guards, and Zhang and Lu [20] use improved simulated annealing to determine where to place the guards. However, grid terrains are less suited for visibility problems than polyhedral terrains [4,23]. Furthermore, polyhedral terrains allow a more compact representation of the data, which may allow us to avoid heuristics when working in external memory [21].

Informally speaking, the viewshed of a guard is the part of the terrain that it can see. An example is shown in Fig. 1. Computing viewsheds of guards at fixed locations is itself useful in many applications, e.g. bird behavioral studies [3,5]. There are efficient

M. Duckham et al. (Eds.): GIScience 2014, LNCS 8728, pp. 318–332, 2014.

algorithms to compute the viewshed for a given guard [17], and even computing the part of the terrain visible by a set of guards can be done efficiently [14]. Unfortunately, it is NP-hard to determine *where* to place a minimum number of guards such that they can together see the entire terrain [7]. Moreover, Eidenbenz et al. [10] showed that there is no polynomial time algorithm that can approximate the number of guards required to cover the whole terrain consisting of n triangles within a factor $O(\ln(n))$ of the optimum unless some complexity classes collapse, which is very unlikely.

Partial Covers. The digital model of the terrain that we work with is often imprecise, and even if we have the true heights for all points on the terrain, there other factors, such as vegetation, that impact visibility. So, instead of requiring that the guards see the entire terrain it may be sufficient if they see a large portion of the terrain. This raises the question if we can efficiently solve or approximate this so-called ε-guarding problem.

To define the ε-guarding problem precisely, we need some definitions. A *guard* g is a point (tower) at height h above a polyhedral terrain $\mathcal{T}$. It can see, or *cover*, a point $p \in \mathcal{T}$, if the open ended line segment $\overline{gp}$ lies entirely above the terrain. We also say that p is *visible* from g. The maximal set of points on $\mathcal{T}$ that is visible from g, denoted $\mathcal{V}(g)$, is the *viewshed* of g. The viewshed of a set of guards $\mathcal{G}$ is the maximal set of points on $\mathcal{T}$ visible from at least one guard in $\mathcal{G}$, that is, $\mathcal{V}(\mathcal{G}) = \bigcup_{g \in \mathcal{G}} \mathcal{V}(g)$. We can measure the *size* $[\![\mathcal{T}']\!]$ of a part $\mathcal{T}'$ of $\mathcal{T}$, and thus the size of a viewshed, in two ways. Either we consider the *area* of $\mathcal{T}'$, or we consider the *number of terrain vertices* in $\mathcal{T}'$. Our algorithms can be used in both cases—only the running time changes through a different viewshed computation.

Definition 1 (ε-cover, ε-Guarding Problem). *Given a polyhedral terrain $\mathcal{T}$, a height h, and a value $\varepsilon > 0$, an ε-cover of $\mathcal{T}$ is a set of guards $\mathcal{G}$ that can together see at least a fraction of $(1 - \varepsilon)$ of the terrain, i.e., a set of guards for which $[\![\mathcal{V}(\mathcal{G})]\!] \geq (1 - \varepsilon)\,[\![\mathcal{T}]\!]$. In the ε-guarding problem, the goal is to find a minimum sized ε-cover.*

Our Contribution. In Section 2, we generalize Eidenbenz et al.'s [10] result and show that the same lower bound holds for ε-guarding problem for any $\varepsilon \leq 1 - 1/n^x$ with

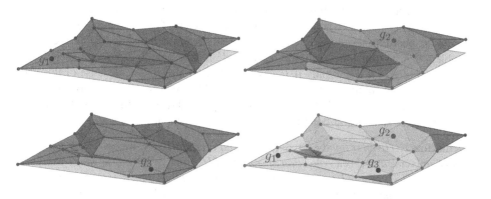

Fig. 1. The viewshed $\mathcal{V}(g_i)$ for each guard g_i, and the viewshed $\mathcal{V}(\{g_1, g_2, g_3\})$

$x < 1$. Next, we make two main observations in this paper, which we experimentally validate. In Section 3 we analyze a practical approach to compute an ε-cover: a greedy algorithm that simply places the guard that covers the largest possible unguarded area, and continues to place more guards until a desired fraction of the terrain is covered. By using ideas from the approximation techniques known for the so-called SET-COVER problem [15] we show that the number of required guards can be related to the optimal number of guards required to cover the whole terrain. We also implemented the approach, and show that in real-world terrains, the number of required guards to cover 95% of a mountainous region of roughly 150km^2 is typically less than 20.

In Section 4, we introduce the concept of *dominating guards*: a potential guard tower *dominates* another potential guard tower if it can see at least the same part of the terrain. Since dominated guards are never necessary in an optimal solution, the computation of a good set of guards can be sped up by precomputing and deleting the dominated guards. We can also relax the concept and say a guard is dominated if another guard sees *almost* everything it sees. We show experimentally that in real-world terrains, computing dominated guards typically reduces the problem size by 15-40%, depending on the resolution of the terrain. For the relaxed version, this percentage drops drastically, reducing the problem size to as little as 10% of the original, but comes at the cost of potentially not allowing every solution anymore. However, we show that in practice, the greedy approach still finds a solution of the same size as without preprocessing.

The idea of reducing a problem's input size by transforming it to another instance, while preserving theoretical guarantees on the solution, is common in theoretical computer science. Agarwal et al. [1] introduce the concept of *core-sets*: a subset P' of a set of objects P (classically, points in a d-dimensional space) with the property that for some function f that one is interested in, $f(P')$ differs from $f(P)$ by at most a factor $(1 - \varepsilon)$. A related concept in complexity theory is *kernalization* [9].

2   Lower Bound

In this section we show a lower bound on the computational complexity of the ε-guarding problem. We first focus on the case where we measure viewsheds and terrain sizes $[\![\mathcal{T}']\!]$ with respect to the number of visible terrain vertices. Eidenbenz et al. [10] have shown in Lemmata 6 and 7, which are part of the proof of the non-approximability result for guarding-terrain problem, that an instance I for the so-called SET-COVER problem can be transformed into an instance I' for the guarding problem such I has a solution of size k if and only if I' has a solution of size $k + 4$. Because of this and because optimal solutions of the SET-COVER problem can be of arbitrarily large size k we only need to show a lower bound for the so-called α-weak SET-COVER problem and obtain the same lower bound for the ε-guarding problem where $\varepsilon = 1 - \alpha$.

Definition 2. *Given a number $0 \leq \alpha \leq 1$ and a tuple $(U, \mathcal{C})$, where U is a finite set called* universe *and $\mathcal{C}$ is a collection of subsets of U with $\bigcup_{X \in \mathcal{C}} X \subseteq U$, an α-weak set cover for $(U, \mathcal{C})$ is a collection $S \subseteq \mathcal{C}$ such that the union of all sets in S contains at least $\alpha|U|$ elements. In the special case $\alpha = 1$, we also call S a set cover for $(U, \mathcal{C})$. The size of S is its cardinality. In the (α-weak) SET-COVER problem, we have to find an (α-weak) set cover of minimal size.*

Lemma 1. *The α-weak* SET-COVER *problem has no polynomial-time approximation algorithm of ratio $c \ln n$ unless* $P = NP$, *where n is the size of the universe, $\alpha \geq 1/n^d$ with $d < 1$, and $c > 0$ is an appropriately chosen constant.*

Proof. Alon, Moshkovitz, and Safra [2] showed that, for an appropriately chosen constant $c'' > 0$, there is no polynomial-time approximation algorithm of ratio $c' \ln |U'|$ for the SET-COVER problem with universe U' unless $P = NP$. Since we can figure out in polynomial time if a set-cover instance has no solution, there can not exist a polynomial-time approximation of approximation ratio $c' \ln |U'|$ for the SET-COVER problem restricted to solvable instances.

We now show the lemma by a reduction from the SET-COVER problem to the α-weak SET-COVER problem. Let (U', C') be a solvable SET-COVER instance. Take $k = \lceil 1/\alpha \rceil$, $U = U' \cup \{x_1, \ldots, x_{k|U'|}\}$ where $x_1, \ldots, x_{k|U'|}$ are new elements, and $C = C' \cup \{\{x_1\}, \ldots, \{x_{d|U'|}\}\}$.

Note that each solution of the set-cover instance (U', C') is also a solution for the α-weak-set-cover instance (U, C). For the reverse direction, assume that there is a solution $S \subseteq C$ for the α-weak-set-cover instance, and let $V := \bigcup_{X \in S} X$ be the elements covered by S. Intuitively speaking, for each element of U' that is not covered, there is an element of $U \setminus U'$ covered by the solution. More exactly, $|U' \setminus V| \leq |(U \setminus U') \cap V|$. To obtain a solution S' for the set-cover instance (U', C') from S, for each $u \in U' \setminus V$, we replace a set in $S \cap (C \setminus C')$ by a set $C' \in C'$ with $u \in C'$. Note that set C' must exist since (U', C') has a solution.

To sum up, each solution of (U', C') can be turned into a solution of the same size for (U, C) and vice versa. Then (U, C) has no polynomial-time approximation algorithm of ratio $c' \ln |U'| = c' \ln(|U|/(k+1)) \leq c' \ln(n/(2n^d)) \leq c'(1-d) \ln(n/2) \leq c \ln n$ if c is chosen appropriately.   □

Corollary 1. *The ε-guarding problem has no polynomial-time approximation algorithm of ratio $c \ln n$ where n is the size of the universe, i.e., the number of the terrain vertices, $c > 0$ is an appropriately chosen constant, and $\varepsilon = 1 - \alpha \leq 1 - 1/n^d$ with $d < 1$.*

If we want to measure terrain sizes $[\![T']\!]$ with respect to the visible area, we can first decompose the terrain into maximal regions such that each region is visible from the same set of guards. We then take the size of the universe to be the number of such regions. If we have $m \leq n$ (potential) guards, there are at least $\Omega(m^2 n^2)$ and at most $O(m^3 n^2)$ such regions, and they can easily be computed in polynomial-time [14]. Thus, also in the case with terrain sizes measured by the visible area, we obtain a $\Omega(\ln n)$ lower bound on the approximation ratio of an ε-cover, for small enough ε.

Note that, if $\varepsilon = 1 - 1/n^x$ with $x = 1$, we need to see one vertex of the terrain to solve the vertex viewsheds ε-guarding problem. Clearly, such a solution can be found easily.

3 Greedy Approach

3.1 Algorithm and Analysis

Consider the following simple and straightforward algorithm GREEDYGUARD that given a terrain $\mathcal{T}$, a parameter ε, and a set of potential guard locations $\mathcal{P}$, computes an ε-cover of $\mathcal{V}(\mathcal{P})$. So, when the guards in $\mathcal{P}$ can together see the entire terrain $\mathcal{T}$, GREEDYGUARD computes an ε-cover of $\mathcal{T}$.

Algorithm. GREEDYGUARD($\mathcal{T}, \varepsilon, \mathcal{P}$)
1. Compute the viewsheds for all guards in $\mathcal{P}$.
2. Let $\mathcal{G} = \emptyset$ and $\mathcal{R} = \mathcal{P}$.
3. **while** $[\![\mathcal{V}(\mathcal{G})]\!] \leq (1 - \varepsilon) \, [\![\mathcal{V}(\mathcal{P})]\!]$ **and** $\mathcal{R} \neq \emptyset$ **do**
4.   Select a guard $g \in \mathcal{R}$ that maximizes the size $[\![\mathcal{V}(g) \setminus \mathcal{V}(\mathcal{G})]\!]$, i.e., the size of the region it can cover but is not covered by $\mathcal{G}$ yet.
5.   Remove g from $\mathcal{R}$ and add it to $\mathcal{G}$.
6. **return** $\mathcal{G}$

We now show that the selected set of guards $\mathcal{G}$ has size at most $O(k/\varepsilon)$, where k is the number of guards required in an optimal solution to cover $\mathcal{V}(\mathcal{P})$.

Lemma 2. *Let $\mathcal{T}$ be a terrain, let $\mathcal{P}$ be a set of potential guard locations, and let $\varepsilon \in (0, 1]$. GREEDYGUARD computes an ε-cover of $\mathcal{T}' = \mathcal{V}(\mathcal{P})$ of at most $O(k/\varepsilon)$ guards, where k is the size of an optimal 0-cover of $\mathcal{T}'$.*

Proof. Consider the guards $g_1, .., g_\ell$ picked, in that order, by the greedy algorithm, and let ρ_i denote the fraction of $\mathcal{T}'$ that remains uncovered by the first i guards, that is, $\rho_i = [\![\mathcal{R}_i]\!] / [\![\mathcal{T}']\!]$, where $\mathcal{R}_i = \mathcal{T}' \setminus \bigcup_{j=1}^{i} \mathcal{V}(g_j)$.

The greedy algorithm stops once the fraction of $\mathcal{T}'$ that remains uncovered is at most ε. That is, $\rho_\ell \leq \varepsilon$. We claim that this is the case for $\ell = ck/\varepsilon$, for some $c \in \mathbb{R}$.

For any i, with $0 \leq i < \ell$, the remaining part $\mathcal{R}_i$ of $\mathcal{T}'$ can be covered with k guards. So, there is a guard, say $g^* \in \mathcal{P}$, that covers at least $[\![\mathcal{R}_i]\!] /k$. It follows the next guard g_{i+1} picked by the greedy algorithm covers at least that much. Thus, we have $\rho_{i+1} [\![\mathcal{T}']\!] = [\![\mathcal{R}_{i+1}]\!] \leq [\![\mathcal{R}_i]\!] - [\![\mathcal{R}_i]\!] /k = \rho_i [\![\mathcal{T}']\!] - \rho_i [\![\mathcal{T}']\!] /k$, and therefore $\rho_{i+1} \leq \rho_i - \rho_i/k = \rho_i(1 - \frac{1}{k})$. It follows that $\rho_i \leq (1 - \frac{1}{k})^i$.

We thus have to show that $\rho_\ell \leq (1 - \frac{1}{k})^\ell \leq \varepsilon$. In case $k = 1$ this is trivially true. In case $k \geq 2$, we have that $(1 - \frac{1}{k})^k \leq \frac{1}{e}$. So, we have to show that

$$\left(1 - \frac{1}{k}\right)^\ell = \left(1 - \frac{1}{k}\right)^{ck/\varepsilon} = \left(\left(1 - \frac{1}{k}\right)^k\right)^{c/\varepsilon} \leq (1/e)^{c/\varepsilon} = 1/(e^{c/\varepsilon}) \leq \varepsilon.$$

Using basic calculus we can reduce $1/(e^{c/\varepsilon}) \leq \varepsilon$ to $\varepsilon \ln(1/\varepsilon) \leq c$. This holds for all $c \geq \frac{1}{e}$. The lemma follows. $\square$

We now analyze the running time of this algorithm. If we are in the case where terrain sizes are measured by the number of terrain vertices, it is easy to compute the viewshed of a guard g in $O(n^2)$ time by simply checking if the line segment between g and each terrain vertex lies above the terrain. Thus, if we are given m potential guards in $\mathcal{P}$, this takes $O(mn^2)$ time in total. From Lemma 2 it follows that the algorithm selects at most $O(k/\varepsilon)$ guards. Finding a guard $g \in \mathcal{R}$ that maximizes $[\![\mathcal{V}(g) \setminus \mathcal{V}(\mathcal{G})]\!]$ takes $O(mn)$ time, so selecting all $O(k/\varepsilon)$ guards takes $O(kmn/\varepsilon)$ time in total. Thus, our algorithm can be implemented in $O(mn^2 + kmn/\varepsilon)$ time.

We can improve on this by preprocessing our terrain for visibility queries. In $O(n\alpha(n)\log n)$ time, where α is the extremely slow growing inverse of Ackermann's function, we construct a data structure that can answer visibility queries from a fixed viewpoint in $O(\log n)$ time [7]. Using this data structure we can compute all viewsheds in $O(mn\alpha(n)\log n + mn\log n) = O(mn\alpha(n)\log n)$ time. Thus, we conclude:

Theorem 1. *Let $\mathcal{T}$ be a terrain, let $\mathcal{P}$ be a set of potential guard locations, and let $\varepsilon \in (0,1]$. GREEDYGUARD computes an ε-cover of $\mathcal{T}' = \mathcal{V}(\mathcal{P})$ of at most $O(k/\varepsilon)$ guards in $O(mn\alpha(n)\log n + kmn/\varepsilon)$ time, where k is the size of an optimal 0-cover of $\mathcal{T}'$, and m is the number of potential guards in $\mathcal{P}$.*

We remark that the GREEDYGUARD algorithm is essentially the well-known greedy algorithm for the SET-COVER problem. Hence, our results immediately transfer to the SET-COVER problem as well. Indeed, when we compute the required guards/sets by our algorithm to *completely* cover the terrain, we get only an $O(\log n)$ approximation. The same applies if we compare our solution to an optimal ε-cover. However, in our approach we simultaneously approximate the amount of terrain covered, and the number of guards required. This allows us to find a decent approximation ratio. We further remark that the minimum number of guards required for an ε-cover may differ significantly from the minimum number of guards to completely cover the terrain.

Finally, in case we measure the size of a viewshed by its area we can compute an ε-cover of $\mathcal{T}'$ in $O(n^2(m\log n + m^4 + k^4/\varepsilon^4))$ time using the results of Hurtado et al. [14]. The $O(n^2m^4)$ term originates from computing $[\![\mathcal{V}(\mathcal{P})]\!] = [\![\mathcal{T}']\!]$. If we know the size of $\mathcal{T}'$ in advance, for example because $\mathcal{T}' = \mathcal{T}$, then we can improve these results to $O(n^2(m\log n + k^4/\varepsilon^4))$.

Experimental Evaluation

We experimentally evaluate results from the greedy algorithm on five terrain models in California, obtained from the U.S. Geological Survey[1] and shown in Table 1. Each terrain model spans approximately 11.5km × 14km. For each terrain we have a coarse and a fine version with approximately 1800 and 16 000 vertices, respectively. In all our experiments we choose the set $\mathcal{P}$ of potential guards to be the set of vertices of the terrain.

To keep the implementation of our algorithms simple, we consider only the case where the size of the viewsheds is measured by the number of terrain vertices. For the same reason we use the naive implementation for the point-to-point visibility rather than building the data structure for visibility queries.

---

[1] http://www.usgs.gov

Table 1. The terrains we use to evaluate our approach. The terrains are shaded by height: higher vertices and faces are lighter.

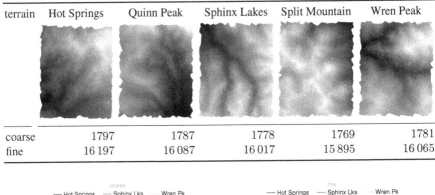

terrain	Hot Springs	Quinn Peak	Sphinx Lakes	Split Mountain	Wren Peak
coarse	1797	1787	1778	1769	1781
fine	16 197	16 087	16 017	15 895	16 065

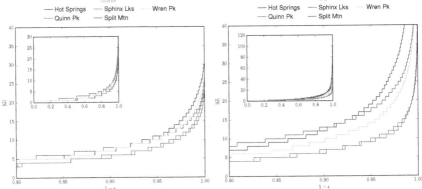

Fig. 2. The number of selected guards in $\mathcal{G}$ as a function of ε. The insets show the results for full domain of ε.

We first investigate the number of guards selected, that is, the size of the set $\mathcal{G}$, as a function of ε. For these results we fix the height h of the guards on 15 meters. The results are shown in Fig. 2 and 3. For both the coarse and fine terrain models we see that the required number of guards rapidly increases when $1 - \varepsilon$ approaches 1, that is, when we attempt to cover almost the complete terrain. However, eleven guards are sufficient for an 0.05-cover on each coarse terrain model. Generally speaking, the same number of guards covers a smaller fraction of the terrain in the fine terrain models than in

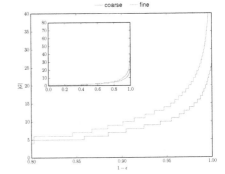

Fig. 3. A comparison between the number of guards needed in the coarse and fine versions of the Wren Peak terrain

the coarse terrain models, indicating that the added detail may be significant for visibility studies. For the coarse terrain models eleven guards can cover at least 88% of the terrain; for an 0.05-cover we need between nine and nineteen guards, depending on the terrain.

Fig. 4 shows the locations of the guards placed by GREEDYGUARD for an 0.05-cover on the coarse Wren Peak terrain (again for height $h = 15$m). The fraction of the terrain covered by those guards is shown in Fig. 3.

The height of the guards influences the number of guards required, see Fig. 5. However, the differences are small. The height has a larger influence on the fine version of the terrains. Even so, covering one of the terrains with guards at height 1m requires only four more guards than guards at 30m.

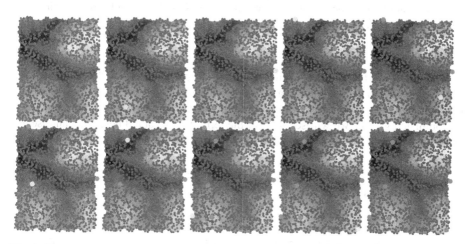

Fig. 4. The ten guards placed by GREEDYGUARD for a 0.05-cover of the coarse Wren Peak terrain (green). Each figure shows the vertices covered so far in blue, and the vertices that remain uncovered in red. The newly selected guards are shown in yellow.

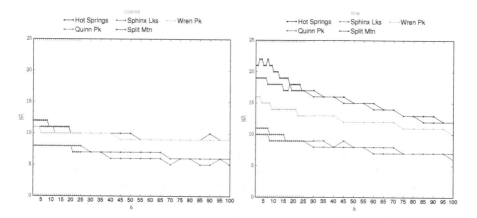

Fig. 5. The number of required guards for an 0.05-cover as a function of the heights h

4   Dominating Guards

4.1   Simple Domination

A guard g *dominates* another guard h if (and only if) the viewshed of h is contained in the viewshed of g, that is, $\mathcal{V}(h) \subseteq \mathcal{V}(g)$. We say g *strictly* dominates h if $\mathcal{V}(h) \subset \mathcal{V}(g)$. We now observe:

Observation 1. *Let $\mathcal{P}$ be a set of potential guards. There is an optimal (minimum size) ε-cover $\mathcal{G}$ of $\mathcal{V}(\mathcal{P})$ such that no guard in $\mathcal{G}$ is strictly covered by any guard in $\mathcal{P}$.*

Proof. Let $\mathcal{G}^*$ be an optimal ε-cover of $\mathcal{V}(\mathcal{P})$. Let h be any guard $h \in \mathcal{G}^*$ that is strictly dominated by another potential guard $g \in \mathcal{P}$. Replace h by g. Repeat this procedure until there are no more guards that satisfy the criterion. Note that this process terminates since strict domination defines a partial order on $\mathcal{P}$ (a guard g can only strictly dominate a guard h if $[\![\mathcal{V}(g)]\!] > [\![\mathcal{V}(h)]\!]$). Let $\mathcal{G}$ be the set of guards we obtain. Clearly, $\mathcal{G}$ contains at most the same number of guards as $\mathcal{G}^*$, and since we have that $\mathcal{V}(\mathcal{G}^*) \subseteq \mathcal{V}(\mathcal{G})$, $\mathcal{G}$ is also an ε-cover of $\mathcal{V}(\mathcal{P})$. □

It now follows that we never have to choose a guard that is strictly dominated by another guard. Hence, when finding an ε-cover of the terrain, we can simply ignore all dominated guards. Although there is no useful theoretical lower bound on the number of dominated guards, we will see that for the terrains considered here, we can discard between 10% and 40% of the potential guards, depending on the resolution of the terrain.

Experimental Evaluation

In Fig. 6 shows the percentage of the potential guards that is dominated as a function of the height h, and Fig. 7 shows the dominating and dominated potential guards for $h = 15$m. For the coarse terrains, between forty and fifty percent of the potential guards are dominated by another potential guard, and can thus be discarded. For the fine terrains the percentage of dominated vertices is lower: between eight and twenty percent. Surprisingly, the percentage of dominated vertices in the fine Hot Springs terrain is well over twenty percent. This behavior differs from the coarse Hot Springs ter-

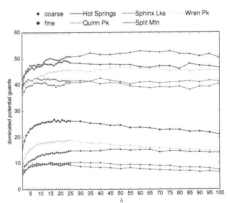

Fig. 6. The percentage of the potential guards that are dominated by another guard as a function of their height

rain, where the percentage of dominated guards is comparable to the other terrains.

The height at which we place the guards only mildly influences the number of dominated vertices. One might expect that the number of dominated vertices increases monotonically as the height grows (since the individual viewsheds grow when the height increases). However, for the heights considered this is not the case: the number of dominated vertices increases at first, but then slightly decreases.

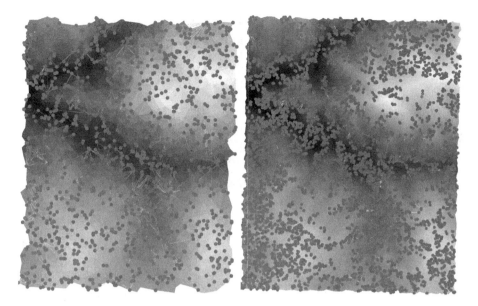

Fig. 7. Dominating potential guards in blue and the potential guards they dominate in red on the Wren Peak terrain. On the left the coarse model and on the right the fine model.

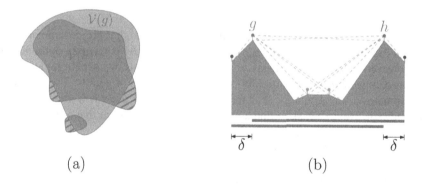

(a)   (b)

Fig. 8. (a) The red guard g δ-dominates the blue guard h if and only if the fraction of $\mathcal{V}(h)$ that is not covered by g (the hatched area) is at most δ. (b) A cross section of a terrain in which all potential guards are δ-dominated by an other guard.

4.2 δ-Domination

Since it is sufficient if the guards cover a fraction of $1 - \varepsilon$ of the terrain, we can consider slightly relaxing the definition of domination. Instead of requiring that guard g should see everything that h sees, it may be sufficient if g sees a large enough portion of h. More formally, guard g *δ-dominates* guard h if (and only if) the fraction of $\mathcal{V}(h)$ that is not covered by g is at most δ, that is, if and only if $[\![\mathcal{V}(h) \setminus \mathcal{V}(g)]\!] / [\![\mathcal{V}(h)]\!] \leq \delta$. See Fig. 8(a) and 9.

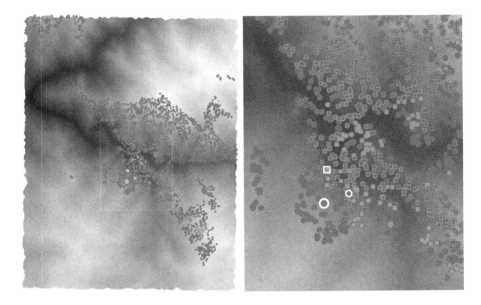

Fig. 9. Three guards on a fine grained Wren Peak terrain (left). The blue guard 0.02-dominates the red and green guards (guards are delineated in white). Only very few vertices in the red and green viewsheds are not visible from the blue guard (right).

In this setting we can no longer just discard all guards that are δ-dominated, as shown in Fig. 8(b). For small values of ε, the set $\mathcal{G} = \{g, h\}$ is the only optimal solution. However, guard g δ-dominates h, and vice versa. Thus, we cannot remove them both. Instead, we extend the notion of δ-domination to sets of guards: a set of guards $\mathcal{G}$ δ-dominates a set of guards $\mathcal{H}$ if for every guard in $\mathcal{H}$ there is a guard in $\mathcal{G}$ that δ-dominates it. We now wish to find a minimum size set of guards that δ-dominate $\mathcal{P}$.

It turns out that computing such a minimum size set is NP-hard as well [13]. However, we can easily find a *maximal* set $\mathcal{D}$ with a simple greedy algorithm. Unfortunately, we cannot provide a good lower bound on the fraction of the terrain that is visible from $\mathcal{D}$, so instead we evaluate δ-domination experimentally.

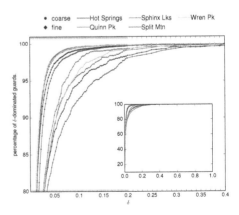

Fig. 10. The percentage of potential guards in $\mathcal{P}$ for which there is an other potential guard that δ-dominates it

Experimental Evaluation

In Fig. 10 we can see the percentage of the guards for which there is another guard that δ-dominates it (for $h = 15$ meters). This figure shows that on the fine terrains and $\delta \geq 0.05$ for practically every vertex there is another vertex that

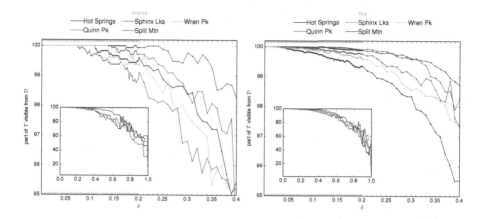

Fig. 11. The fraction of $\mathcal{T}$ that it still coverable with a minimal set $\mathcal{D}$ of δ-dominating guards

δ-dominates it. For the coarse terrains we need slightly higher values of δ. This is to be expected, since the vertices, and thus the locations for the potential guards, are spread further apart than in the coarse terrains. Even so, we note that for $\delta = 0.05$ already 90% of the potential guards are δ-dominated by another guard.

As noted before, we can no longer remove *all* guards that are δ-dominated, so instead consider a minimal size set $\mathcal{D}$ of δ-dominating guards. Fig. 11 shows the percentage of the terrain that is still coverable by $\mathcal{D}$ (again for $h = 15$ meters). For values of δ up to 0.2—so 20% of the viewshed of a guard h may not be visible from a guard g that δ-dominates it—a minimal set $\mathcal{D}$ of δ-dominating guards can still see more than 99% of the terrain.

5 The Greedy Approach with δ-Domination

We can now use the δ-domination to limit the number of potential guards we have to consider when computing an ε-cover. As a preprocessing step we determine which guards are δ-dominated, so we can discard some of the potential guards. We then use the greedy algorithm from Section 3 to compute an ε-cover on the remaining guards. More specifically, consider the following algorithm DOMINATINGGUARD, that computes an ε-cover using the δ-domination, if possible. When the fraction of the terrain $\hat{\delta}$ that is no longer visible from $\mathcal{D}$ is larger than ε we (obviously) cannot obtain an ε-cover any more.

Algorithm. DOMINATINGGUARD($\mathcal{T}, \varepsilon, \delta, \mathcal{P}$)
1. Compute the viewsheds for all guards in $\mathcal{P}$.
2. Compute a minimal set of guards $\mathcal{D}$ that δ-dominates $\mathcal{P}$.
3. Let $\hat{\delta} = [\![\mathcal{V}(\mathcal{D})]\!] / [\![\mathcal{V}(\mathcal{P})]\!]$ be the fraction of $\mathcal{V}(\mathcal{P})$ covered by $\mathcal{D}$.
4. Let $\gamma = (\varepsilon - \delta)/(1 - \hat{\delta})$ and let $\hat{\mathcal{T}} = \mathcal{V}(\mathcal{D})$.
5. **return** GREEDYGUARD($\hat{\mathcal{T}}, \gamma, \mathcal{D}$)

We note, however, that the terrain can be preprocessed by removing δ-dominated vertices, irrespective of the algorithm used to then place the guards. Since the data size can be reduced drastically, it may also be feasible to use more computation-intensive solutions than the simple greedy heuristic presented in this study.

Experimental Evaluation

We choose $\varepsilon = 0.05$ and $h = 15$m and run DOMINATINGGUARD for various values of δ. Fig. 12 shows the number of guards in the set $\mathcal{D}$, as a function of δ. The number of potential guards to be given to the greedy algorithm rapidly decreases when δ increases. On the fine grained terrains and for $\delta \geq 0.04$ we can discard well over 90% of all potential guard locations due to the δ-domination. For the coarse terrains this number is roughly 80%.

We observed that the number of guards selected is independent of δ: for all values the algorithm returns the same number of guards. Hence, removing δ-dominated potential guards does not seem to affect the final solution.

6   Concluding Remarks

We investigated practical approaches to compute an ε-cover of a polyhedral terrain $\mathcal{T}$; that is, a set of guards that together can see at least a fraction $1 - \varepsilon$ of the terrain. We showed that, for any ε of interest, no constant-approximation algorithm exists to compute an ε-cover. In addition, we provided a theoretical analysis of a straightforward greedy algorithm to compute an ε-cover. This analysis shows that we need at most $O(k/\varepsilon)$ guards, where k is the minimum number of guards required to completely cover the terrain. Through experiments we show the algorithm gives a reasonable number of guards in practice. Furthermore, we introduced the notions of dominated and

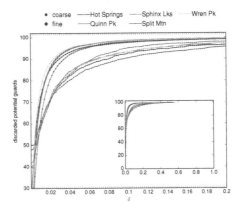

Fig. 12. The percentage of guards that we can discard in step two of DOMINATINGGUARD. When the selected minimal set of dominating guards no longer cover at least $1 - \varepsilon$ of the terrain the algorithm does not yield a solution.

δ-dominated guards. Our experiments show that these can greatly reduce the number of potential guard locations we have to consider.

In our experiments we use viewsheds that only consider the vertices of the terrain. It would be very interesting to see if we obtain equally good numbers for viewsheds that also incorporate the (interior of) the faces of the terrain. A different direction of future work entails a theoretical investigation of the concept of δ-domination. For our current definition, we cannot state any theoretical guarantees, though it performs well in practice.

Acknowledgments. This work has been partially supported by the Netherlands Organisation for Scientific Research (NWO) under grants 639.021.123 and 612.001.022.

References

1. Agarwal, P.K., Har-Peled, S., Varadarajan, K.R.: Geometric approximation via coresets. In: Combinatorial and Computational Geometry. MSRI, University Press, pp.1–30 (2005)
2. Alon, N., Moshkovitz, D., Safra, S.: Algorithmic construction of sets for k-restrictions. In: ACM Transactions on Algorithms, pp. 153–177 (2006)
3. Aspbury, A.S., Gibson, R.M.: Long-range visibility of greater sage grouse leks: A gis-based analysis. Animal Behaviour 67(6), 1127–1132 (2004)
4. Burrough, P.A., McDonnell, R., Burrough, P.A., McDonnell, R.: Principles of geographical information systems, vol. 333. Oxford University Press, Oxford (1998)
5. Camp, R.J., Sinton, D.T., Knight, R.L.: Viewsheds: A complementary management approach to buffer zones. Wildlife Society Bulletin 25(3), 612–615 (1997)
6. Catry, F.X., Rego, F.C., Santos, T., Almeida, J., Relvas, P.: Fires prevention in Portugal - using GIS to help improving early fire detection effectiveness. In: 4th Int. Wildland Fire Conf. (2007)
7. Cole, R., Sharir, M.: Visibility problems for polyhedral terrains. Journal of Symbolic Computation 7(1), 11–30 (1989)
8. De Floriani, L., Marzano, P., Puppo, E.: Line-of-sight communication on terrain models. International Journal of Geographical Information Systems 8(4), 329–342 (1994)
9. Downey, R.G., Fellows, M.R.: Parameterized Complexity, p. 530. Springer (1999)
10. Eidenbenz, S., Stamm, C., Widmayer, P.: Positioning guards at fixed height above a terrain – an optimum inapproximability result. In: Bilardi, G., Pietracaprina, A., Italiano, G.F., Pucci, G. (eds.) ESA 1998. LNCS, vol. 1461, pp. 187–198. Springer, Heidelberg (1998)
11. Floriani, L.D., Magillo, P.: Algorithms for visibility computation on terrains: A survey. Environ. Plann. B 30(5), 709–728 (2003)
12. Franklin, W.: Siting observers on terrain. In: Richardson, D., Oosterom, P. (eds.) Advances in Spatial Data Handling, pp. 109–120. Springer, Heidelberg (2002)
13. Garey, M.R., Johnson, D.S.: Computers and Intractability: A Guide to the Theory of NP-completeness. WH Freeman & Co., New York (1979)
14. Hurtado, F., Löffler, M., Matos, I., Sacristán, V., Saumell, M., Silveira, R.I., Staals, F.: Terrain visibility with multiple viewpoints. In: Cai, L., Cheng, S.-W., Lam, T.-W. (eds.) ISAAC2013. LNCS, vol. 8283, pp. 317–327. Springer, Heidelberg (2013)
15. Johnson, D.S.: Approximation algorithms for combinatorial problems. Journal of Computer and System Sciences 9(3), 256–278 (1974)
16. Jones, N.L., Wright, S.G., Maidment, D.R.: Watershed delineation with triangle-based terrain models. Journal of Hydraulic Engineering 116(10), 1232–1251 (1990)

17. Katz, M.J., Overmars, M.H., Sharir, M.: Efficient hidden surface removal for objects with small union size. Computational Geometry 2(4), 223–234 (1992)
18. Kim, Y.H., Rana, S., Wise, S.: Exploring multiple viewshed analysis using terrain features and optimisation techniques. Computers & Geosciences 30, 1019–1032 (2004)
19. Kreveld, M.: Digital elevation models and tin algorithms. In: van Kreveld, M., Roos, T., Nievergelt, J., Widmayer, P. (eds.) CISM School 1996. LNCS, vol. 1340, pp. 37–78. Springer, Heidelberg (1997)
20. Lv, P., Zhang, J.-f., Lu, M.: An optimal method for multiple observers sitting on terrain based on improved simulated annealing techniques. In: Ali, M., Dapoigny, R. (eds.) IEA/AIE 2006. LNCS (LNAI), vol. 4031, pp. 373–382. Springer, Heidelberg (2006)
21. Magalhaes, S.V.G., Andrade, M.V.A., Franklin, W.R.: An optimization heuristic for siting observers in huge terrains stored in external memory. In: 2010 10th International Conference on Hybrid Intelligent Systems (HIS), August 2010, pp. 135–140 (2010)
22. Silfer, A.T., Kinn, G.J., Hassett, J.M.: A geographic information system utilizing the triangulated irregular network as a basis for hydrologic modeling. In: Proc. Auto-Carto., vol. 8, pp. 129–136 (1987)
23. Silveira, R.I.: Optimization of polyhedral terrains. ASCI Disseratation Series (178) (2009)
24. Stephen, B.K., Wicker, S.B.: Experimental analysis of local search algorithms for optimal base station location (2000)

Oriented Regions for Linearly Conceptualized Features

Joshua A. Lewis and Max J. Egenhofer

School of Computing and Information Science, University of Maine, USA

Abstract. The typical phenomena in geographic space are 2-dimensional or 3-dimensional in nature, yet people often conceptualize some of them as 1-dimensional entities embedded in a 2-dimensional space—rivers have widths and depths, and extent across the surface of the Earth, but for some tasks they are thought of as linear objects; likewise, roads as travel paths have widths as they wind through the landscape, but in some scenarios the extent is ignored and only connectivity between points along the path is considered. A critical property that makes these features special is the *orientation* that is attached (e.g., through the flow of the water or the traffic directions imposed by an authority). Contemporary spatial models capture such features either 1-dimensionally as networks of lines or directed lines, or 2-dimensionally simply as regions, each abstracting away one key property—in the case of the network the features' extents and connections to neighboring areas, and in the case of regions their orientations. This paper introduces *oriented regions* as a model that preserves the key properties from both abstractions. Key properties of this approach are the sequences in which the boundaries of oriented regions interact, and the placement of objects with respect to the topological hull of a set of oriented regions. This model, dubbed *hull+i*, is based on topological hulls and the *i-notation*, a systematic method to capture boundary interactions between oriented regions, and provides a means for representing entire spatial scenes with an arbitrary number of objects, separations, and instances where ensembles of objects surround other objects.

1   Introduction

Abstractions of spatial phenomena to geometric representations have been at the core of qualitative spatial reasoning. While regions as models for extended objects in a 2-dimensional embedding space are ubiquitous in spatial reasoning [6], spatial-relation reasoning about linear representations are found less frequently. Yet the geographic domain offers enough examples of spatial phenomena that favor, for specific tasks, a graph-like representation with edges and nodes at their intersections.

Fig. 1 shows a street scene from Washington, DC with Streets and Avenues, with multiple lanes, their driving directions, and such entities as buildings and parks, which are related to the roads. It provides topological information about connectivity, neighborhood, and inclusion. For instance, parks are on either side of Pennsylvania Ave; the World Bank is surrounded by Pennsylvania Ave, G St., 18th St, and 19th St, but the same network does not surround the IMF; and driving from Pennsylvania Ave to the intersection of G St and 18th St (within the selected area) is only possible via 19th St and G St.

M. Duckham et al. (Eds.): GIScience 2014, LNCS 8728, pp. 333–348, 2014.

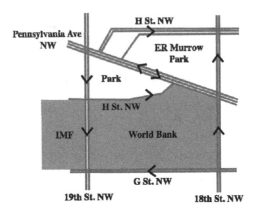

Fig. 1. A network of roads in Washington, DC. One-way, two-way, and multi-lane roads form a complex scene as they surround various parks and buildings.

Contemporary approaches to modeling such a scenario either capture roads or lanes as 2-dimensionally extended regions that are embedded in $\mathbb{R}^2$, or abstract the roads or lanes to directed edges within a graph. This paper pursues an alternative approach to modeling linearly conceptualized spatial features by representing them as *oriented regions*, essentially combining an inherent property of the 2-dimensional model (the extent of objects and their embedding in $\mathbb{R}^2$) with an inherent property of a directed graph (the orientation of the roads or lanes) into a single model. Ideally, each object should be represented with its proper extent and orientation, and relations between objects should be expressed independent of how many objects are involved, allowing entire spatial scenes to be captured completely.

When describing such a feature as a road, limiting the description to simple connectivity grossly oversimplifies people's real-world experiences with it. Simply driving in one direction, versus the opposite, is a potentially unique experience, as different intersections are encountered and landmarks are oriented differently. Therefore, the development of a theory that allows for the representation of these features, while capturing these additional semantic distinctions, is desirable, and has the potential to change how GISs represent linear features within spatial scenes.

The remainder of this paper is structured as follows: Section 2 reviews related work in the field of line and region representations and their topological relations. Section 3 introduces *oriented regions* to represent linearly conceptualized items as regions. Section 4 introduces *hull+i*, the new model for oriented regions that is based on topological hulls and the *i-notation*, a systematic method to capture boundary interactions between oriented regions. Section 5 shows with a series of examples how *hull+i* models some complex spatial scenes of roads and other linearly conceptualized features. Section 6 concludes with a summary and a discussion of future work.

2   Representing Linearly Conceptualized Features and Their Topological Relations

Current approaches to modeling binary topological relations generate similar sets of eight relations between two regions (Fig. 2), either with a connection-based approach [5] or an intersection-based approach [13,14]. The latter also has been shown to identify 19 relations between a region and a simple line [16], and 33 relations between two simple lines [9]. In order to support similarity reasoning, conceptual neighborhood graphs have been derived for these region-region [7,10], line-region [15], and line-line relations [32]. The relations' conceptual neighborhoods have provided a rationale for expressing the semantics of natural-language-like spatial predicates [20,26,35].

These relations are embedded in $\mathbb{R}^2$, but classical work has also explored the thirteen relations between line intervals in $\mathbb{R}^1$[2].

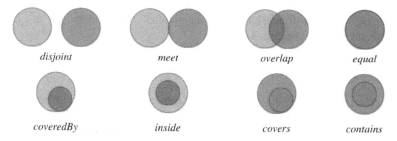

Fig. 2. The eight relations between two regions in $\mathbb{R}^2$ with their 9-intersection labels

The 9-intersection [14] has been expanded for relations between complex regions, such as those with holes and separations [34], and those with multiple touching or crossing configurations for regions [12], as well as for lines [4] and their metric refinements [30]. Alternative approaches for specific relations [17] focus on binary topological relations exclusively. Further research examined alternative line relations, such as directed line relations [22–24] and line-line relations with uncertainty [31], as well as dipole relations [27,28], which are also bound to a binary approach to describing topological relations.

When considering the problems of wayfinding and cartographic abstraction, any of these approaches can provide meaningful, qualitative descriptions of space within an appropriate context. However, much work has been tailored to problems specific to this area. For instance, the level of abstraction must be determined and might vary from the level of an observer to that of a planner, depending on how the space needs to be conceptualized to complete a task [36]. As such, human cognition plays a role; if the goal of a system is to generate driving directions, for instance, it is useful to understand how a person perceives decision points within a road network [19,21]. The choice of which features to represent, and how they respond to transformations, such as changes in scale [3,29,33], has also been considered. Many of these considerations have been brought together in comprehensive projects, such as OpenStreetMap [18], where a combination of crowdsourcing and a set of tools allow users to build up a cartographic representation of their space. Other approaches use natural-language descriptions to construct detailed topological and relative position information over a set of objects [37].

These approaches also rely on binary topological relations, or a graph representation of space. However, emerging research has led to the description of relations between more than two relations, allowing entire scenes to be captured. MapTree [38] uses combinatorial maps to represent multiple regions, capturing a hierarchy of containment, whereas hull+o [25] captures the relations between multiple regions through the boundary intersections between them, and provides a means of capturing separations and holes within objects through the use of a topological-hull operator.

3   From Lines to Regions

Abstracting a phenomenon linearly vs. regionally is not only a choice of detail, but also one about inferences. After a discussion of some of line-line vs. region-region relations (Section 3.1), we introduce the concept of *oriented regions* as a means for capturing with 2-dimensional representations the orientation of linearly conceptualized features (Section 3.2).

3.1   Lines as Regions

When representing a line as a region the first requirement is to settle on an interpretation. Fig. 3 illustrates how a simple line-line relation (Fig. 3a) may be represented in three different ways as a region-region relation (Figs. 3b-d), each capturing a different semantics.

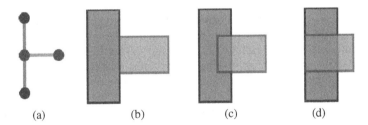

(a)   (b)   (c)   (d)

Fig. 3. Various interpretations of a scene: (a) a line-line relation, and (b-d) three different versions using regions to depict the same relation with endpoints maintained for the purpose of comparison

Other options, such as placing one region completely into the other's interior or placing them such that they have no connectivity between them, would miss some critical properties about the line-line relation. None of the three choices is, however, a rigid choice that is preserved across all objects and their semantics, as each has a different impact on how such properties as traversability (Section 4.5) are captured. Also the spatial-reasoning power differs whether one maps the line-line relation onto *meet* (Fig. 3b) vs. *overlap* (Figs. 3c-d) between the two regions, as *meet* and *overlap* have different compositions [8]. For this reason the choice to model a relation one way or another is highly dependent on the scene and what is to be described.

3.2   Oriented Regions

Regions with an orientation are plentiful in geographic space (Fig. 4). Each day people drive about roadways on a specific side, rotate a screw one way to loosen it, and another to tighten, and race about a track (hopefully) in an established direction. This quality is not restricted to human-made features, however. Wind flows about a wing in a specific fashion, water can swirl in a persistent whirlpool in the middle of a river, and planetary bodies spin about their axes.

(a)   (b)

Fig. 4. Regions with orientations: (a) skaters in a rink usually adopt a counter-clockwise rotation and (b) the interaction of gears or directed fields rotating toward a central point

Orientation is a fundamental property in algebraic topology [1], which has been applied in topological data models, based on the ordering of vertices of a simplex and a simplicial complex [11]. For a particular dimension $n>0$, an ordered n-simplex S_n may be represented by its vertices v_i (Eqn. 1).

$$S_n = < v_0, ..., v_n > \tag{1}$$

The orientation fixes the vertices to lie in a specific sequence, defining an order for the associated ordered simplex. The orientations of a 1-simplex can be interpreted as the direction from vertex v_0 to vertex v_1, and reverse from vertex v_1 to vertex v_0. The orientations of a 2-simplex can be interpreted as *clockwise* and *anticlockwise* (also referred to as counterclockwise) ordering of its three vertices, either v_0, v_1, v_2 or reverse v_2, v_1, v_0. For a 2-simplex the start vertex does not matter, that is, v_0, v_1, v_2 is the same as v_1, v_2, v_0 as well as v_2, v_0, v_1.

We distinguish five types of *oriented regions*:

- A *clockwise-oriented region* is homeomorphic to a 2-simplex with a clockwise orientation.
- An *anticlockwise-oriented region* is homeomorphic to a 2-simplex with a counter-clockwise orientation.
- A *monodirectional region*, which has either clockwise or counterclockwise orientation, but not both.
- A *bidirectionally oriented region* is an oriented region with both a clockwise and an anticlockwise orientation.
- For an *orientation-free region* the orientation is immaterial.

4   *Hull+i* for Oriented Regions

In order to model spatial scenes with oriented regions, we introduce *hull+i*, a variation of hull+o [25] that accommodates orientation. It consists of two parts: (1) the concepts of a *topological hull* and an *aggregate topological hull* (Section 4.1) and (2) the *i-notation* (Section 4.2), an account for the boundary interactions.

4.1   Topological Hull and Aggregate Topological Hull

Topological hull and aggregate topological hull follow immediately from [25], but for the sake of being self-contained, the two definitions are repeated.

Definition 1. Let A be a closed, path-connected set in $\mathbb{R}^2$ with co-dimension 0 within the standard topology, and let B be the smallest closed set homeomorphic to an n-disk such that $A \subseteq B$. B is called the *topological hull of A*, denoted as $[A]$.

To represent relations with holes, those holes need to be accounted for (Fig. 5a). To this end the topological hull acts as a hole-filling operation (Fig. 5b), allowing objects representative of any holes within an object to be defined by subtracting the original object from its hull (Fig. 5c).

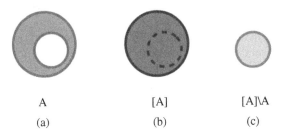

A	[A]	[A]\A
(a)	(b)	(c)

Fig. 5. The process of identifying holes with the topological hull: (a) A closed disk A, with a path-connected interior in $\mathbb{R}^2$, (b) the object's topological hull, such that $\partial A \subseteq [A]$, and (c) the hole of A, $[A] \setminus A$

Definition 2. Let A be a closed set in $\mathbb{R}^2$ with co-dimension 0 within the standard topology. Consider the collection of path-connected subsets P such that $\bigcup_{p \in P} p = A°$ and $|P|$ is minimized. The *aggregate topological hull* $[A]$ is the set $B = \bigcup_{p \in P}[p]$ [25].

The topological hull is a special case of the aggregate topological hull, and maintains the same notation. As the aggregate topological hull of an object A is the union of A's path-connected subsets, the aggregate topological hull allows separations—or satellites—of objects to be identified, in addition to any holes within each subset.

4.2   Basic *i-Notation*

While the *o-notation* [25] delivers a powerful means for describing the *boundary interactions* between multiple objects, it suffers from redundancy. When considering a spatial scene with many objects it is often the case that most *o-notation* strings will have large and similar sets of objects listed in the *outside* component of the string. Since the vast majority of objects are *disjoint* in geographic space, the *o-notation* becomes cumbersome.

The *i-notation* takes the reverse approach, considering the collection of regions for which the boundary intersection is currently *inside*—not *outside*—of, symbolized by the letter *i*. As such, the *i-notation* is defined as all boundary interaction—dimension and sequence—within a spatial scene *S*. It distinguishes between both 1-dimensional (Fig. 6a) and 0-dimensional touching (Fig. 6c) relations, as well as 1-dimensional (Fig. 6b) and 0-dimensional crossing (Fig. 6d) relations.

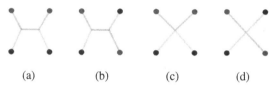

(a)   (b)   (c)   (d)

Fig. 6. Examples of line interactions: (a) 1-dimensional touching, (b) 1-dimensional crossing, (c) 0-dimensional touching, and (d) 0-dimensional crossing

The sequence of the interaction is captured by traversing the boundary of each object and recording each interaction in order (Eqn. 2).

$$\partial A_{comp} : i_s(dimension, T, C) \qquad (2)$$

The symbol ∂A_{comp} represents the boundary of an oriented region A, where S is the collection of regions for which the boundary interaction is currently *inside* of. The qualitative length of the interactions—either 0 or 1—is their *dimension*, followed by T, the collection of region boundaries that are undergoing a *touch* interaction, and finally C, which captures the collection of regions that undergo a cross *interaction*.

4.3   Binary Topological Relations with *i-Notation*

With the *i-notation* it is possible to represent the eight topological relations between two regions in $\mathbb{R}^2$ (Figs. 2).

Certain relations are subject to stacking (i.e., repeated sequences), indicated with an addition symbol. Such stacking allows for capturing increasingly complex configurations for the relations *overlap*, *meet*, *covers*, and *coveredBy*, each of which may have multiple boundary interactions.

While each *i-notation* string can represent a relation between two objects, it can also be expanded to include a larger set of objects, since an individual intersection can accommodate any number of objects simultaneously. Only *overlap* needs two *i-notation* strings. As the boundaries of both objects are simple closed curves, any crossing-into an object must be followed (not necessarily immediately) by a crossing-out-of that object. For *overlap*, this results in an additional *i-notation* string.

When representing multiple intersections along the boundary of a single object, one needs to differentiate intersections that appear immediately back-to-back (Fig. 8b) from intersections that appear after a break (Fig. 8a). To remove such ambiguity in the *i-notation* between immediate and delayed intersections, we introduce three markers that punctuate an *i-notation* string: (1) a plus sign (+) to signify the immediate addition of a following intersection, (2) a comma (,) to signify a pause before the start of the next intersection, and (3) optionally a period (.) in combination with one of the other symbols to simulate the endpoints of a line segment, which can be used to distinguish between different sections of an object, such as the sides of a street.

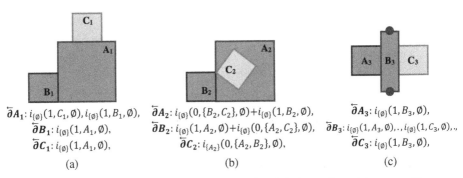

$\vec{\partial}A_1$: $i_{\{\emptyset\}}(1, C_1, \emptyset), i_{\{\emptyset\}}(1, B_1, \emptyset),$
$\vec{\partial}B_1$: $i_{\{\emptyset\}}(1, A_1, \emptyset),$
$\vec{\partial}C_1$: $i_{\{\emptyset\}}(1, A_1, \emptyset),$

$\vec{\partial}A_2$: $i_{\{\emptyset\}}(0, \{B_2, C_2\}, \emptyset) + i_{\{\emptyset\}}(1, B_2, \emptyset),$
$\vec{\partial}B_2$: $i_{\{\emptyset\}}(1, A_2, \emptyset) + i_{\{\emptyset\}}(0, \{A_2, C_2\}, \emptyset),$
$\vec{\partial}C_2$: $i_{\{A_2\}}(0, \{A_2, B_2\}, \emptyset),$

$\vec{\partial}A_3$: $i_{\{\emptyset\}}(1, B_3, \emptyset),$
$\vec{\partial}B_3$: $i_{\{\emptyset\}}(1, A_3, \emptyset), ., i_{\{\emptyset\}}(1, C_3, \emptyset), .,$
$\vec{\partial}C_3$: $i_{\{\emptyset\}}(1, B_3, \emptyset),$

(a)   (b)   (c)

Fig. 7. Scene between multiple oriented regions, with their *i-notations*: (a) scene with a gap between intersections, (b) scene between three objects without a gap between objects, and (c) scene between a road and a simple region on each side

The *o-notation* [25] treats *equal* as a 1-dimensional *touch* relation, and the only way to differentiate *equal* from any other case involving a single 1-*touch* is through the use of the *topological hull*. In the *i-notation*, *equal* does not need the topological hull for this distinction, because the object (or set of objects) that *equal* the target object are recorded along with the collection of regions for which the boundary intersection is currently *inside* of, and they are accented with an underline to represent equality similar to the subset notation (Fig. 7).

binary topological relation	$\vec{\partial}A$	$\vec{\partial}B$
disjoint	$i_{\{\emptyset\}}(\emptyset, \emptyset, \emptyset)$	$i_{\{\emptyset\}}(\emptyset, \emptyset, \emptyset)$
*meet**	$i_{\{\emptyset\}}(x, B, \emptyset)$	$i_{\{\emptyset\}}(x, A, \emptyset)$
*overlap**	$i_{\{\emptyset\}}(x, \emptyset, B), i_{\{B\}}(x, \emptyset, B)$	$i_{\{\emptyset\}}(x, \emptyset, A), i_{\{A\}}(x, \emptyset, A)$
equal	$i_{\{\underline{B}\}}(1, B, \emptyset)$	$i_{\{\underline{A}\}}(1, A, \emptyset)$
*coveredBy**	$i_{\{B\}}(x, B, \emptyset)$	$i_{\{\emptyset\}}(x, A, \emptyset)$
inside	$i_{\{B\}}(\emptyset, \emptyset, \emptyset)$	$i_{\{\emptyset\}}(\emptyset, \emptyset, \emptyset)$
*covers**	$i_{\{\emptyset\}}(x, B, \emptyset)$	$i_{\{A\}}(x, A, \emptyset)$
contains	$i_{\{\emptyset\}}(\emptyset, \emptyset, \emptyset)$	$i_{\{A\}}(\emptyset, \emptyset, \emptyset)$

Fig. 8. The *i-notation* strings corresponding to the eight binary relations between regions. An *x* denotes a wildcard, which can be either 0 or 1. An asterisk signifies a relation that can be subject to stacking.

4.4   Orientation into *i-Notation*

In order to account for an oriented region's orientation, the *i-notation* captures it for each object explicitly. Such recording gives rise to such scenarios as modeling road networks with different driving directions, as in the U.S. and the U.K., one-way roads, and regions of roads that allow motorists to freely merge between multiple lanes. In the U.S. and most continental regions drivers travel along the right side of the road, giving it an anticlockwise orientation. Depending on locale it makes sense to alter the order in which strings are recorded, so that the ordering of intersections more easily maps onto the experience of a driver, for instance. To this end an appropriately directed arrow is added to the boundary component, $\vec{\partial}A$ signifying a clockwise orientation and $\overleftarrow{\partial}A$ signifying a counterclockwise orientation.

In order to accommodate orientation, the *i-notation* is expanded with a fourth argument (Eqn. 3).

$$\partial A_{comp}(O): i_S(dimension, T, C) \tag{3}$$

The symbol O represents the orientation of the object. Five values signify different orientations (Fig. 9), and each object possesses a single orientation.

orientation	value
no orientation	N
clockwise	C
anticlockwise	A
monodirectional	M
bidirectional	B

Fig. 9. The five *i-notation* values for orientation between intersections

Semantically, roadways usually have a start and an end, which is well captured with a directed line segment. When representing lines as regions, this explicit representation of start and endpoints needs to be preserved, in order to distinguish between one side of a road and its opposite side. The period (.) string punctuation allows the sequence to be divided into multiple components. When recording the intersection string for a monodirectional object care should be taken to start at the beginning of the segment on the dominant side of the road, that is, starting with the right side in the U.S., or the left side in the U.K. and similar areas.

Both edges of such a region, which is used to model a one-way street, travel in the same direction. Since a normal driver (in the U.S. example) will always transit along their subjective right side, regardless of road direction, such a careful recording of *i-notation* strings allows for not just differences between one side and the opposite to be explored, but can be used to distinguish between the driver's left and right.

4.5   Traversability into *i-Notation*

The final addition to the *i-notation* is the incorporation of *traversability* in order to identify valid connections within a network of objects. To consider this, imagine two roadway components: a 4-way intersection and an over/underpass. From the standpoint of a driver, it is sound to travel from one road to another in the four-way intersection, by making a left or a right turn. However, such a transition is not desirable when traveling on an overpass.

Since the objects are embedded in the plane, distinctions such as these, which typically result from components existing at different elevations, are generally not expressible through the usual planar topological relations. Thus far with *i-notation* such overpasses and actual intersections would both look the same. The lack of connectivity between the two roads in an overpass, however, needs to be represented explicitly to signify valid transitions within a network. Thus, the final component to be built in to the model of oriented regions is traversability (Eqn. 5).

$$\partial A_{comp}(O): i_S(dimension, T, C, \boldsymbol{R}) \tag{5}$$

Here R is the set of objects that can be reached from a given intersection. The value of R is relative to the object currently being traversed—this is not necessarily symmetric across all objects involved within that intersection. Once the extent of an object has been accounted for, internal components of the object, such as additional lanes in a road, can be expressed with additional regions, depending on what is desirable to model. The objective is not necessarily to convert all linear features to regions, but to preserve the qualities of the real-world objects, such as extent, while providing a rich descriptive framework that includes features, such as connectivity. To this end, there are two further distinctions that can be made when representing lines as regions: (1) explicit line segments and (2) implicit connections. *Explicit line segments* are the divisions within a roadway, such as additional lanes. Additionally, explicit line segments complement the visual representation of the scene at the highest levels of detail, but all of them would not necessarily be represented at more distant scales. *Implicit connections* are those connections that are added exclusively to represent some aspect of the internal connectivity within an object, such as an area spanning multiple lanes of a highway where cars can merge left or right. These connections might not ever appear in a visual representation of a scene. By representing these connections as regions with a bidirectional orientation, areas where motorists can safely switch lanes can be modeled.

When approaching a T-intersection (Fig. 10a), a driver has multiple choices, some illegal (Fig. 10b), others legal (Fig 10c). While the *i-notation* for the two scenarios would be nearly identical, the key difference between the two paths is a single character within one string in A's boundary: $i_{\{B\}}(1, B, \emptyset, \boldsymbol{B})$. If the final symbol, B, was expanded to $\{A, B\}$, a driver would have the ability to make a U-turn (Fig. 10b). The legality of a route is subject to local convention.

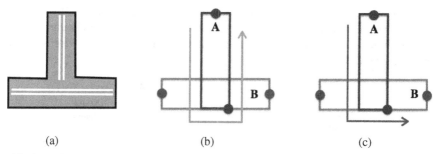

(a)   (b)   (c)

Fig. 10. A road scene (a) with a T-intersection, (b) an invalid path for a vehicle, and (c) a vehicle's valid path with its *i-notation* $\overleftrightarrow{\partial}A(A)$: $i_{\{B\}}(0,\emptyset,B,A),.,i_{\{\emptyset\}}(0,\emptyset,B,\{A,B\}),i_{\{B\}}(1,B,\emptyset,B)+.,$ and $\overleftrightarrow{\partial}B(A)$: $i_{\{\emptyset\}}(0,\emptyset,A,\{A,B\}),i_{\{A\}}(0,\emptyset,A,B),.,i_{\{\emptyset\}}(1,A,\emptyset,\{A,B\}),.,$

5 *Hull+i* for Scenes with Oriented Regions

With the ability to append orientation and traversability to regional objects within a spatial scene, it is now possible to model the geometry of complex features such as roads, highways, and rivers—along with the other components that exist within a space, such as buildings. Each object can be represented with its proper extent, and relations between objects can now be expressed independent of how many objects are involved, allowing scenes to be captured completely. As such, the *i-notation* provides a toolset, with which it is possible to explore different scenarios.

To start, consider the intersection of two branches of a river that *meet* at a downstream location (Fig. 11). This scenario includes two objects with both a spatial extent and a specific orientation (i.e., the flow direction of each river).

When considering the river, all water flows downward with respect to the image. When A *crosses* B, the flow must join that of region B, taking on its orientation. Since both rivers have a monodirectional orientation, the segment that is recorded first determines the relative direction of flow (Section 4.3). In a true-to-life representation it would not be expected that the flow of the river from A to B hugs the first shoreline as the two objects intersect, thus the extent of A *crosses* through B to the opposing shore. This is a choice where choosing how to represent a line-line relation as an extended feature requires an understanding of the scene. With such a representation it is shown that the two flows meet at an intersection, which itself has an extent, and proceed downward across that space. By looking at the *i-notation* strings Orono, River A, and Marsh Island are all clearly represented on one shore of B, with Bradley located on the opposite shore—based on the punctuation defined for River B. Similar determinations can be made with regard to River A.

Fig. 11b has the following *i-notation* strings:

$\vec{\partial}A(M)$: $i_{\{\emptyset\}}(1,MarshIsland,\emptyset,A)+i_{\{\emptyset\}}(0,MarshIsland,B,\{A,B\}),i_{\{B\}}(1,\{B,Bradly\},\emptyset,\{A,B\})$
    $+.,i_{\{B\}}(0,Orono,B,B)+i_{\{\emptyset\}}(1,Orono,\emptyset,A),.,$
$\vec{\partial}B(M)$: $i_{\{\emptyset\}}(1,Bradly,\emptyset,B)+i_{\{\emptyset\}}(1,\emptyset,\{A,Bradly\},\{A,B\})+i_{\{\emptyset\}}(1,Bradly,\emptyset,B),.,$
$i_{\{\emptyset\}}(1,Orono,\emptyset,B)+i_{\{\emptyset\}}(0,Orono,A,\{A,B\}),i_{\{A\}}(0,MarshIsland,A,\{A,B\}),i_{\{\emptyset\}}(1,MarshIsland,\emptyset,B),.,$
$\vec{\partial}Orono(N)$: $i_{\{\emptyset\}}(1,A,\emptyset,\emptyset)+i_{\{\emptyset\}}(0,\{A,B\},\emptyset,\emptyset)+i_{\{\emptyset\}}(1,B,\emptyset,\emptyset),$
$\vec{\partial}MarshIsland(N)$: $i_{\{\emptyset\}}(1,B,\emptyset,\emptyset)+i_{\{\emptyset\}}(0,\{A,B\},\emptyset,\emptyset)+i_{\{\emptyset\}}(1,B,\emptyset,\emptyset),$
$\vec{\partial}Bradley(N)$: $i_{\{\emptyset\}}(1,B,\emptyset,\emptyset)+i_{\{\emptyset\}}(1,\{A,B\},\emptyset,\emptyset)+i_{\{\emptyset\}}(1,B,\emptyset,\emptyset),$

The next example is the extent of a multilane road (Fig. 12a). Thus far roads have been described using the anticlockwise orientation, however this only corresponds to the highest level of abstraction. To capture more detail the anticlockwise road can be seen to contain two monodirectional lanes (Fig. 12b).

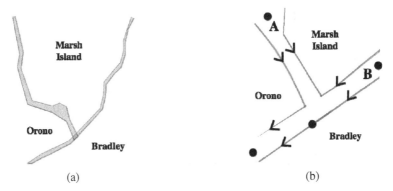

Fig. 11. Two branches of a river in Maine: (a) a representation of the scene similar to a traditional map, and (b) a representation that can be expressed in terms of the *i-notation*, modeling the preservation of flow across the various intersections

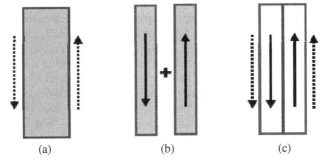

Fig. 12. A simple road: (a) the extent an entire road object, (b) the driving lanes within the road object, and (c) the features combined

Just before an intersection one of these lanes might be further divided to accommodate a lane for straight travel, and a lane for drivers making a right turn (Fig. 13a). Drivers need to be able to merge between these new lanes, so a bidirectional region is used to create a passing area (Fig. 13b).

When all of these components are added together (Fig. 14) they form a complete model of a street scene spanning different levels of detail. The complete extent of the object is represented by the anticlockwise region—this object captures the general sense that the road travels in two directions. Contained within the parent object are the major lane divisions, providing an intermediate level of detail. At the highest level of detail additional elements are inserted, such as passing and merging areas, allowing the model to functionally represent not only the scene, but how a driver might interact with it. Thus, *i-notation* is compatible with different levels of scale and different modeling needs.

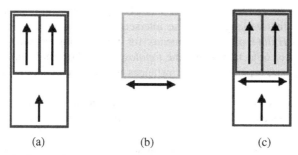

Fig. 13. The area before an intersection: (a) two different turning lanes, (b) the addition of a bidirectional region to allow for merging, and (c) the features combined

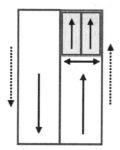

Fig. 14. The complete scene created by Fig. 12 and Fig. 13 merged together

The final example reexamines an aspect of the Washington, DC scene (Fig. 15a). By taking the topological hull (Fig. 15c) of a block of roads (Fig. 15b), any objects they surround, which might not meet the road, can be identified. In this scenario the road network contains a set of holes, p_1 *and* p_2 (Fig. 15d), which themselves contain a small park, and the World Bank on the other side of H Street, respectively.

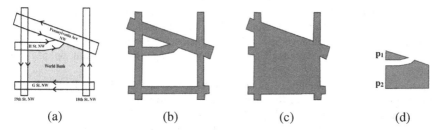

(a)   (b)   (c)   (d)

Fig. 15. A simplified scene in Washington, DC near the World Bank: (a) the building and the roads that surround it, (b) the union of the roads, (c) the topological hull of the local road network, and (d) objects representing the cavities within the union of roads

6   Conclusions and Future Work

To conclude, *hull+i* allows the representation of spatial scenes through a description of the boundary intersections between sets of objects. This approach preserves the ordering and qualitative length of these intersections, allowing a finer level of detail to be captured. *Hull+i* provides a means for representing spatial scenes with an arbitrary number of objects, through the *topological hull* operator, in order to capture holes within objects, separations, and instances where ensembles of objects *surround* other objects.

Future work aims to normalize *i-notation*, so that a single intersection is recorded only one time, instead of once per object involved in a given intersection. A normalized *i-notation* also will provide the framework for automatic scene generation—allowing a computer to recreate a scene based on a *hull+i* description.

The development of *hull+i* will also include an exploration of *i-notation* strings for scenes with two and three objects, conceptual neighborhood graphs for those relations, and an exploration of the rich composition that is possible with *i-notation*.

Furthermore, *hull+i* can be expanded to handle instances where scenes change over time. Scenarios include the closure and opening of lanes during different times of the day, flow changes due to construction, and other dynamic elements that could be explored in an implementation of *hull+i*.

Acknowledgments. This work was partially supported by NSF Grant IIS-1016740 (PI: Max Egenhofer).

References

1. Alexandroff, P.: Elementary Concepts of Topology. Dover, Mineola (1961)
2. Allen, J.: Maintaining Knowledge about Temporal Intervals. Communications of the ACM 26(11), 832–843 (1983)
3. Barkowsky, T., Latecki, L.J., Richter, K.-F.: Schematizing Maps: Simplification of Geographic Shape by Discrete Curve Evolution. In: Habel, C., Brauer, W., Freksa, C., Wender, K.F. (eds.) Spatial Cognition II. LNCS (LNAI), vol. 1849, pp. 41–53. Springer, Heidelberg (2000)
4. Clementini, E., Di Felice, P.: Topological Invariants for Lines. IEEE Transactions on Knowledge and Data Engineering 10(1), 28–45 (1998)
5. Cohn, A., Bennett, B., Gooday, J., Gotts, N.: Qualitative Spatial Representation and Reasoning with the Region Connection Calculus. Geoinformatica 1(3), 275–316 (1997)
6. Cohn, A., Renz, J.: Qualitative Spatial Representation and Reasoning. In: van Hermelen, F., Lifschitz, V., Porter, B. (eds.) Handbook of Knowledge Representation, pp. 551–596 (2008)
7. Egenhofer, M.J.: The Family of Conceptual Neighborhood Graphs for Region-Region Relations. In: Fabrikant, S.I., Reichenbacher, T., van Kreveld, M., Schlieder, C. (eds.) GIScience 2010. LNCS, vol. 6292, pp. 42–55. Springer, Heidelberg (2010)
8. Egenhofer, M.: Deriving the Composition of Binary Topological Relations. Journal of Visual Languages and Computing 5(2), 133–149 (1994)
9. Egenhofer, M.: Definitions of Line-Line Relations for Geographic Databases. IEEE Data Engineering Bulletin 16(3), 40–45 (1993)

10. Egenhofer, M., Al-Taha, K.: Reasoning about Gradual Changes of Topological Relationships. In: Frank, A.U., Formentini, U., Campari, I. (eds.) GIS 1992. LNCS, vol. 639, pp. 196–219. Springer, Heidelberg (1992)

11. Egenhofer, M., Frank, A., Jackson, J.: A Topological Data Model for Spatial Databases. In: Buchmann, A.P., Smith, T.R., Wang, Y.-F., Günther, O. (eds.) SSD 1989. LNCS, vol. 409, pp. 271–286. Springer, Heidelberg (1990)

12. Egenhofer, M., Franzosa, R.: On the Equivalence of Topological Relations. International Journal of Geographical Information Systems 9(2), 133–152 (1995)

13. Egenhofer, M., Franzosa, R.: Point-Set Topological Spatial Relations. International Journal of Geographical Information Systems 5(2), 161–174 (1991)

14. Egenhofer, M., Herring, J.: Categorizing Binary Topological Relationships Between Regions, Lines, and Points in Geographic Databases, Department of Surveying Engineering, University of Maine, Orono, ME (1991)

15. Egenhofer, M., Mark, D.: Modeling Conceptual Neighborhoods of Topological Line-Region Relations. International Journal of Geographical Information Systems 9(5), 555–565 (1995)

16. Egenhofer, M., Sharma, J., Mark, D.: A Critical Comparison of the 4-Intersection and 9-Intersection Models for Spatial Relations: Formal Analysis. In: McMaster, R., Armstrong, M. (eds.) Autocarto-Conferance, Minneapolis, MN, vol. 11, pp. 1–11 (1993)

17. Galton, A.: Modes of Overlap. Journal of Visual Languages and Computing 9(1), 61–79 (1998)

18. Haklay, M., Weber, P.: OpenStreetMap: User-Generated Street Maps. IEEE Pervasive Computing 7(4), 12–18 (2008)

19. Klippel, A.: Wayfinding Choremes. In: Kuhn, W., Worboys, M.F., Timpf, S. (eds.) COSIT 2003. LNCS, vol. 2825, pp. 301–315. Springer, Heidelberg (2003)

20. Klippel, A., Li, R., Yang, J., Hardisty, F., Xu, S.: The Egenhofer-Cohn Hypothesis or, Topological Relativity? In: Raubal, M., Mark, D., Frank, A. (eds.) Cognitive and Linguistic Aspects of Geographic Space, pp. 195–215. Springer (2013)

21. Klippel, A., Tappe, H., Kulik, L.: Wayfinding Choremes—A Language for Modeling Conceptual Route Knowledge. Journal of Visual Languages and Computing 16(4), 311–329 (2005)

22. Kurata, Y.: The 9+-Intersection: A Universal Framework for Modeling Topological Relation. In: Cova, T.J., Miller, H.J., Beard, K., Frank, A.U., Goodchild, M.F. (eds.) GIScience 2008. LNCS, vol. 5266, pp. 181–198. Springer, Heidelberg (2008)

23. Kurata, Y., Egenhofer, M.: The Arrow Semantic Interpreter. Spatial Cognition and Computing 8(4), 306–332 (2008)

24. Kurata, Y., Egenhofer, M.: The Head-Body-Tail Intersection for Spatial Relations Between Directed Line Segments. In: Raubal, M., Miller, H.J., Frank, A.U., Goodchild, M.F. (eds.) GIScience 2006. LNCS, vol. 4197, pp. 269–286. Springer, Heidelberg (2006)

25. Lewis, J.A., Dube, M.P., Egenhofer, M.J.: The Topology of Spatial Scenes in R^2. In: Tenbrink, T., Stell, J., Galton, A., Wood, Z. (eds.) COSIT 2013. LNCS, vol. 8116, pp. 495–515. Springer, Heidelberg (2013)

26. Mark, D., Egenhofer, M.: Modeling Spatial Relations between Lines and Regions: Combining Formal Mathematical Models and Human Subjects Testing. Cartography and Geographic Information Systems 21(3), 195–212 (1994)

27. Moratz, R., Lücke, D., Mossakowski, T.: A Condensed Semantics for Qualitative Spatial Reasoning about Oriented Straight Line Segments. Artifical Intelligence 175(16-17), 2099–2127 (2011)

28. Moratz, R., Renz, J., Wolter, D.: Qualitative Spatial Reasoning about Line Segments. In: Horn, W. (ed.) Proceedings of the 14th European Conference on Artificial Intelligence, ECAI 2000, Berlin, pp. 234–238 (2000)

29. Morehouse, S.: GIS-Based Map Compilation and Generalization. In: Müller, J.-C., Lagrange, J., Weibel, R. (eds.) GIS and Generalization: Methodology and Practice, pp. 21–30. Taylor&Francis, Bristol (1995)

30. Nedas, K., Egenhofer, M., Wilmsens, D.: Metric Details of Topological Line-Line Relations. International Journal of Geographical Information Science 21(1), 21–48 (2007)

31. Reis, R., Egenhofer, M., Matos, J.: Topological Relations Using Two Models of Uncertainty for Lines. In: Caetano, M., Painho, M. (eds.) 7th International Symposium on Spatial Accuracy Assessment in Natural Resources and Environmental Sciences, Lisbon, Portugal, pp. 5–7 (2006)

32. Reis, R., Egenhofer, M., Matos, J.: Conceptual Neighborhoods of Topological Relations between Lines. In: Ruas, A., Gold, C. (eds.) Headway in Spatial Data Handling, pp. 557–574. Springer (2008)

33. Ruas, A., Lagrange, J.: Data and Knowledge Modelling for Generalization. In: Müller, J., Lagrange, J., Weibel, R. (eds.) GIS and Generalization: Methodology and Practice, pp. 73–90. Taylor&Francis, Bristol (1995)

34. Schneider, M., Behr, T.: Topological Relationships Between Complex Spatial Objects. ACM Transactions on Database Systems 31(1), 39–81 (2006)

35. Schwering, A.: Evaluation of a semantic similarity measure for natural language spatial relations. In: Winter, S., Duckham, M., Kulik, L., Kuipers, B. (eds.) COSIT 2007. LNCS, vol. 4736, pp. 116–132. Springer, Heidelberg (2007)

36. Timpf, S., Volta, G., Pollock, D., Egenhofer, M.: A Conceptual Model of Wayfinding Using Multiple Levels of Abstraction. In: Frank, A.U., Formentini, U., Campari, I. (eds.) GIS 1992. LNCS, vol. 639, pp. 348–367. Springer, Heidelberg (1992)

37. Vasardani, M., Timpf, S., Winter, S., Tomko, M.: From Descriptions to Depictions: A Conceptual Framework. In: Tenbrink, T., Stell, J., Galton, A., Wood, Z. (eds.) COSIT 2013. LNCS, vol. 8116, pp. 299–319. Springer, Heidelberg (2013)

38. Worboys, M.: The Maptree: A Fine-Grained Formal Representation of Space. In: Xiao, N., Kwan, M.-P., Goodchild, M.F., Shekhar, S. (eds.) GIScience 2012. LNCS, vol. 7478, pp. 298–310. Springer, Heidelberg (2012)

RCC*-9 and CBM*

Eliseo Clementini[1] and Anthony G. Cohn[2]

[1] University of L'Aquila, Information Engineering, L'Aquila, Italy
[2] University of Leeds, School of Computing, Leeds, UK

Abstract. In this paper we introduce a new logical calculus of the Region Connection Calculus (RCC) family, RCC*-9. Based on nine topological relations, RCC*-9 is an extension of RCC-8 and models topological relations between multi-type geometric features: therefore, it is a calculus that goes beyond the modeling of regions as in RCC-8, being able to deal with lower dimensional features embedded in a given space, such as linear features embedded in the plane. Secondly, the paper presents a modified version of the Calculus-Based Method (CBM), a calculus for representing topological relations between spatial features. This modified version, called CBM*, is useful for defining a reasoning system, which was difficult to define for the original CBM. The two new calculi RCC*-9 and CBM* are introduced together because we can show that, even if with different formalisms, they can model the same topological configurations between spatial features and the same reasoning strategies can be applied to them.

1   Introduction

The modeling of topological relations in Geographical Information Systems (GISs) and spatial databases has been a central topic of research since the early 90s. Three models have played a very important role, both in terms of theoretical developments and practical applications: the 9-intersection model (9IM) [14], RCC-8 [11], and CBM [5]. Regarding their modeling capabilities, RCC-8 is able to represent topological relations between regions, while 9IM and CBM are able to represent topological relations between spatial features of any dimensionality. With respect to reasoning capabilities, composition tables were defined for RCC-8 and 9IM [12] (for regions only), while composition tables for the CBM were never developed. Having composition tables for all kinds of spatial data types is essential for several tasks, e.g., for spatial query optimization [1]: applying the constraints of the tables, it is possible to discover contradictions in the query expression before the real processing of the query actually starts.

The RCC family of calculi [8] uses a logical approach for the representation of qualitative topological relations. The calculi were developed with regions as the primitive spatial entity and the connection relation as the primitive topologic relation between regions, from which other relations can be defined. RCC-8, the most representative calculus of the family, can model eight topological relations between regions of the plane: there is a one-to-one correspondence with the eight topological relations that are definable with the 9IM between 2D simple regions. As remarked in

M. Duckham et al. (Eds.): GIScience 2014, LNCS 8728, pp. 349–365, 2014.

[18], there exist few attempts to express topological relations between features of lower dimensions than the embedding space, such as lines in R^2, due to the difficulties of dealing with different types. In [19], Galton introduced an axiomatic system for multidimensional mereotopology, using primitives for 'part' (P) and 'boundary' (B). Gott's "INCH" calculus dealt with closed sets of points of uniform dimensionality [21] using a single primitive binary relation INCH ("includes a chunk of"). See further analysis in [9].

CBM [5] is a model for expressing topological relations between regions, lines, and points. It was especially defined for expanding the querying capabilities of database query languages towards spatial data. The operators of CBM have been adopted by the Open GeoSpatial Consortium (OGC) [23] and implemented in all spatial database systems. CBM relations can find an equivalent expression in terms of Egenhofer matrix-based methods [15] and vice versa. In particular, as it was shown in [3], CBM is more expressive than 9IM and equivalent to the Dimensionally-Extended 9-Intersection Model (DE+9IM) [3]. Despite its success in spatial databases and in the standardization process, CBM had little impact in the Qualitative Spatial Reasoning (QSR) community, due to the absence of a strong logical formulation and in particular its lack of composition tables. As pointed out in [16, 22], CBM is difficult to compare to logical calculi such as the RCC and no reasoning rules have been defined for it. The definitions of CBM were dependent on the dimension of the features participating in the relation. For example, a *cross* between a line and a region had a different definition from a *cross* between two lines. This meant it was not possible to find a single composition table for the calculus: at best, it would have been possible to find composition tables for each group of relations, that is, for region/region relations, for line/region relations, and so on, as proposed in [22].

In this paper, we aim at establishing a bridge between RCC and CBM, by defining an extension of RCC-8 that is capable of modeling topological relations between spatial features of any dimensionality and an extension of CBM that is capable of reasoning. To achieve this goal, a new calculus of the RCC family is defined, called RCC*-9, able to deal with features of various dimensions, not just regions[1]. A modification of CBM, called CBM*, is introduced that maps straightforwardly onto calculi of the RCC family and allows a composition table for reasoning to be found. Finally, it is shown that the two new calculi, RCC*-9 and CBM*, are able to model the same topological configurations, even though they are defined in different ways, and they are compared with 9IM.

In Section 2, we recapitulate the definition of the geometric data model on which we base our work. In Section 3, we briefly recall the definitions of CBM. In Section 4, we introduce CBM* and discuss the changes between CBM and CBM*. In Section 5, we introduce the logical calculus RCC*-9. In Section 6, we define the spatial reasoning system of both RCC*-9 and CBM*. In Section 7, we discuss how to express CBM* relations and RCC*-9 relations in terms of 9IM. In Section 7, we make some concluding remarks.

---

[1]   The reason for the asterisk in the name is that it is not just a change in the number of relations. There is also a substantial change in the spatial primitives the new calculus is able to deal with. Further, we will also need in the paper to introduce a coarser calculus which we called RCC*-7, since an RCC-7 already exists [20].

2   Definition of Geometric Features

In this paper, we will adopt the same terminology of the OGC where various point-sets of the plane R^2 are called features, distinguishing between simple features and complex features [23]. The OGC simple feature model definitions were in turn taken from [4]. In the following, we briefly recall those definitions. First of all, features are classified with respect to their dimension: regions of dimension 2, lines of dimension 1, and points of dimension 0.

Let x be a two-dimensional point-set.
Def. 1. The interior $x°$ of x is defined as the union of all open sets contained in x.
Def. 2. The closure $\bar{x}$ of x is defined as the intersection of all closed sets containing x.
Def. 3. The boundary ∂x of x is defined as the set difference between its closure and its interior, i.e., $\bar{x} - x°$.
Def. 4. The exterior x^- of x is defined as the set difference $R^2 \setminus \bar{x}$.
Def. 5. x is regular closed if $x=\bar{x°}$.
Def. 6. A simple region is a regular closed (non-empty) two-dimensional point-set x with a connected interior and connected exterior.

Def. 6 implies that a simple region is homeomorphic to the closed unit disk. A simple region does not have holes and is connected. If we remove the constraint of connected exterior from the definition, we obtain regions with holes [13]. In OGC simple feature specifications, regions with holes are implemented with the Polygon spatial data type. If we remove the constraint of connected interior, we obtain complex regions, that is, regions with holes and separations. Complex regions are implemented in OGC feature model with the MultiPolygon spatial data type.

Def. 7. A simple line is a closed (non-empty) one-dimensional point-set x defined as the image of a continuous mapping $f:[0,1] \rightarrow R^2$, such that $\forall t_i,t_j \in [0,1]$, $t_i \neq t_j$, $f(t_i) \neq f(t_j)$.

In other words, a simple line is the mapping of the unit interval in the plane with no self-intersections. A simple line can be described as the trace of a pencil on a sheet of paper without detaching the pencil and by not passing twice on the same position. The initial and final point of the simple line, defined as $f(0)$ and $f(1)$, are called the end-points of the line.

Topologically, a simple line embedded in R^2, being a one-dimensional set, has an empty interior. As common practice in GIS [15] and in OGC standards as well, the boundary ∂x of a line x is considered to be the set of its endpoints and the interior of the line the difference, $x°=x \setminus \partial x$. In this paper, we will adopt these definitions of boundary and interior of a line feature. In OGC feature model, simple lines are implemented with the Polyline spatial data type.

From Def. 7, if we remove the constraint of no self-intersections, we obtain lines with self-intersections. A particular case of a line with self-intersections is the closed ring, where $f(0)=f(1)$. If a one-dimensional point set can be obtained as the union of several mappings from the unit interval to the plane, then we obtain the concept of a complex line. A complex line can be made of several disjoint components. A complex line in OGC feature model is implemented with the MultiPolyline spatial data type.

A simple point is a zero-dimensional element of the embedding space. A complex point is the union of a finite number of simple points. Following the OGC convention, we assume that point features have an empty boundary. Simple and complex point features are implemented in OGC standards with the Point and Multipoint spatial data types, respectively.

3   CBM

One of the basic ideas behind CBM [5] was to provide an easy spatial extension of the tuple relational calculus [7] to express queries such as:

$$\{x| \exists y \ [River(x) \wedge Region(y) \wedge cross(x,y) \wedge y= \text{'Abruzzo'}]\}$$

The above CBM expression corresponds to the query "Retrieve all the rivers that cross the Abruzzo region." The topological relations of CBM can be applied not only to simple variables but to the boundaries of geometric features. Boundaries are extracted by the three operators b (boundary – the closed line representing the boundary of a simple region), f (from – the first endpoint of a line), t (to – the second endpoint of a line)[2]. For example, the following queries can be expressed in CBM:

$$\{x| \exists y \exists z \ [River(x) \wedge Mountain(y) \wedge Sea(z) \wedge in(f(x),y) \wedge y= \text{'Apennines'} \wedge touch(t(x),z) \wedge z= \text{'Adriatic'}]\}$$

$$\{x| \exists y \ [Road(x) \wedge Region(y) \wedge cross(x,y) \wedge overlap(x,b(y)) \wedge y= \text{'Abruzzo'}]\}$$

The above expressions correspond to "Retrieve all the rivers that rise in the Apennines mountains and flow into the Adriatic sea" and "Retrieve all the roads that cross the Abruzzo region and have a part of the road along the region's boundary."

The five topological relations of CBM are named *disjoint, touch, in, cross, overlap*. The definition of these relations are (the 'dim' operator evaluates to 0, 1, 2 depending whether the argument is a 0-, 1-, or 2-dimensional point set):

Def. 8. *disjoint*$(x,y) =_{\text{def}} x \cap y = \varnothing$
Def. 9. *touch*$(x,y) =_{\text{def}} x° \cap y° = \varnothing \ \wedge \ x \cap y \neq \varnothing$
Def. 10. *in*$(x,y) =_{\text{def}} x \cap y = x \ \wedge \ x° \cap y° \neq \varnothing$
Def. 11. *cross*$(x,y) =_{\text{def}} \dim(x° \cap y°) < \max(\dim(x°),\dim(y°)) \ \wedge \ x \cap y \neq x \ \wedge \ x \cap y \neq y$
Def. 12. *overlap*$(x,y) =_{\text{def}} \dim(x° \cap y°) = \dim(x°) = \dim(y°) \ \wedge \ x \cap y \neq x \ \wedge \ x \cap y \neq y$

The relations can be applied to all geometric types, either simple or complex [4]. They were implemented by the OGC feature model as a set of functions with names Disjoint, Touches, Within, Crosses, Overlaps. Additionally, the converse function of Within was called Contains and the function Equals was defined

---

[2]   These boundary extraction operators were specifically defined to extract the first and last endpoint of a directed line. More generally, when the direction is not known or there are more than two endpoints (such as in the case of complex lines) or no endpoints (such as in the case of closed rings), a generic boundary operator b is used that extracts the boundary of the feature.

as Within and Contains at the same time [23]. An expression of the relational tuple calculus extended with the five topological relations and the three boundary operators can be expressed by the Egenhofer matrix-based methods. Conversely, any instance of the DE+9IM can be expressed by an expression of CBM [3].

4   CBM*

In this section, we introduce a modification of CBM, called CBM*, for which it is easier to find an equivalence in terms of calculi of the RCC family and to find a composition table for reasoning. The basic relations of CBM* have a slightly different meaning from the corresponding relations of CBM. We assume the following definitions (we adopt the same names followed by a *) accompanied by a qualitative explanation of the meaning:

Def. 13. $disjoint^*(x,y)$, the two features are disjoint:

$disjoint^*(x,y) =_{\text{def}} x \cap y = \varnothing$

Def. 14. $touch^*(x,y)$, the two features intersect, but their interiors are disjoint (and it excludes containment):

$touch^*(x,y) =_{\text{def}} x° \cap y° = \varnothing \ \wedge \ x \cap y \neq \varnothing \ \wedge \ x \cap y \neq x \ \wedge \ x \cap y \neq y$

Def. 15. $in^*(x,y)$, feature x is part of feature y (it excludes equality):

$in^*(x,y) =_{\text{def}} x \cap y = x \ \wedge x \neq y$

$in^{*-1}(x,y) =_{\text{def}} x \cap y = y \ \wedge x \neq y$

$equal^*(x,y) =_{\text{def}} x = y$

Def. 16. $cross^*(x,y)$, the interiors of the two features intersect, but at least one feature's boundary does not intersect the other feature:

$cross^*(x,y) =_{\text{def}} x° \cap y° \neq \varnothing \ \wedge \ (\partial x \cap y = \varnothing \ \vee \ x \cap \partial y = \varnothing)$

Def. 17. $overlap^*(x,y)$: the interiors of the two features intersect and also each feature's boundary intersects the other feature (and it excludes containment).

$overlap^*(x,y) =_{\text{def}} x° \cap y° \neq \varnothing \ \wedge \ \partial x \cap y \neq \varnothing \ \wedge \ x \cap \partial y \neq \varnothing \ \wedge \ x \cap y \neq y \ \wedge x \cap y \neq x$

The proof that the relations of CBM* make a jointly exhaustive and pairwise disjoint (JEPD) set is readily obtained from the decision tree (see Fig. 1). Let us comment in more detail upon the differences between CBM and CBM* definitions. The *disjoint* and *disjoint** relations are the same. The *touch** relation is more restrictive than the *touch* relation, since cases where one feature is entirely contained inside the boundary of another one are excluded and are instead classified as *in** (see Fig. 2). The *in** relation takes over the cases ruled out by *touch** and excludes the case of equality between the two features: therefore, an explicit *equal** relation is needed in CBM*. The *cross** and *overlap** relations take into account the remaining cases with a different criterion for partitioning the cases with respect to the original *cross* and *overlap* relations. In Fig. 2, we can see the differences with some representative configurations. The *overlap* relation between a region and a line was not possible in CBM, while the relation *overlap** between a region and a line corresponds to a real case.

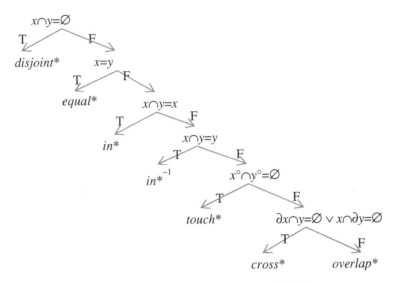

Fig. 1. Decision tree for the relations of CBM*

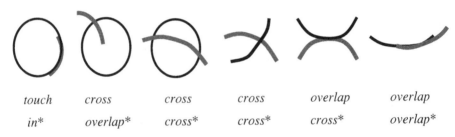

touch	cross	cross	cross	overlap	overlap
in*	overlap*	cross*	cross*	cross*	overlap*

Fig. 2. Some differences between CBM and CBM* relations

5   Definition of RCC*-9

In Cohn and his coauthors' work, the spatial primitive entities of the calculus are regions [8, 11]. The primitive spatial entities of the proposed calculus RCC*-9 are instead generic spatial features, without forcing an interpretation in terms of regions, lines, or points. As discussed in Section 2, in topology a feature of co-dimension bigger than zero (such as a line or a point in R^2) does not have an interior. One consequence is that a line in R^2 cannot have a non-tangential proper part (see also Galton's work [18]). The RCC definitions work when the universe of discourse contains regions of dimension R^n, for any $n>0$. But the definitions do not work for points or for universes of discourse containing regions of mixed dimensionality[3].

---

[3] Different mereotopologies (such as RCC) take a different semantic stance as to what kinds of spatial entities are allowed. See Cohn and Varzi [10] for an extended discussion and analysis on this issues and a comparison of the different approaches, as well as axiomatisations of merotopologies allowing boundaries (though the *cross* relation considered in this paper is not defined there).

The boundary of an interval is made up of its two endpoints. A non-tangential proper part of an interval is another interval that is inside the first one and that does not connect with the endpoints of the first one. Adopting the "usual" GIS definitions [4, 15], non-tangential proper parts of lines embedded in R^2 can be defined as a mapping from one-dimensional intervals to the plane. In this way, we can find RCC*-9 definitions of topological relations that apply to all kinds of spatial features.

Analogously to RCC-8, we consider a primitive *connected* relation between two features $C(x,y)$. There are several models for RCC in the literature; here, for consistency with CBM*, we take our universe of discourse to be closed regions (possibly disconnected), closed lines (also possibly disconnected), and sets of isolated points. $C(x,y)$ is interpreted as being true when x and y have at least one point in common. The *connected* relation enjoys two axioms:

$C(x,x),$
$C(x,y) \rightarrow C(y,x).$

From the primitive *connected* relation, other relations are consequently defined. These are as in RCC8 except as noted. The *disconnected* relation is defined as:

Def. 18. $DC(x,y) =_{def} \neg C(x,y)$

The *part* relation between x and y is defined by saying that the feature x cannot be connected to features disconnected from y:

Def. 19. $P(x,y) =_{def} \forall z [C(z,x) \rightarrow C(z,y)]$

The *proper part* relation excludes the case of equality between the two features:

Def. 20. $PP(x,y) =_{def} P(x,y) \wedge \neg P(y,x)$

The *equal* relation is defined as:

Def. 21. $EQ(x,y) =_{def} P(x,y) \wedge P(y,x)$

In the original RCC, the *overlap* relation was defined as: $O(x,y) = \exists z [P(z,x) \wedge P(z,y)]$. Such a definition sufficed to refine the *connected* relation and make a distinction between the *overlap* and the *externally connected* relation. In RCC*-9, when we remove the limitation that features are regions only, the fact that there is a common part belonging to the two features x and y would not suffice to identify a new relation. In essence, the $O(x,y)$ relation would coincide with the $C(x,y)$ relation, since the common part could be a line or a point. Therefore, we need to find another definition for the *overlap* relation. The *externally connected* relation in RCC-8 was defined as $EC(x,y) = C(x,y) \wedge \neg O(x,y)$. This means that the EC relation cannot be defined simply by negating O. Further, in RCC-8, the *non-tangential proper part* relation needed the EC relation for its definition, which was $NTPP(x,y) = PP(x,y) \wedge \neg \exists z[EC(z,x) \wedge EC(z,y)]$.

To overcome the above issues, we need to introduce a new topological primitive and we choose the *boundary* relation $B(x,y)$, expressing the fact that feature x is the *boundary* of feature y. The type of x must be of different type to that of y. For a line y,

x is the set of its endpoints[4]. If y is a simple region, then x is the closed line representing y's boundary; if y is a complex region (holed or multipiece), then x *is a set of lines*. This effectively also introduces several kinds of spatial entities, so that our intended universe of discourse now consists of *regions* (2D entities), 1D lines (such as *boundaries of regions*), and sets of isolated points (boundaries of lines). The *boundary* relation obeys the following axiom:

$B(x,y) \rightarrow PP(x,y)$.

Hence, we give a new definition of the *non-tangential proper part* relation:

Def. 22. $NTPP(x,y) =_{def} PP(x,y) \land \forall y_1 [B(y_1, y) \rightarrow DC(x, y_1)]$

The definition is illustrated in Fig. 3. The feature x is a proper part of y and does not touch the boundary of y. Such a definition of NTPP, though it is different, has exactly the same semantics as the original RCC definition in the case of regions.

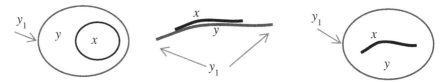

Fig. 3. Illustrations of the NTPP definition of RCC*-9. (Note that in the middle illustration, x actually is part of y, but it is drawn alongside it for clarity of illustration; we use the same convention in later figures as well).

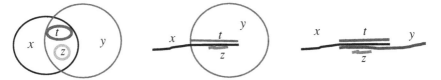

Fig. 4. Illustration of the O relation

The new definition for the tangential proper part relation is:

Def. 23. $TPP(x,y) =_{def} PP(x,y) \land \neg NTPP(x,y)$

We can now give a new definition of the *overlap* relation, which is more restrictive than the corresponding definition of RCC-8:

Def. 24. $O(x,y) =_{def} \exists z[NTPP(z,x) \land NTPP(z,y)] \land \exists t[TPP(t,x) \land TPP(t,y)]$

The above definition of *overlap* expresses the fact that there is a common non-tangential proper part belonging to the two features and a common tangential proper

---

[4] Since B is a relation rather than a functor in RCC*-9, if y is a closed ring and its boundary is empty, it means that there is no value x for which $B(x,y)$ is true. Similarly $B(x,y)$ is never true when y is a point.

part as well. The second part of the rule would not be necessary for regions, but it is necessary for lines (see Fig. 4), otherwise also cases of *cross* (see later on) would be regarded as *overlap*.

As in RCC-8, as a refinement of $O(x,y)$, the *partially overlap* relation corresponds to excluding the inclusion of one feature into the other one:

Def. 25. $PO(x,y) =_{\text{def}} O(x,y) \wedge \neg P(x,y) \wedge \neg P(y,x)$

Considering the new domain of spatial features instead of only regions, there are two other kinds of connection that are not included in the overlap definition, namely, the *externally connected* and the *cross* relations. We use the following definition for the *externally connected* relation (which differs from the RCC8 one):

Def. 26. $EC(x,y) =_{\text{def}} C(x,y) \wedge \neg O(x,y) \wedge \forall z \ [[P(z,x) \wedge P(z,y)] \rightarrow [TPP(z,x) \vee TPP(z,y)]]$

Fig. 5 depicts the EC definition in case of two regions, a region and a line, and two lines. The whole common part z needs to be a tangential proper part of x or y (this is ensured through the universal quantifier $\forall z$). In the case of a line x and a region y in Fig. 5, the common part z is a tangential proper part of y. Also in the case of the two lines, the common part z is a tangential proper part of y. The EC relation maintains the same semantics as RCC-8 for (2D) regions.

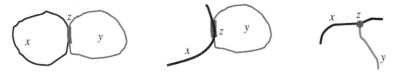

Fig. 5. Illustrations of the EC relation

Finally, we add the definition of *cross*, which corresponds to the remaining kind of connection and is not included in the previous ones (see Fig. 6)[5]:

Def. 27. $CR(x,y) =_{\text{def}} C(x,y) \wedge \neg O(x,y) \wedge \neg EC(x,y)$

Fig. 6. Cases of the CR relation

---

[5] Note that the *cross* relation is between a region and a line or a pair of lines; one could also imagine a scenario where two regions "cross" each other (so that they form a kind of "fat cross"; this is not an instance of the *cross* relation, but just of the PO relation – but see Galton [17] for definitions of relations specialising PO in this way.

The inverse relations of the asymmetric *part* relation and its specializations are defined as:

Def. 28. $Pi(x,y) =_{def} P(y,x)$
Def. 29. $PPi(x,y) =_{def} PP(y,x)$
Def. 30. $NTPPi(x,y) =_{def} NTPP(y,x)$
Def. 31. $TPPi(x,y) =_{def} TPP(y,x)$

For completeness with respect to the original RCC family of calculi, a DR relation (*discrete*) is defined as:

Def. 32. $DR(x,y) =_{def} EC(x,y) \vee DC(x,y)$

The 9 relations DC, EC, PO, TPP, NTPP, TPPi, NTPPi, EQ, and CR form a provably JEPD set of relations and are the base relations of RCC*-9. A hierarchical implication structure of all the relations defined above is given in Figure 7. To show that Figure 7 correctly reflects the implication hierarchy of the relations is mostly straightforward from the definitions. The only cases which are not trivial are the subsumption of O by C, and of P and Pi by O. We also define the JEPD set DC, EC, PO, PP, PPi, EQ, and CR, which we name RCC*-7 and, as we shall see below, corresponds to CBM*.

It is important to stress the fact that the changes we have made to some definitions of RCC-8 to obtain RCC*-9 are alternative definitions of RCC-8 relations to accommodate multi-type features. There is no change of meaning for these relations if we apply them to (2D) regions. RCC*-9 introduces the new CR relation, which can only hold when one of the entities is a 1D entity.

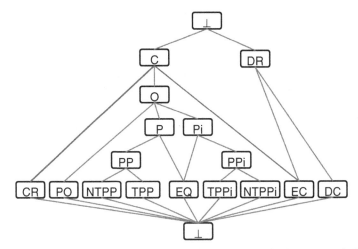

Fig. 7. The subsumption hierarchy of RCC*-9 relations. The lines indicate semantic inclusion – i.e., whenever two relations are linked, the lower one implies the upper one.

6   Spatial Reasoning

We introduced in Section 4 a modified CBM, called CBM*, and in Section 5 an extension of RCC-8, called RCC*-9. The latter is logically defined in FOPC in terms of a single primitive relation $C(x,y)$ whereas the former's relations are all taken as primitive and each have their own semantic definitions. It is provable that the two systems can express the same topological relations, as illustrated in Table 1. We can observe a direct correspondence of the base relations of CBM* with the 7 relations of RCC*-7. In CBM*, there are no base relations expressing RCC*-9 relations NTPP and TPP, which can be expressed by a logical formula involving boundaries of features.

Table 1. Correspondence between CBM* and RCC*-9

RCC*-9	CBM*
$DC(x,y)$	$disjoint^*(x,y)$
$EC(x,y)$	$touch^*(x,y)$
$PP(x,y)$	$in^*(x,y)$
$CR(x,y)$	$cross^*(x,y)$
$PO(x,y)$	$overlap^*(x,y)$
$PPi(x,y)$	$in^{*-1}(x,y)$
$EQ(x,y)$	$equal^*(x,y)$
$NTPP(x,y)$	$in^*(x,y) \wedge disjoint^*(x,b(y))$
$TPP(x,y)$	$in^*(x,y) \wedge \neg disjoint^*(x,b(y))$
$NTPPi(x,y)$	$in^{*-1}(x,y) \wedge disjoint^*(b(x),y)$
$TPPi(x,y)$	$in^{*-1}(x,y) \wedge \neg disjoint^*(b(x),y)$
$C(x,y)$	$\neg disjoint^*(x,y)$
$P(x,y)$	$in^*(x,y) \vee equal^*(x,y)$
$Pi(x,y)$	$in^{*-1}(x,y) \vee equal^*(x,y)$
$O(x,y)$	$overlap^*(x,y) \vee in^*(x,y) \vee$ $in^{*-1}(x,y) \vee equal^*(x,y)$
$DR(x,y)$	$disjoint^*(x,y) \vee touch^*(x,y)$

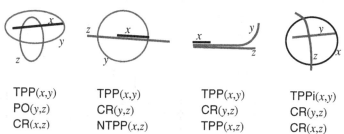

$TPP(x,y)$	$TPP(x,y)$	$TPP(x,y)$	$TPPi(x,y)$
$PO(y,z)$	$CR(y,z)$	$CR(y,z)$	$CR(y,z)$
$CR(x,z)$	$NTPP(x,z)$	$TPP(x,z)$	$CR(x,z)$

Fig. 8. Some new cases of composition involving the CR relation

Given the correspondence between the CBM* and RCC*-9, we proceed to find the composition tables for these calculi. The composition tables contain the basic rules to perform qualitative spatial reasoning with such calculi (see, for example, [9]). Given the relation $r_1(x,y)$ and the relation $r_2(y,z)$, the composition is the relation $r_3(x,z)$. The composition table gives all the possible results of composition for each combination of relations. Such results are expressed as disjunctions of the basic relations. For RCC*-9, the results of compositions are those reported in Table 2. Such a table is a direct extension of the composition table of RCC-8 [11], that is, if we restricted Table 2 to regions, we would re-obtain the composition table of RCC-8: the CR relation cannot hold between two regions.

In general, the proof of composition tables is difficult, especially when the semantics of the calculus depends on higher-order constructs such as sets [24]. There are two aspects to proving that a composition table is correct: (1) showing that each disjunct in each cell is necessary; (2) showing that there are no missing disjunctions. The former is usually achieved by demonstrating (i.e., providing a model such as a figure) of each combination of r_1, r_2 and a disjunct from r_3[6]. Showing that there are no missing disjuncts, given an axiomatic theory of the calculus, can be achieved by proving a theorem that r_1 and r_2 imply r_3 for each cell[7]. If we consider that the RCC*-9 composition table is an extension of the RCC-8 table, one way of finding a proof is 'by difference', that is, limiting the analysis to the new cases involving the CR relation only. We found in total 89 new compositions that can be instantiated in R^2 involving the CR relation: see Fig. 8 for a sample of them. The fact that no other cases with CR are possible can be proved with a theorem for each entry, but by using redundancy elimination techniques as in [2], the actual number of entries that need to be proved can be reduced significantly. Alternatively, a proof could be developed with a semi-automatic reasoner as proposed in [24]. Besides formal proofs, in future work we also plan to apply heuristics such as in [6], where composition tables can be filled up by running tests on random data sets made up of points, polygons, and polylines.

Table 2. Composition table for RCC*-9

r_2 / r_1	DC	EC	PO	TPP	NTPP	TPPi	NTPPi	EQ	CR
DC	no info	DR, PO, PP, CR	DR, PO, PP, CR	DR, PO, PP, CR	DR, PO, PP, CR	DC	DC	DC	DR, PO, PP, CR
EC	DR, PO, PPi, CR	DR, PO, TPP, EQ, TPPi, CR	DR, PO, PP, CR	EC, PO, PP, CR	PO, PP, CR	DR	DC	DC	DR, PO, PP, CR
PO	DR, PO, PPi, CR	DR, PO, PPi, CR	no info	PO, PP, CR	PO, PP, CR	DR, PO, PPi, CR	DR, PO, PPi, CR	PO	DR, PO, PP, PPi, CR
TPP	DC	DR	DR, PO, PP, CR	PP	NTPP	DR, PO, TPP, EQ, TPPi, CR	DR, PO, PPi, CR	TPP	DR, PP, PO, CR
NTPP	DC	DC	DR, PO, PP, CR	NTPP	NTPP	DR, PO, PP, CR	no info	NTPP	DR, PP, PO, CR
TPPi	DR, PO, PPi, CR	EC, PO, PPi, CR	PO, PPi, CR	PO, TPP, EQ, TPPi	PO, PP, CR	PPi	NTPPi	TPPi	PO, PPi, CR
NTPPi	DR, PO, PPi, CR	PO, PPi, CR	PO, PPi, CR	PO, PPi, CR	O, CR	NTPPi	NTPPi	NTPPi	PO, PPi, CR
EQ	DC	EC	PO	TPP	NTPP	TPPi	NTPPi	EQ	CR
CR	DR, PO, PPi, CR	DR, PO, PPi, CR	DR, PO, PP, PPi, CR	PP, PO, CR	PP, PO, CR	DR, PPi, PO, CR	DR, PPi, PO, CR	CR	no info

---

[6] This is what we actually did to find the RCC*-9 composition table, that is, finding configurations like those in Fig.8 satisfying each result of the table.

[7] An automatic proof of RCC-8 composition table based on encoding RCC-8 in an intuitionistic propositional calculus has been proposed in [2].

From the composition table of RCC*-9, we can infer the composition table of CBM*. First, we need to find an intermediate result: the composition table of RCC*-7. This is just a reduced version of the composition table of RCC*-9 that is obtained by making the union of relations TPP and NTPP and of relations TPPI and NTPPi. From the composition table of RCC*-7, and from the correspondences between CBM* and RCC-7 (Table 1), we can obtain as an almost immediate result the composition table for CBM* (Table 3) by simple renaming of the relations.

Table 3. Composition table for CBM*. Adopted abbreviations: *di*=*disjoint*, *to*=*touch*, *ov*=*overlap*, *eq*=*equal*, *cr*=*cross*.

r_1 \ r_2	di*	to*	ov*	in*	in*⁻¹	eq*	cr*
di*	no info	di*, to*, ov*, in*, cr*	di*, to*, ov*, in*, cr*	di*, to*, ov*, in*, cr*	di*	di*	di*, to*, ov*, in*, cr*
to*	di*, to*, ov*, in*⁻¹, cr*	no info	di*, to*, ov*, in*, cr*	to*, ov*, in*, cr*	di*, to*	di*	di*, to*, ov*, in*, cr*
ov*	di*, to*, ov*, in*⁻¹, cr*	di*, to*, ov*, in*⁻¹, cr*	no info	ov*, in*, cr*	di*, to*, ov*, in*⁻¹, cr*	ov*	di*, to*, ov*, in*, in*⁻¹, cr*
in*	di*	di*, to*	di*, to*, ov*, in*, cr*	in*	no info	in*	di*, to*, in*, ov*, cr*
in*⁻¹	di*, to*, ov*, in*⁻¹, cr*	to*, ov*, in*⁻¹, cr*	ov*, in*⁻¹, cr*	ov*, in*, in*⁻¹, eq*, cr*	in*⁻¹	in*⁻¹	ov*, in*⁻¹, cr*
eq*	di*	to*	ov*	in*	in*⁻¹	eq*	cr*
cr*	di*, to*, ov*, in*⁻¹, cr*	di*, to*, ov*, in*⁻¹, cr*	di*, to*, ov*, in*, in*⁻¹, cr*	in*, ov*, cr*	di*, to*, in*⁻¹, ov*, cr*	cr*	no info

7   Comparison with 9-Intersection

In this section, we compare the calculi CBM* and RCC*-9 with 9IM [15] and DE+9IM [3]. This is useful for practical reasons to easily implement the relations of the proposed calculi in OGC-compliant systems. We use the `Relate` function defined in the OGC Simple Features Specification [23]. The function returns true if the two features satisfy the topological relation corresponding to the string parameter. Such a string represents a set of values for 9IM matrix by rows: characters allowed in the string are 'F' for an empty intersection, 'T' for a non-empty intersection, and '*' for 'don't care'. The value 'T' in the string of the `Relate` function can be specialized to the values 0, 1, 2 to express the dimension of the intersection set: this corresponds to the DE+9IM matrix introduced in [3]. Table 4 summarizes the correspondence between CBM* and 9IM. The equivalent expressions of 9IM can be easily inferred from CBM* definitions.

Table 4. Correspondence between CBM* relations and 9IM relations

CBM*	9IM
disjoint*(x,y)	Relate($x,y,$ "FF*FF****")
touch*(x,y)	Relate($x,y,$ "FTT***T**") ∨
	Relate($x,y,$ "F*TT**T**") ∨
	Relate($x,y,$ "F*T*T*T**")
in*(x,y)	Relate($x,y,$ "**F**F***") ∧
	¬ Relate($x,y,$ "TFFFTFFFT")
cross*(x,y)	Relate($x,y,$ "T**FF****") ∨
	Relate($x,y,$ "TF**F****")
overlap*(x,y)	Relate($x,y,$ "TTTT**T**") ∨
	Relate($x,y,$ "T*T*T*T**")
$in*^{-1}(x,y)$	Relate($x,y,$ "******FF*") ∧
	¬ Relate($x,y,$ "TFFFTFFFT")
equal*(x,y)	Relate($x,y,$ "TFFFTFFFT")

Table 5. Correspondence between RCC*-9 and 9IM

RCC*-9	9IM
DC(x,y)	Relate($x,y,$ "FF*FF****")
EC(x,y)	Relate($x,y,$ "FTT***T**") ∨
	Relate($x,y,$ "F*TT**T**") ∨
	Relate($x,y,$ "F*T*T*T**")
NTPP(x,y)	Relate($x,y,$ "*FF*FF***")
TPP(x,y)	Relate($x,y,$ "*TF**F***") ∨
	Relate($x,y,$ "**F*TF***") ∧
	¬ Relate($x,y,$ "TFFFTFFFT")
CR(x,y)	Relate($x,y,$ "T**FF****") ∨
	Relate($x,y,$ "TF**F****")
PO(x,y)	Relate($x,y,$ "TTTT**T**") ∨
	Relate($x,y,$ "T*T*T*T**")
NTPPi(x,y)	Relate($x,y,$ "***FF*FF*")
TPPi(x,y)	Relate($x,y,$ "***T**FF*") ∨
	Relate($x,y,$ "****T*FF*") ∧
	¬ Relate($x,y,$ "TFFFTFFFT")
EQ(x,y)	Relate($x,y,$ "TFFFTFFFT")

We can see in Table 4 that CBM* relations do not need the dimension of the intersection set to find equivalent expressions. Therefore, it is possible to find equivalent expressions of CBM* queries in terms of 9IM without the need to resort to the much expressive DE+9IM. The CBM relations needed the dimension to find equivalent expressions in the DE+9IM matrices, being 9IM matrix alone not sufficient. In [3], it was proved that CBM is equivalent to DE+9IM in terms of the number of topological configurations that the models are able to distinguish. Though it is out of the scope of this paper, it is provable that CBM* is equivalent to 9IM in terms of number of topological configurations. In this sense, CBM* can be considered weaker than CBM because CBM* does not include the possibility of checking the dimension of intersections. Of course, this is not a real weakness of CBM* since an operator to check dimension could be easily added to the calculus to recuperate the ability of checking set dimension. Given the correspondence between CBM* and RCC*-9 (Table 1), we can express RCC*-9 relations in terms of 9IM matrices by using Table 4 to obtain Table 5.

8   Conclusions and Further Work

An extension towards multidimensional mereotopology [18] has been advocated for a long time. RCC*-9 is our contribution to address this issue. We defined RCC*-9 by modifying the definition of the basic relations of RCC-8 and adding two new relations, namely, a new primitive $B(x,y)$ to express that x is *boundary* of y and $CR(x,y)$ for the defined *cross* relation. The variables of RCC*-9 no longer range just over regions, but features (or sets of features) of dimension 2, 1, or 0, embedded in R^2. These changes extend rather than change[8] the semantics of RCC-8, since if we consider only regions, then RCC*-9 collapses to RCC-8. The composition table of RCC*-9 with respect to the composition table of RCC-8 presents the relation CR as an additional possible result of composition, but it does not affect the already present results – i.e., each entry in the composition table for RCC*-9 is either the same, or a superset of the corresponding RCC-8 composition table entry (except for the rows and columns labelled by CR, which are new).

In this paper, we also introduced the CBM*, a modified version of CBM where we lose the possibility of distinguishing the dimension of set intersections. CBM* definitions do not depend on the type of features, e.g., a *cross* between two lines has the same definition of the *cross* between a line and a region. With the new definitions, it is possible to obtain a single composition table for all features.

Finally, we provided the usual basis for a reasoning system for a qualitative calculus, i.e., a composition table, for the new calculi, extending the earlier composition tables from simple regions to the case of generic spatial features. Another interesting aspect that we discussed is how to find equivalent expressions of both calculi in terms of 9IM, which is essential to enabling a straightforward implementation in OGC-compliant systems.

Further work is needed to provide a formal proof of the correctness of the composition tables. Another issue that is not covered in this paper is the study of the cognitive adequacy of the group of relations inside CBM* and RCC*-9 models. It would be interesting to find out the differences in subjective perceptions especially of the previous CBM calculus versus the new CBM* calculus. Finally, an assessment of how the calculi behave for complex features and for higher dimensional spaces remains to be done.

---

[8] Strictly, it only extends RCC8 if we consider the 2D interpretation of RCC8: RCC8 can be interpreted in any dimension ≥ 2; in principle the definitions here may apply to regions of other dimensions but we have not investigated this yet.

Acknowledgments. The authors wish to express their gratitude to the referees for their useful suggestions and to Paolo Fogliaroni for his comments to a previous version of the manuscript. The financial support of EU projects RACE (FP7-ICT-287752) and STRANDS (FP7-ICT-600623) is gratefully acknowledged.

References

1. Aref, W.G., Samet, H.: Optimization Strategies for Spatial Query Processing. In: 17th International Conference on Very Large Databases, Barcelona, Spain, pp. 81–90 (1991)
2. Bennett, B.: Logical Representations for Automated Reasoning about Spatial Relationships. Ph.D. Thesis, School of Computer Studies, University of Leeds (1997)
3. Clementini, E., Di Felice, P.: A Comparison of Methods for Representing Topological Relationships. Information Sciences 3, 149–178 (1995)
4. Clementini, E., Di Felice, P.: A Model for Representing Topological Relationships Between Complex Geometric Features in Spatial Databases. Information Sciences 90, 121–136 (1996)
5. Clementini, E., Di Felice, P., van Oosterom, P.: A Small Set of Formal Topological Relationships Suitable for End-User Interaction. In: Abel, D.J., Ooi, B.-C. (eds.) SSD 1993. LNCS, vol. 692, pp. 277–295. Springer, Heidelberg (1993)
6. Clementini, E., Skiadopoulos, S., Billen, R., Tarquini, F.: A reasoning system of ternary projective relations. IEEE Transactions on Knowledge and Data Engineering 22, 161–178 (2010)
7. Codd, E.F.: A relational model for large shared data banks. Communications of the ACM 13, 377–387 (1970)
8. Cohn, A.G., Bennett, B., Gooday, J., Gotts, N.: Qualitative Spatial Representation and Reasoning with the Region Connection Calculus. GeoInformatica 1, 275–316 (1997)
9. Cohn, A.G., Renz, J.: Qualitative Spatial Representation and Reasoning. In: Harmelen, F.V., Lifschitz, V., Porter, B. (eds.) Handbook of Knowledge Representation, vol. 1, pp. 551–596. Elsevier (2007)
10. Cohn, A.G., Varzi, A.C.: Mereotopological connection. Journal of Philosophical Logic 32, 357–390 (2003)
11. Cui, Z., Cohn, A.G., Randell, D.A.: Qualitative and Topological Relationships in Spatial Databases. In: Abel, D.J., Ooi, B.-C. (eds.) SSD 1993. LNCS, vol. 692, pp. 296–315. Springer, Heidelberg (1993)
12. Egenhofer, M.J.: Deriving the composition of binary topological relations. Journal of Visual Languages and Computing 5, 133–149 (1994)
13. Egenhofer, M.J., Clementini, E., Di Felice, P.: Topological relations between regions with holes. International Journal of Geographical Information Systems 8, 129–142 (1994)
14. Egenhofer, M.J., Franzosa, R.D.: Point-Set Topological Spatial Relations. International Journal of Geographical Information Systems 5, 161–174 (1991)
15. Egenhofer, M.J., Herring, J.R.: Categorizing Binary Topological Relationships Between Regions, Lines, and Points in Geographic Databases. Department of Surveying Engineering, University of Maine, Orono, ME Technical Report (1990)
16. Gabrielli, N.: Investigation of the Tradeoff between Expressiveness and Complexity in Description Logics with Spatial Operators. Ph.D. Thesis: University of Verona (2009)
17. Galton, A.: Modes of overlap. Journal of Visual Languages and Computing 9, 61–79 (1998)

18. Galton, A.: Multidimensional Mereotopology. In: Dubois, D., Welty, C., Williams, M.-A. (eds.) Proceedings of the Ninth International Conference on Principles of Knowledge Representation and Reasoning (KR 2004), Whistler, BC, Canada, June 2-5, pp. 45–54. American Association for Artificial Intelligence (2004)

19. Galton, A.P.: Taking dimension seriously in qualitative spatial reasoning. In: Proceedings of the Twelfth European Conference on Artificial Intelligence (ECAI 1996), Budapest, Hungary, August 11-16 (1996)

20. Gerevini, A., Renz, J.: Combining topological and size information for spatial reasoning. Artificial Intelligence 137, 1–42 (2002)

21. Gotts, N.M.: Formalizing Commonsense Topology: The INCH Calculus. pp. In: Proceedings of the Fourth International Symposium on Artificial Intelligence and Mathematics (1996)

22. Isli, A., Cabedo, L.M., Barkowsky, T., Moratz, R.: A Topological Calculus for Cartographic Entities. In: Habel, C., Brauer, W., Freksa, C., Wender, K.F. (eds.) Spatial Cognition 2000. LNCS (LNAI), vol. 1849, pp. 225–238. Springer, Heidelberg (2000)

23. OGC Open Geospatial Consortium Inc. OpenGIS Simple Features Implementation Specification for SQL. vol. OGC 99–049 (1999)

24. Wölfl, S., Mossakowski, T., Schröder, L.: Qualitative constraint calculi: Heterogeneous verification of composition tables. In: 20th International FLAIRS Conference (FLAIRS 2007), pp. 665–670 (2007)

Author Index